This authoritative volume is the first to present a comprehensive and up-to-date review of our new understanding of accretion disks around black holes.

Interest in black hole accretion disks has undergone a renaissance in recent years because of developments in three complementary areas: theoretical modelling of relativistic plasmas, numerical simulations with supercomputers, and observational tests now possible using new observatories such as the Japanese X-ray satellite, ASCAR. This volume presents review papers on all these topics from leading world authorities who gathered at an international conference in Reykjavik, Iceland. The paper authors are M. Abramowicz, P. Artymowicz, G. Björnsson, A. Brandenburg, P. Charles, A. Fabian, J. Krolik, J.-P. Lasota, G. Madejski, R. Narayan, I. Novikov, J. Papaloizou, J. Poutanen, J. Pringle, M. Rees, E. Spiegel, R. Svensson, and P. Witta.

This timely volume provides an up-to-date review of the theory of black hole accretion disks for graduate students and researchers in astrophysics and theoretical physics.

T0206315

CAMBRIDGE CONTEMPORARY ASTROPHYSICS

Theory of Black Hole Accretion Disks

CAMBRIDGE CONTEMPORARY ASTROPHYSICS

Series editors
José Franco, Steven M. Kahn, Andrew R. King and Barry F. Madore

Theory of
Black Hole
Accretion Disks

Edited by
MAREK A. ABRAMOWICZ
Göteborg University, Sweden

GUNNLAUGUR BJÖRNSSON
University of Iceland, Reykjavik

JAMES E. PRINGLE
University of Cambridge

CAMBRIDGE
UNIVERSITY PRESS

CAMBRIDGE UNIVERSITY PRESS
Cambridge, New York, Melbourne, Madrid, Cape Town, Singapore,
São Paulo, Delhi, Dubai, Tokyo

Cambridge University Press
The Edinburgh Building, Cambridge CB2 8RU, UK

Published in the United States of America by Cambridge University Press, New York

www.cambridge.org
Information on this title: www.cambridge.org/9780521152952

© Cambridge University Press 1998

This publication is in copyright. Subject to statutory exception
and to the provisions of relevant collective licensing agreements,
no reproduction of any part may take place without the written
permission of Cambridge University Press.

First published 1998
This digitally printed version 2010

A catalogue record for this publication is available from the British Library

ISBN 978-0-521-62362-9 Hardback
ISBN 978-0-521-15295-2 Paperback

Cambridge University Press has no responsibility for the persistence or
accuracy of URLs for external or third-party internet websites referred to in
this publication, and does not guarantee that any content on such websites is,
or will remain, accurate or appropriate.

Contents

Stability of accretion discs

Coherent structures

Summary

Preface

Accretion discs play a fundamental role in many astronomical objects: stellar binaries, active galactic nuclei, and as nurseries of planetary formation. Those in quasars are the most powerful and efficient energy sources in the Universe. Accretion discs are governed by non linear phenomena and often experience internal motions approaching the speed of light, ultra high temperatures, and extreme strengths of gravitational and magnetic fields.

Understanding of accretion discs progresses through a multidisciplinary union of observational, theoretical and computational methods. High tech observations are performed by ground based radio, and adaptive optics telescopes, and Hubble, X-ray and gamma-ray telescopes in space. They cover the whole electromagnetic spectrum, from radio waves to TeV gamma rays. Theory investigates profound interrelations between atomic physics, magnetohydrodynamics, black holes physics, radiation theory, and chaotic phenomena such as turbulence. It is supported by advanced supercomputer simulations.

The Midsummer Symposium on *Non-Linear Phenomena in Accretion Discs around Black Holes* that took place in Laugarvatn, near Reykjavik in Iceland on June 18-21, 1997, was devoted to discuss the significant recent progress in understanding theory of black hole accretion discs, in particular in understanding the origin of turbulent viscosity, radiative processes in very hot, optically thin plasma, and the role of advective cooling. The invited speakers at the Symposium played a key role in all these developments. The list of participants as well as details of the Symposium are available at http://fy.chalmers.se/~marek/Reykjavik.html

The Symposium was sponsored by several Nordic institutions: Royal Swedish Academy of Sciences through its Nobel Institute of Physics (Stockholm), The Swedish Natural Science Research Council (Stockholm), Chalmers University of Technology (Göteborg), Theoretical Astrophysics Centre (Copenhagen), University of Iceland (Reykjavik), Göteborg University (Göteborg), Nordita (Copenhagen), and NorFa (Oslo).

This book is not just a Proceedings volume of the Midsummer Symposium. We have compiled a monograph, summarizing the remarkable recent progress, achieved in understanding astronomical objects powered by the black hole accretion, and covering both theory and observations. For this reason, the book opens with two observational reviews, written by experts who did not attended the purely theoretical Midsummer Symposium.

All articles in this book are written by authors who made fundamental contributions to the subject. As the subject is rapidly expanding, the articles describe not only results that have been obtained and progress that has been made, but also unresolved issues, difficulties, and fundamental questions that need to be answered. Articles in this book are personal and individual: they reflect thoughts, ideas and preferences of their authors. Often, they are critical, and polemic, and display the characteristic style and sense of humour of their authors.

The Editors have not attempted in any way to standardize the style of the articles. We have only added a few cross-references to help the reader navigate within this book. Colour versions of many of the figures in the book may be retrieved from http://raunvis.hi.is/~gulli/Reykjavik.html

M. A. Abramowicz, G. Björnsson, J. E. Pringle
Nordita, Copenhagen
June, 1998

Preface

Acknowledgements

We would like to express our sincere thanks to all institutes and organizations that provided financial support for the meeting in Iceland. These are listed in the Preface. In particular, we would like to thank Nordita, Copenhagen, it's director and staff for providing a stimulating environment, and for continuous support through the *Nordic Project: Non-linear phenomena in accretion disks around black holes.*

Black holes in our Galaxy: observations

By PHIL CHARLES

Department of Astrophysics, Oxford University, Keble Road, Oxford OX1 3RH, UK

This paper reviews the X-ray, optical, radio and IR observations of galactic X-ray binaries suspected to contain black hole compact objects, with particular emphasis on the supporting dynamical evidence.

1. Introduction

From just one object (Cyg X-1) 25 years ago, the number of strong black-hole candidates in the Galaxy has increased dramatically in the last decade. This is due to two main factors: (i) the provision of all-sky monitoring capabilities on the current generation of X-ray observatories, and (ii) the discovery of the class of low-mass X-ray binary (LMXB) known as *soft X-ray transients* (SXT). These are short-period (5hrs – 6days) binaries with mostly K–M type secondaries, and many similarities with dwarf novae. Their rare, dramatic X-ray outbursts (typically separated by decades) can reach very high luminosities ($> 10^{38}$erg s^{-1}) and are considered to be accretion disc instability events. Remarkably, a very high fraction of SXTs are black-hole candidates. This is largely because they are the only LMXBs for which accurate mass determinations are possible, through the ability to undertake dynamical studies of the secondary star during the long periods of quiescence. While Cyg X-1 has been joined by other high-mass X-ray binaries (HMXB) also believed to contain black-holes, they all suffer from the serious limitation that an accurate mass is required for the early-type mass donor before it is possible to precisely determine the compact object mass. Given their complex binary evolutionary path this is difficult to obtain.

It has been more than 30 years since the first X-ray binary was optically identified (Sco X-1), but a detailed knowledge of their binary parameters only started to come in the 1970s with the identification and study of Cyg X-1, Her X-1 and Cen X-3. However, these all have early-type companions, observable in spite of the presence in the binary of luminous X-ray emission (for a review see van Paradijs & McClintock 1995). Apart from Cyg X-1, most of these high-mass X-ray binaries (HMXRBs) had pulsating neutron star compact objects, thereby providing the potential for a full solution of the binary parameters since they were essentially double-lined spectroscopic binaries. From this has come the detailed dynamical mass measurements of neutron star systems which have recently been collated by Thorsett & Chakrabarty (1998), showing that they are all consistent with a mass of 1.35±0.04M$_\odot$.

However, when HMXRBs are suspected of harbouring much more massive compact objects (as is the case for Cyg X-1), the mass measurement process runs into difficulties. By definition, there will be no dynamic features (such as pulsations) associated with the compact object that can be observed. Hence all mass information must come from the mass-losing companion, and the mass of the compact object cannot be determined unless the companion's mass is accurately known.

The situation for low-mass X-ray binaries (LMXBs), such as Sco X-1, is completely different, in that their short orbital periods require their companion stars to be of low mass. This can be demonstrated quite simply as follows (see King 1988). Since these are interacting binaries in which the companion fills its Roche lobe, then we may employ the useful Paczynski (1971) relation

$$R_2/a = 0.46(1+q)^{-1/3}, \qquad (1.1)$$

where the mass ratio $q = M_X/M_2$. Combining this with Kepler's 3rd Law yields the well-known result that the mean density, ρ (in g cm^{-3}) of the secondary,

$$\rho = 110/P_{hr}^2. \qquad (1.2)$$

And if these stars are on or close to the lower main sequence, then $M_2 = R_2$ and hence $M_2 = 0.11P_{hr}$. Therefore short period X-ray binaries must be LMXBs and so the companion star will be faint. The major observational problem with this is that the optical light will then be dominated by reprocessed X-radiation from the disc (or heated face of the companion star; see van Paradijs & McClintock 1994). This is why the optical spectra of LMXBs are hot, blue continua (U–B typically -1) with superposed broad hydrogen and helium emission lines, the velocities of which indicate that they largely arise in the inner disc region, thereby denying us access to the dynamical information that is essential if accurate masses are to be determined. Hence, the evidence for the nature of the compact object in most bright LMXBs has come from indirect means, usually X-ray bursting behaviour (as few are X-ray pulsars) or the fast flickering first seen in Cyg X-1 (and hence used as a suggestion of the presence of a black hole). [Note, however, that while it is useful to employ the Paczynski relation in this way, it is only valid for $q > 1$, and there is a more accurate algorithm due to Eggleton (1983) which is valid for all q.]

To make real progress in determining the nature of compact objects in our Galaxy requires dynamical mass measurements of the type hitherto employed on neutron stars in HMXRBs. But without velocity information associated with the compact object, all that can be measured (in the case of Cyg X-1, and the other two HMXRBs suspected of harbouring black holes, LMC X-1 and LMC X-3) is the mass function

$$f(M) = \frac{PK^3}{2\pi G} = \frac{M_X^3 \sin^3 i}{(M_X + M_2)^2}, \qquad (1.3)$$

where P is the orbital period and K is the radial velocity amplitude. And since $M_2 \geq M_X$, then M_X is not accurately known because M_2 has a wide range of uncertainty ($\sim$12–20M$_\odot$) given the unusual evolutionary history of the binary. The compact object in Cyg X-1 almost certainly is a black hole, but an accurate mass determination is not possible from the available data which simply constrain it to be >3.8M$_\odot$ (Herrero et al 1995). This is close to the canonical maximum mass of a neutron star, based on the oft-quoted Rhoades-Ruffini Theorem (1974). However, there are a number of assumptions built into this which need careful examination in the light of the masses of the compact objects reviewed here (see e.g. Miller 1998 and Miller et al 1998).

For the LMXBs we clearly need to find systems in which the companion star *is* visible, which requires sources where the X-ray emission switches off for some reason. This is the basis of the new field of study of the *soft X-ray transients*, hereafter SXTs (and sometimes referred to as *X-ray novae*). Remarkably, of the $\sim$23 currently known, only 6 (i.e. $\sim$25%) are confirmed neutron star systems (they display type I X-ray bursts), the remainder are all black-hole candidates, the highest fraction of any class of X-ray source.

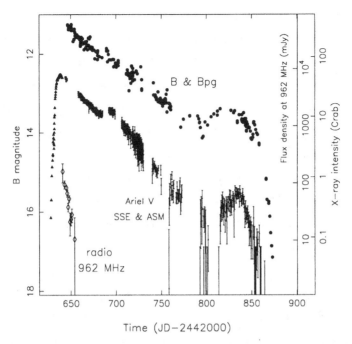

FIGURE 1. Optical, X-ray and radio outburst light curves of the prototype SXT A0620-00 (=Nova Mon 1975), adapted from Kuulkers (1998).

2. X-ray/Optical Behaviour

2.1. *Outburst*

The SXTs typically outburst every ∼10–20 years. The first one (Cen X-2) was found by early rocket flights (Harries et al 1967), but the prototype of the class (due to its proximity and detailed multi-wavelength study) is widely considered to be A0620-00 (Nova Mon 1975), for several months the brightest X-ray source in the sky and peaking at 11th mag in the optical (Elvis et al 1975; see Figure 1, taken from Kuulkers 1998). Their light curves tend to show a fast rise followed by an exponential decay (see Chen et al 1997 for a compendium of all SXT light curves), the optical amplitude of which has been shown by Shahbaz & Kuulkers (1998) to be related to the orbital period, and the precise form of the decay is related to the peak X-ray luminosity at outburst (King & Ritter 1997; Shahbaz et al 1998a).

It takes ∼1 year for SXTs to reach optical/X-ray quiescence after an outburst, but note that there have been subsequent *mini*-outbursts in some systems (e.g. GRO J0422+32, Kuulkers 1998, and references therein) and erratic re-brightenings in others (e.g. GRO J1655-40). The observed properties of the 9 SXTs for which full dynamical analyses have been performed are summarised in Table 1 and listed in order of orbital period, apart from separating the two SXTs which display a much earlier spectral type companion star.

At the time of outburst most (but not all) exhibit *ultra-soft* X-ray spectra with black-body colour temperatures of $kT \sim 0.5 - 1\text{keV}$ superposed on a hard power-law extending to much higher energies (see Tanaka & Lewin 1995). It is this characteristic that gives the SXTs their name, and effectively distinguishes them from the much harder Be X-ray transients that appear to be almost exclusively long-period neutron star systems. [Note also that the ultra-soft X-ray spectrum of SXTs is very different from the *super-soft*

TABLE 1. Optical/IR Properties of Soft X-ray Transients

Source	Outbursts	P (hrs)	Sp. Type	E_{B-V}	V (quiesc)	K	$v\sin i$ (km/s)	K_2 (km/s)
J0422+32	1992	5.1	M2V	0.3	22	16.2	$\leq$80	381
A0620-00	1917,75	7.8	K5V	0.35	18.3	6	83	433
GS2000+25	1988	8.3	K5V	1.5	21.5	17	86	518
GRS1124-68	1991	10.4	K0-4V	0.29	20.5	16.9	106	399
H1705-25	1977	12.5	K	0.5	21.5	-	$\leq$79	448
Cen X-4	1969,79	15.1	K7IV	0.1	18.4	15.0	45	146
V404 Cyg	1938,56,89	155.3	K0IV	1	18.4	12.5	39	208.5
4U1543-47	1971,83,92	27.0	A2V	0.5	16.6	-	-	124
J1655-40	1994+	62.9	F3-6IV	1.3	17.2	-	-	228

designation applied (mostly) to the (very) much cooler accreting white dwarf systems in the LMC and SMC (see Kahabka & van den Heuvel 1997).]

Additionally the SXTs (e.g. GS2023+338) can show extremely erratic variability which is very similar to that displayed by Cyg X-1. Hence the X-ray spectrum and variability are used as key discriminators to hunt for black holes. However, it must be noted that, in certain circumstances, neutron star systems can mimic these properties (e.g. Cir X-1 and X0331+53), and so we must use only dynamical evidence in the final analysis as to the nature of the compact objects (McClintock 1991).

2.2. *Quiescence*

Even in quiescence, optical studies (see section 3) show that mass transfer continues in the SXTs, and indeed many have been detected by X-ray observatories (Einstein, EX-OSAT, ROSAT) as very weak sources (e.g. Verbunt et al 1994). However, the observed luminosities are substantially lower than expected for the continuing accretion rate, and this has led various groups (see e.g. Abramowicz et al 1995; Narayan et al 1997a) to propose that *advective accretion* is taking place. The inner disc at low accretion rates evaporates due to the X-radiation into a very hot low density corona. Such hot gas cannot radiate efficiently and transports most of its thermal energy onto the compact object (the advection process). (Such models can also account for the spectral shapes during outburst, see e.g. Chen et al 1995; Chakrabarti & Titarchuk 1995; Esin et al 1997 and Chakrabarti 1998). If it is a black hole, then that energy is lost! But if it is a neutron star then the energy will be radiated from the neutron star's surface. The model therefore predicts that black-hole SXTs will be X-ray fainter in quiescence (relative to outburst) than neutron-star systems, and there is some evidence for this (see discussion in McClintock 1998).

2.3. *X-ray Spectroscopy of BHXRBs*

With Cyg X-1 as the first BH candidate, its X-ray properties were not surprisingly proposed as key indicators to help search for similar systems. Cyg X-1 exhibits 2 X-ray "states", a low, hard state with a power law spectrum extending to very high energies, and a high, soft state where the low energy data are well represented by a (multi-colour or disk) black-body spectrum at a temperature of $kT \sim 1$ keV. The power law can extend to hundreds of keV, sometimes to MeV. This is usually attributed to Comptonisation, in which case the highest energy photons require more scatterings, and hence an energy-dependent time delay would be expected. Such an effect is seen in Cyg X-1 (see van Paradijs 1998 and references therein) but it is complex.

The black-body component is explained as arising from the inner accretion disc around the compact object. Assuming that this could be approximated by a "multi-colour disc" (MCD) model which incorporates the temperature variation with disc radius, Mitsuda et al (1984) fitted the Ginga X-ray spectra to show that all those suspected (on other grounds) of being BHC had inner disc radii of $\sim 3r_S$, appropriate for stellar mass black hole candidates. r_S is the Schwarzschild radius, and ~ 3 times this value is the last stable orbit for matter around such an object. Interestingly, correcting the MCD model to allow for the effects of GR made no difference to the fitting of the continuum spectrum. However, it does affect the profiles of the spectral lines, and this is believed to have been seen in AGN (Tanaka et al, 1995; see accompanying article by Madejski), but not yet in XRBs (see Ebisawa 1997, although Życki et al (1997) have shown that such profiles are consistent with the X-ray outburst Fe line spectra obtained by Ginga). It has also recently been pointed out (Zhang et al 1997) that the BH spin can affect the size of r_S, from ~ 1–9 times the radius of the event horizon, and this will be discussed further below.

2.4. *X-ray Spectral/Variability States*

The presence of type-I X-ray bursts or pulsations immediately identifies the compact object as a neutron star, and detailed modelling of their light curves leads to constraints on the physical properties of the compact object. However, the discovery of rapid but non-periodic variability (*quasi-periodic oscillations* or QPOs) by EXOSAT in GX5-1 and other LMXBs (see Lewin, van Paradijs & van der Klis 1988) opened up new avenues of investigation into the physical processes occurring close to the interface between the accretion disc and compact object. Simultaneous analysis of the X-ray spectral and temporal variability led to the description of the behaviour of the source in terms of source "states" which could be understood as functions of the mass accretion rate (see review by van Paradijs 1998).

The 2-state behaviour of Cyg X-1 (with its low state, hard X-ray spectrum contrasting with the soft, high state) has been well-studied since the 1970s and is even considered as a possible black-hole diagnostic. Similar studies of the BHXRBs have extended these concepts and shown interesting correlations between their X-ray spectra and temporal variability. The presence of QPOs also reinforces their interpretation as a property of the inner accretion disc rather than the compact object. The SXT outbursts, when studied from peak to quiescence, cover a very large range in accretion rate onto the compact object, and detailed X-ray observations of N Mus 1991 by Ginga (van der Klis 1994, 1995; see Figure 2) suggested that the change in its X-ray spectral and temporal behaviour was indeed a simple function of mass accretion rate. Table 2 summarises the key X-ray states observed so far in SXTs as defined by the temperature of the X-ray spectrum and variability characteristics displayed by the power density spectrum (PDS).

However, the fact that the IS has properties similar to the VHS indicate that the problem must be more complex than just a function of mass transfer rate. Care must also be taken in drawing comparisons between observations made in different energy bands, as can be seen by examining the decay light curves of Nova Mus 1991 by BATSE (see Ebisawa et al 1994).

It should also be noted that:

• the hard power law component in Cyg X-1 is very stable, it is the US component that can change on timescales of $\sim$ day(s);

• the US component is *anti*-correlated with the radio emission;

BLACK—HOLE—CANDIDATE POWER SPECTRA

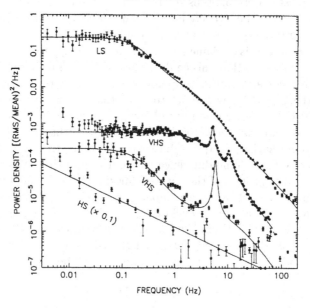

FIGURE 2. Power spectra of BHXRBs in various spectral states as observed by *Ginga*. The low state is of Cyg X-1, whereas the others are of Nova Mus 1991. From van der Klis (1995).

TABLE 2. X-ray Spectral and Temporal Properties of SXTs

Source State	X-ray Spectrum	Temporal Characteristics
Low (LS)	no ultra-soft (US) component	power law PDS, substantial variability
Intermediate (IS)	US + steeper power law at high E	Lorentzian noise
High (HS)	US dominates (MCD model), very weak power law component	very little variability
Very high (VHS)	strong US + PL component	strong QPOs at $\sim$ 10Hz, Lorentzian noise

• Barret et al (1996) proposed that observations are consistent with only the BHXRBs exhibiting both a hard power law component and a high L_X (but note the energy range concern);

• LS BHXRBs and neutron star systems in the atoll state are very similar. Hence the presence of a solid surface and strong magnetic field makes little difference in this situation!

2.5. *Light Curve Shapes*

The X-ray light curves are reviewed by Tanaka & Shibazaki (1996), but are most comprehensively presented by Chen et al (1997). In spite of the similarity of GS2023+338, GS2000+25 and N Mus 1991, they do in fact exhibit a wide variety of outburst behaviour.

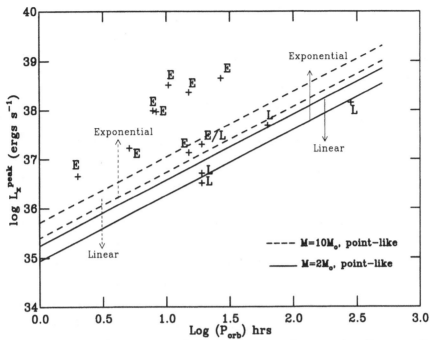

FIGURE 3. The critical luminosity needed to ionise the entire accretion disc, according to the calculations of Shahbaz et al (1998a). Results are shown for total masses of 2 and $10 M_\odot$ corresponding to neutron star and black hole SXTs respectively. These luminosities are a factor 2 smaller for exponential decays compared to linear ones which is due to the difference in the circularisation and tidal radii of the disc. The SXTs shown are SAX J1808.4-3658, GRO J0422+32, A0620-00, GS2000+25, GS1124-68, Cen X-4, Aql X-1, 4U1543-47, GRO J1655-40 and GRO J1744-28. Observed decay shapes are designated E (exponential) and L (linear).

Nevertheless the basic shape is that of a fast rise followed by an exponential decay, the latter being interpreted as a disc instability (see Cannizzo 1998, King & Ritter 1997). However, more recently Shahbaz et al (1998a) have shown that SXTs can exhibit both exponential and linear decays. As pointed out by King & Ritter (1998), the exponential decay is a natural consequence of the X-ray irradiation maintaining the disc in a hot (viscous) state, thereby producing outbursts that last much longer than in their dwarf nova analogues. If the SXT outburst is not bright enough to ionise the outer edge of the disc, then a linear decay will result. This is shown in figure 3 for those systems with known orbital period and whose decay light curve shape can be determined. Furthermore, Shahbaz et al show that the radius of the hot disc at peak of outburst is related to the time at which the secondary maximum is seen in the light curve, opening up the possibility of using the light curve as form of standard candle.

As mentioned earlier Shahbaz & Kuulkers (1998) have shown that the optical outburst amplitude ΔV is related to the system's orbital period (for periods <1 day) according to:

$$\Delta V = 14.36 - 7.63 \log P \qquad (2.4)$$

where P is given in hours. This relation is shown along with the observed points in figure 4, and is essentially due to the brighter secondaries (which must be Roche-lobe filling) in longer period systems. This now provides the valuable function of being able to predict the orbital period providing the peak and quiescent magnitudes are known.

However, a key observational problem concerns SXTs' quiescent properties. Optical

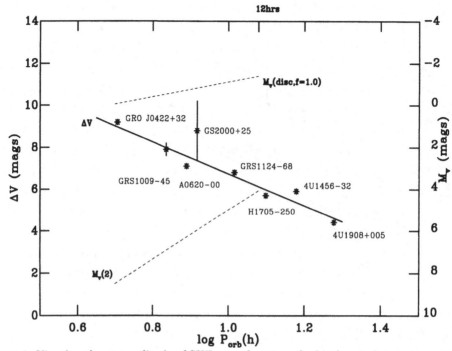

FIGURE 4. Visual outburst amplitude of SXTs as a function of orbital period, together with the least-squares fit (taken from Shahbaz & Kuulkers 1998). The dashed lines show the calculated M_V of the disc in an outbursting SXT and the M_V of the secondary star according to Warner (1995).

spectroscopy shows, in all cases strong Hα emission, doppler tomography of which indicates continuing mass transfer into the accretion disc. However, the implied rate is not consistent with the very low observed quiescent X-ray fluxes. This led Narayan et al (1997b) to propose ADAFs as a means of accounting for this difference between the BH and NS systems. At low mass transfer rates, the temperature and density of gas in the inner disc region would be such as to produce a cooling time for the gas that exceeded the radial infall time. Hence the thermal energy of the gas would be advected into the black hole, and thereby lost to the observer. And while similar flows would occur in NS systems, the thermal energy would eventually be radiated from the NS surface, hence producing an apparently much higher L_X from the same rate of accreting matter.

2.6. *Effect of Black Hole Spin*

In spite of their observational title of SXT, there are two transients (GS2023+338 and GRO J0422+32) that do not show US components in their bright state, nor do Cyg X-1 and GX339-4 when in their hard (low) state. Both optical and X-ray studies show that this cannot be due to a high inclination in any of these. Zhang et al (1997) have calculated the disc emission from both Schwarzschild and Kerr BHs, allowing the specific angular momentum ($a_* = a/r_g$, where $r_g = GM/c^2$) to take values of 0 (non-spinning hole) and ± 1 (for maximal spin) where a negative value implies a retrograde spin. From this they inferred that we would observe a black-body (US) component whose colour temperature was a function of a_*.

Given the great distances of the SXTs (they are rare and spread throughout the galaxy), a key requirement of the US component is that it has $kT_{col} \geq 0.5$–1 keV, in order for the emission not to be obscured by interstellar absorption. The calculations show that the highest kT_{col} (and most luminous US component) occur for $a_* = +1$, and hence that GRO J1655-40 and GRS1915+10 must both have this value. And they are both jet sources!

Interestingly it can already be asserted that GRO J1655-40 *cannot* be a Schwarzschild BH as the dynamically determined $M_X = 7M_\odot$ requires the radius of the last stable orbit to be $2.3r_g$ and yet theoretically the minimum value required is $6r_g$.

3. Mass Measurements

3.1. *Radial Velocity Curves*

It is when they reach quiescence that the SXTs become such valuable resources for research into the nature of LMXBs. Their optical brightness has typically declined by a factor of 100 or more, with all the known SXTs having quiescent magnitudes in the range 17–23. The quiescent light is now dominated by the companion star and, while technically challenging, presents us with the opportunity to determine its spectral type, period and radial velocity curve (whose amplitude is the K-velocity). From the latter two we can calculate the mass function (equation 1.3) and the results for the same 9 systems (this time listed in order of their mass functions) are summarised in table 3, again separating out the two early spectral type systems as well as the single neutron star SXT, Cen X-4. Hence the enormous importance of the SXTs, since all are LMXBs which have $M_X > M_2$. The mass functions in table 3 represent the *absolute minimum* values for M_X since (for all of them) $i < 90^o$ and $M_2 > 0$, both of which serve only to *increase* the implied value of M_X. That is why the work of McClintock & Remillard (1986) on A0620-00 and Casares et al (1992) on V404 Cyg has generated so much interest.

It should also be noted that it can be possible to derive some dynamical information about the system even during outburst, providing spectroscopic data of sufficient resolution is obtained. Casares et al (1995) observed GRO J0422+32 during one of its subsequent "mini-outbursts" and found intense Balmer and HeIIλ4686 emission that was modulated on what was subsequently shown to be the orbital period. Furthermore, a sharp component of HeII displayed an S-wave that was likely associated with the hotspot.

However, to determine the actual value of M_X we need additional constraints that will allow us to infer values for M_2 and i.

3.2. *Rotational Broadening*

Since the secondary is constrained to corotate with the primary in short period interacting binaries, we can exploit our knowledge of its size by making *assumption 1* that R_2 is given by equation (1). Hence the result (Wade and Horne, 1988)

$$v_{rot} \sin i = \frac{2\pi R_2}{P} \sin i = K_2 \times 0.46 \frac{(1+q)^{2/3}}{q} \qquad (3.5)$$

from which q can be derived if v_{rot} is measurable. Typical values are in the range 40–100 km s^{-1} and clearly require high resolution and high signal-to-noise spectra of the secondary.

Figure 5 (second from top) shows the Casares & Charles (1994) WHT summed spectrum of V404 Cyg after doppler correcting all individual spectra into the rest-frame of the secondary. The bottom spectrum is a very high S/N spectrum of a K0IV star which was used as a template, and which clearly has much narrower (actually they are unresolved) absorption lines. The template is broadened by different velocities (together with the effects of limb darkening), subtracted from that of V404 Cyg and the residuals χ^2 tested. This gave $v_{rot} \sin i = 39 \pm 1$ km s^{-1} and hence $q = 16.7\pm1.4$. The full details

TABLE 3. Derived Parameters and Dynamical Mass Measurements of SXTs

Source	$f(M)$ $(M_\odot)$	ρ $(g\ cm^{-3})$	q $(=M_X/M_2)$	i	M_X $(M_\odot)$	M_2 $(M_\odot)$	Ref.
V404 Cyg	6.08±0.06	0.005	17±1	55±4	12±2	0.6	[1–2]
G2000+25	5.01±0.12	1.6	24±10	56±15	10±4	0.5	[3–5]
N Oph 77	4.86±0.13	0.7	>19	60±10	6±2	0.3	[6–9]
N Mus 91	3.01±0.15	1.0	8±2	54^{+20}_{-15}	6^{+5}_{-2}	0.8	[13–15]
A0620-00	2.91±0.08	1.8	15±1	37±5	10±5	0.6	[16–18]
J0422+32	1.21±0.06	4.2	>12	20–40	10±5	0.3	[19–20]
J1655-40	3.24±0.14	0.03	3.6±0.9	67±3	6.9±1	2.1	[10–12]
4U1543-47	0.22±0.02	0.2	-	20–40	5.0±2.5	2.5	[21]
Cen X-4	0.21±0.08	0.5	5±1	43±11	1.3±0.6	0.4	[22–23]

[1] Casares & Charles 1994; [2] Shahbaz et al 1994b; [3] Filippenko et al 1995a; [4] Beekman et al 1996; [5] Harlaftis et al 1996; [6] Filippenko et al 1997; [7] Remillard et al 1996; [8] Martin et al 1995; [9] Harlaftis et al 1997; [10] Orosz & Bailyn 1997; [11, 12] van der Hooft 1997, 1998; [13] Orosz et al 1996; [14] Casares et al 1997; [15] Shahbaz et al 1997; [16] Orosz et al 1994; [17] Marsh et al 1994; [18] Shahbaz et al 1994a; [19] Filippenko et al 1995b; [20] Beekman et al 1997; [21] Orosz et al 1998; [22] McClintock & Remillard, 1990; [23] Shahbaz et al 1993.

can be found in Casares & Charles, and Marsh et al (1994). It should also be noted that while the accretion disc around the compact object might be expected to provide some velocity information, there are serious difficulties with this. The Hα line in figure 5 is extremely broad ($\geq$1000 km s^{-1}) and yet the compact object's motion in such high q systems will be very small (typically $\leq$30 km s^{-1}). Nevertheless such motions have been seen (e.g. Orosz et al 1994), but their interpretation is not straightforward as there is a small phase offset relative to the motion of the companion star, and so they cannot be used as part of the dynamical study.

Having determined q, the range of masses consistent with the observed $f(M)$ is plotted in figure 6, where the only remaining unknown is the orbital inclination i. To date, none of the SXTs is eclipsing (although GRO J1655-40 shows evidence for a grazing eclipse), and so it is the determination of i that leads to the greatest uncertainty in the final mass measurement. Nevertheless there are methods by which i can be estimated.

Note also, that high mass ratios for these systems is also implied by the work of O'Donoghue & Charles (1996) which demonstrates the *superhumps* have been seen in SXT optical light curves during outburst decay. Such features had been seen before during the *superoutbursts* of the SU UMa subclass of dwarf novae and are attributed as arising from tidal stressing of the accretion discs in high mass ratio interacting binaries. Their calculation of system parameters based on this model provides satisfactory confirmation of these values.

3.3. *Ellipsoidal Modulation*

We exploit one more property of the secondary, it's peculiarly distorted shape responsible for the so-called *ellipsoidal modulation* as we view the varying projected area of the secondary around the orbit. This leads to the classical double-humped light curve, as

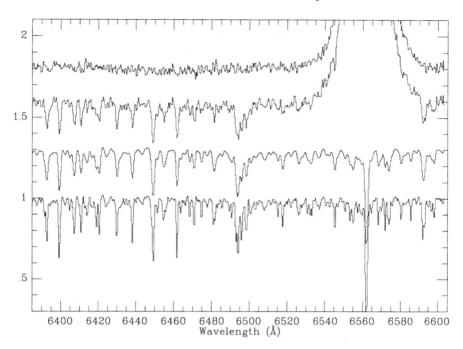

FIGURE 5. Determining the rotational broadening in V404 Cyg. From bottom to top: the K0IV template star (HR8857); the same spectrum broadened by 39 km s^{-1}; doppler corrected sum of V404 Cyg (dominated by intense Hα emission from the disc); residual spectrum after subtraction of the broadened template (from Casares and Charles, 1994).

shown for A0620-00 in figure 7. If the secondary's shape is sufficiently well-determined by theory (i.e. the form of the Roche lobe) then the observed light curve depends on only 2 parameters, q and i. In several cases (as described in the previous section) q is already determined, but in practice the ellipsoidal modulation is largely insensitive to q for values $q > 5$. Details of the light curve modelling can be found in Shahbaz et al (1993), and the collected results are in table 3.

This final stage in the SXT orbital solutions has made 2 key assumptions: *assumption 2*, that the secondary in quiescence fills its Roche lobe; *assumption 3*, that the light curve is not contaminated by any other light sources. It is felt that the former is reasonable since there is strong evidence through doppler tomography (e.g. Marsh et al 1994) for continued mass transfer in quiescence from the secondary. However, the principal (and potentially significant) uncertainty is the problem of any other contaminating light sources. This would mainly be the accretion disk, but residual X-ray heating and starspots on the surface of the secondary might also be present. It is for this reason that this work has been performed in the K band whenever possible. The disc contamination has been measured in the optical around Hα (as a by-product of the spectral type determination by searching for excess continuum light) and is typically $\leq 10\%$. It should therefore be even less in the IR given the blue colour of the disc. However, the outer disc edge has been found to be an IR emitter in CVs (Berriman et al 1985) and the light curves might be contaminated, as was suggested in the case of V404 Cyg by Sanwal et al (1996). This is potentially an important effect, since a contaminating (and presumably steady) contribution will reduce the amplitude of the ellipsoidal modulation, which will lead to a lower value of i being inferred, and hence a higher mass for the compact object.

For this reason Shahbaz et al (1996, 1998b) undertook IR K-band spectroscopy of the

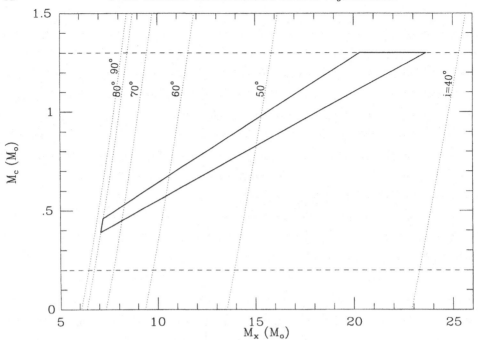

FIGURE 6. Constraints on M_X and M_2 for a range of values of i in V404 Cyg based on the radial velocity curve $(f(M))$ and determination of q (from the rotational broadening). It is the limited constraint on i (absence of eclipses) that leads to a wide range of M_X (from Casares and Charles, 1994).

two brightest and best studied SXTs, V404 Cyg and A0620-00. Only upper limits were derived in both cases, showing that any contamination must be small and hence the masses derived can (at most) be reduced by only small amounts. It is also interesting to note that, in their study of the non-orbital optical variability in V404 Cyg, Pavlenko et al (1996) found that (as first noted by Wagner et al 1992) the ellipsoidal modulation could be discerned underlying the substantial flickering in the light curve. Interpreting the flickering as a completely independent component, Pavlenko et al showed that the *lower envelope* of this light curve (rather than the mean) produced an ellipsoidal light curve which, when fitted as described above, gave essentially identical results to those obtained from the K-band analysis, thereby providing further weight to the significance of the mass determinations.

The results of these analyses are collected together in figure 8, which contains all neutron star and black hole mass measurements. It should also be noted that the value for Cen X-4 (one of only two neutron star SXTs, identified on the basis of its type I X-ray bursts) has been derived exactly by the method outlined here (Shahbaz et al 1993) and yields a value of $1.3M_\odot$, in excellent accord with that expected for a neutron star.

4. Lithium in the Companion Stars

One of the remarkable by-products of our high resolution radial velocity study of V404 Cyg was the discovery of strong LiI $\lambda6707$ absorption in the secondary star (Martín et al 1992). This was, of course, not present in any of the template stars which we were using for the spectral fitting, since Li is a characteristic feature of young, pre-main sequence and T Tau objects. Subsequent convection in late type stars leads to the

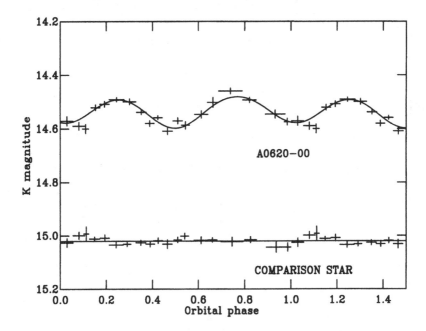

FIGURE 7. IR light curve of A0620-00 showing the classical double-humped modulation (from Shahbaz et al, 1994a).

destruction of lithium, and observed abundances in normal stars (like the Sun) are a thousand times lower. Similar Li enhancements have since been identified in A0620-00, J0422+32, GS2000+25, N Mus 1991 and Cen X-4 (see Martín et al 1996), the latter indicating that this phenomenon is not confined solely to the BHC.

Li is an important element in galactic chemical abundances because galactic material is found to be enriched in Li relative to the halo, and the source of this enrichment is a subject of current research. Since the SXTs are clearly highly evolved objects which are extremely unlikely to have retained such high Li abundances, then we must find mechanisms whihc create Li within the structure of an SXT. Those systems in which Li has been detected cover a wide range of period and secondary size, but they all display high luminosity, recurrent X-ray outbursts. Martín et al (1994) therefore suggested that spallation processes during these outbursts produce Li in large quantities close to the compact object. Subsequent large mass outflows result in the transfer of some of this Li to the secondary, the energetics of which are considered by Martín et al (1994). Support for this suggestion is cited as the interpretation of the 476keV γ-ray line that was observed during the N Mus 1991 outburst (Sunyaev et al 1992; Chen et al 1993). Originally interpreted as the gravitationally redshifted e^--e^+ 511keV line, it might instead be associated with the 478keV line of ^{7}Li. The temporal behaviour of this line (it only lasted for about 12 hours) could give insight into the spallation mechanism. That high luminosities are needed, whatever the mechanism, is indicated by the absence of Li in CVs of comparable spectral type to the SXTs, nor has Li been detected in the nova GK Per (Martín et al 1995).

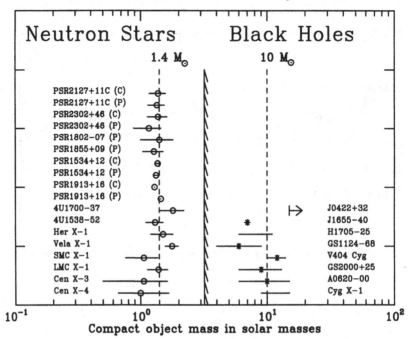

FIGURE 8. The mass distribution of neutron stars and black holes. Note the remarkably narrow spread of neutron star masses, and the large factor by which the BHXRB masses exceed the (canonical) maximum mass of $3.2 M_\odot$.

5. The Superluminal Transients

In 1994 there were two new X-ray transients discovered, GRS1915+105 and GRO J1655-40, that brought an entirely new type of behaviour to this field. As with many of the transient outbursts, they also emitted strongly in the radio, but VLA and VLBI observations showed that these objects also exhibited ejection events that were "superluminal" (Mirabel & Rodríguez 1994; Hjellming & Rupen 1995), the first time that such phenomena had been observed within the Galaxy. Further dynamical studies of GRS1915+105 are severely hampered by (a) its extremely high interstellar extinction ($A_V \sim 26$), leaving only a variable, K$\sim$14 IR counterpart, and (b) its continuing and extremely variable X-ray activity that is totally unlike any of the "classical" SXTs. It is not even clear that GRS1915+105 is an LMXB (see Mirabel et al 1997), and it demonstrates an extraordinarily rich variety of X-ray variability (e.g. Morgan & Remillard 1996, Belloni et al 1997).

GRO J1655-40 (N Sco 1994), on the other hand, is optically the brightest in quiescence of all the SXTs, and so has extremely well-determined photometric light-curves, and is an excellent candidate for a dynamical study. The companion also has one of the earliest (confirmed) spectral types (mid-F) of the SXTs which means that, in quiescence, the effects of the accretion disk are very small, almost negligible. And the high γ-velocity led to the suggestion (Brandt et al 1995) that J1655-40 could be an example of a NS system that had suffered accretion-induced collapse. However, J1655-40's behaviour is not at all typical of other SXTs, with quiescent studies severely hampered by its return to activity in 1996. This return was fortuitously observed by Orosz et al (1997) who found that the optical brightening began $\sim$6 days before the X-ray activity began. They (and Hameury et al 1997; see also Lasota, this volume) interpreted this as an "outside-in" outburst of the accretion disc, with the substantial delay arising due to the ADAF flow (in quiescence) having evaporated the inner disc, and which needed to be re-filled before accretion onto the compact object could take place. Subsequent multi-wavelength (UV/optical/X-ray with HST and RXTE) observations of this period of activity (Hynes et al 1998) demonstrated two interesting properties. They found that the X-ray and optical variations were, at times, correlated, but with the optical variations lagging the X-ray by ~19 secs. With the known size of the binary from its orbital period, this lag is too short to be due to irradiation of the secondary, and hence must be associated with the accretion disc. Furthermore, Hynes et al found that as the outburst progressed, the optical/UV emission declined as the X-rays increased! They suggested that this might arise through the driving of a large corona early in the outburst which can then allow subsequent up-scattering of hard X-rays later on. However, this requires much more extensive and detailed multi-wavelength studies to be undertaken throughout an outburst in order for it to be fully tested.

The orbital system parameters have been derived from several photometric studies of J1655-40 by van der Hooft et al (1997), Orosz & Bailyn (1997) and van der Hooft et al (1998). Figure 9 shows the van der Hooft et al light curve together with the system schematic of Orosz & Bailyn. The values recorded in table 3 are those from the latter paper due to their more conservative error analysis. J1655-40 is unusual in this class in that it has a low mass ratio of $q \sim 3$ (but this has not yet been obtained from a rotational broadening study, due to its return to activity shortly after its initial outburst). At such a value, the ellipsoidal modulation is sensitive to both q and i (the latter also being tightly constrained here as a result of its grazing eclipse). Hence, once it becomes possible to spectroscopically determine the rotational broadening of the F star (when it re-enters an extended period of quiescence) it will then be possible to perform a check of the entire basis on which the quiescent SXT light curves have been modelled and used to determine i. It has also been suggested (Kolb et al 1997) that the secondary star is in a very interesting evolutionary state in which it is crossing the Hertzsprung gap and about to ascend the giant branch. This is what is driving the much higher mass transfer rate than in the other SXTs, but temporary drops in $\dot{M}$ return it to the transient domain.

6. Outburst Mechanisms

Over the last 10 years there has been the same debate over the mechanism for SXT outbursts as had been taking place over the cause of dwarf nova outbursts, with the same two competing models, namely enhanced mass transfer from the secondary star (as a result of X-ray heating) and the thermal (viscous) instability in the accretion disc itself (both are discussed by Lasota 1996). However, as a result of much more sensitive quiescent X-ray observations of SXTs by ROSAT (Verbunt 1996) it is clear that these levels of X-ray emission (as low as 2.5×10^{30} erg s^{-1} for A0620-00) are incapable of heating the secondary star sufficiently to generate the mass transfer necessary to account for the observed outbursts. And whilst there have been models proposed (e.g. Chen et al 1993; Augusteijn et al 1993) that combine elements of both the mass transfer and disc instability explanations, these have not yet been supported by observations.

Strong support for the disc instability model has appeared in papers by van Paradijs (1996) and King et al (1996). They point out that, when calculating whether the disc instability mechanism will occur in an SXT disc (which requires that the temperature somewhere in the disc be below the ionisation temperature of hydrogen, 6500K), it is necessary to take into account the effects of (time-averaged over outbursts) X-ray heating. In this way they derive an expression for the X-ray luminosity (as a function of period)

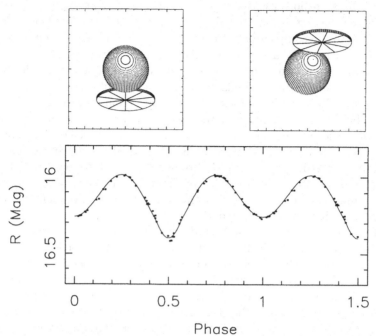

FIGURE 9. R-band light curve of GRO J1655-40 taken by van der Hooft et al (1997) when the source was quiescent. To obtain acceptable fits (shown) it was necessary to add partial obscuration by a cool accretion disc at phase 0.5, as demonstrated in the system schematic (above) taken from Orosz & Bailyn (1997).

that separates the steady and transient sources:

$$\log L_X = 35.8 + 1.07 \log P \tag{6.6}$$

and find that this is very well supported by the observations. Indeed, the continued activity of GRO J1655-40 has brought it very close to this line, indicating that it is in fact "almost" a steady source. The average SXT mass transfer rate from the secondary is found to be:

$$< \dot{M_2} > = -4 \times 10^{-10} P_d^{0.93} M_2^{1.47} M_\odot y^{-1} \tag{6.7}$$

giving values $\sim 10^{-10} M_\odot y^{-1}$ for most SXTs, but about 3× higher for GRO J1655-40.

7. Population Size and Distribution

Important questions for current and future surveys for X-ray transients, such as those likely to be undertaken with AXAF and XMM in nearby galaxies (M31 and M33), concerns the number of quiescent systems, the chance of their being observed if they do outburst, and their distribution through the galaxy. The largest uncertainty in such calculations is the typical recurrence time of an SXT. Following Tanaka & Shibazaki (1996), the current sample's outburst properties (table 1) show that this can range from less than a year to ∼50 years. However, the average is likely to be around 10–20 years, with typically 2 SXTs being observed per year (with regular all-sky coverage by CGRO and RXTE). With ∼90% of SXTs located within a galactic longitude range of ±80°, this implies that they lie within 8kpc of the Galactic Centre, but distance estimates for the current sample suggest they are all <5kpc from us. Hence, we are only detecting ∼10% of the transients that occur within our galaxy (due to a combination of interstellar absorption and sensitivity), and the total number of SXTs is ~200–1000 for assumed recurrence times of 10 and 50 years respectively. Long-term surveys of nearby galaxies will allow these estimates to be more rigorously examined, as well as deriving more accurate luminosities and galactic distribution information.

White & van Paradijs (1996) have examined and compared the distribution in galactic latitude and longitude of the BHC LMXBs and the NS systems. They find that the BHC have a dispersion in z of ~400pc, whereas the NS are ~1kpc, interpreting the difference as an indication that the kick velocities received in BH formation are less than in NS. The corrolary of this is that the BHC LMXBs are not formed by the accretion-induced collapse of a NS (except possibly GRO J1655-40) where the higher NS kick would be inherited by the resultant BH.

8. Nature of the Compact Object

Since the existence of compact objects with masses $\geq 10 M_\odot$ can now be taken as secure, the question arises as to what are they. Or, more importantly, what is the maximum mass of a neutron star? This is usually quoted as $3.2 M_\odot$ on the basis of the Rhoades-Ruffini Theorem (1974). But there are a number of assumptions built into this theorem that, if relaxed, can lead to a very different result (see Miller 1997 and references therein).

In particular, the assumption of causality (which requires the sound speed to be less than c) is only applicable in a non-dispersive medium. Also, the density up to which the equation of state is well-defined may be optimistically high. Both these effects, together with significant rotation of the compact object, can lead to the maximum mass of the neutron star only being constrained to be $< 14 M_\odot$! Nevertheless, it should be recognised that current models of neutron star equations of state are compatible with the original $3.2 M_\odot$ limit.

However, an alternative suggestion that such compact objects may be Q-stars has been made by Bahcall et al (1990). In such objects, the strong force confines neutrons and protons at densities below nuclear density, leading to a very different equation of state (in which the Q stands for conserved Quantity, the baryon number). They can be very compact and hence consistent with our current understanding of the properties of neutron stars. But Miller, Shahbaz & Nolan (1998) have shown that if this model is to be applied to V404 Cyg (i.e. that it is a Q-star of $12 M_\odot$), then it requires that the threshold density for this effect must be ~10× **below** that of nuclear density, which is considered to be too implausible given the results of current experiments. In which case, we conclude that V404 Cyg and related objects must be black holes.

9. Conclusions

The dramatic advances in the field of galactic black-hole studies have come about during this decade for 2 main reasons: (i) the almost continuous monitoring of the X-ray sky that is now provided by all-sky monitors such as those on CGRO and RXTE has provided a steady stream of new X-ray transients for subsequent ground-based observations once they reach quiescence, and (ii) the availability of high performance optical spectrographs with good red sensitivity on 4m and larger telescopes. With these facilities we have obtained more detailed information about the nature of both components of LMXBs than was hitherto possible. In particular, the discovery of the mass function of V404 Cyg has revolutionised attitudes concerning the existence of compact objects that must be heavier than the canonical maximum mass of a neutron star. And while there are useful indicators from X-ray observations as to the possible presence of black holes in X-ray binaries, the ultimate diagnostic has to be the dynamical study that has been described here. The next major advances will come from observing the many quiescent transients that are too faint for current 4m class telescopes and require access to the about to be completed VLT and Gemini telescopes in the southern hemisphere.

10. Future Work

The SXTs are providing a very fertile hunting ground for BH candidates, and with CGRO and RXTE both having all-sky monitors there should be a steady stream of new transients discovered over the next decade. Key questions to be addressed include:

- are there any *low-mass* ($<5M_\odot$) BHs?

i.e. formed in 2 stages via accretion-induced neutron star collapse (see Brown et al. 1996).

- or any very *high-mass* ($>20M_\odot$) BHs?

i.e. from massive stars that succeeded in retaining most of their mass until the time of collapse.

- or are they all formed from He stars at the end of the W-R phase?
- therefore we need at least a dozen *accurate, i.e. $<10\%$* mass determinations, which will require exploiting the new generation of 8–16m class telescopes.
- can advective flows demonstrate the existence of the Event Horizon in the BHXRBs?
- or high resolution X-ray spectroscopy of Fe emission line profiles reveal the distorted shape expected due to General Relativity?

The observations have now securely established compact objects with masses in the 5–$15M_\odot$ range, and if they are *not* BHs then this has profound implications for particle physics (and general relativity). Once thought impossible, SXTs allow perhaps the cleanest study of the BH environment.

I am particularly grateful to Erik Kuulkers and Tariq Shahbaz for their assistance in the preparation of the figures for this manuscript.

REFERENCES

ABRAMOWICZ, M.A., CHEN, X., KATO, S., LASOTA, J.-P. & REGEV, O. 1995, *ApJ*, **438**, L37.

AUGUSTEIJN, T., KUULKERS, E. & SHAHAM, J. 1993, *A&A*, **279**, L13.

BAHCALL, S., LYNN, B.W. & SELIPSKY, S.B. 1990, *ApJ*, **362**, 251.

BEEKMAN, G., SHAHBAZ, T., NAYLOR, T. & CHARLES, P.A. 1996, *MNRAS*, **281**, L1.

BEEKMAN, G., SHAHBAZ, T., NAYLOR, T., CHARLES, P.A., WAGNER, R.M. & MARTINI, P. 1997, *MNRAS*, **290**, 303.

BELLONI, T., ET AL. 1997, *ApJ*, **479**, L145.

BERRIMAN, G., SZKODY, P. & CAPPS, R.W. 1985, *MNRAS*, **217**, 327.

BRANDT, W.N., PODSIADLOWSKI, P.& SIGURDSSON, S. 1995, *MNRAS*, 277, L35.

CANNIZZO, J.K. 1998, *ApJ*, **494**, 366.

CASARES, J., CHARLES, P.A. & NAYLOR, T. 1992, *Nature*, **355**, 614.

CASARES, J. & CHARLES, P.A. 1994, *MNRAS*, **271**, L5.

CASARES, J. ET AL. 1995, *MNRAS*, **274**, 565.

CASARES, J., MARTÍN, E.L., CHARLES, P.A., MOLARO, P. & REBOLO, R. 1997, *New Astron.*, **1**, 299.

CHAKRABARTI, S.K. 1998, *Ind.J.Phys.(Reviews)*, in press.

CHAKRABARTI, S.K. & TITARCHUK, L.G. 1995, *ApJ*, **455**, 623.

CHEN, W., LIVIO, M. & GEHRELS, N. 1993 *ApJ*, **408**, L5.

CHEN, W., SHRADER, C. & LIVIO, M. 1997, *ApJ*, **491**, 312.

CHEN, X., ABRAMOWICZ, M.A., LASOTA, J.-P., NARAYAN, R. & YI, I. 1995, *ApJ*, **443**, L61.

EBISAWA, K. 1997, in *X-ray Imaging and Spectroscopy of Cosmic Hot Plasmas*, eds. F. MAKINO, K. MITSUDA, Tokyo, Universal Academy Press.

EBISAWA, K. ET AL 1994, *PASJ*, **46**, 375.

EGGLETON, P.P. 1983, *ApJ*, **268**, 368.

ELVIS, M., PAGE, C.G., POUNDS, K.A., RICKETTS, M.J. & TURNER, M.J.L. 1975, *Nature*, **257**, 656.

ESIN, A.A., McCLINTOCK, J.E. & NARAYAN, R. 1997, *ApJ*, **489**, 865.

FILIPPENKO, A.V., MATHESON, T. & BARTH, A.J. 1995a, *ApJ*, **455**, L139.

FILIPPENKO, A.V., MATHESON, T. & HO, L.C. 1995b, *ApJ*, **455**, 614.

FILIPPENKO, A.V., MATHESON, T., LEONARD, D.C., BARTH, A.J. & SCHUYLER, D.V. 1997, *PASP*, **109**, 461.

HAMEURY, J.-M., LASOTA, J.-P., McCLINTOCK, J.E. & NARAYAN, R. 1997, *ApJ*, **489**, 234

HARLAFTIS, E.T., HORNE, K. & FILIPPENKO, A.V. 1996, *PASP*, **108**, 762.

HARLAFTIS, E.T., STEEGHS, D., HORNE, K. & FILIPPENKO, A.V. 1997, *AJ*, **114**, 1170.

HARRIES, J.R., McCRACKEN, K.G., FRANCEY, R.J. & FENTON, A.G. 1967, *Nature*, **215**, 38.

HERRERO, A., KUDRITZKI, R.P., GABLER, R., VILCHEZ, J.M. & GABLER, A. 1995, *A&A*, **297**, 556.

HJELLMING, R.M. & RUPEN, M.P. 1995, *Nature*, **375**, 464.

HYNES, R.I. ET AL. 1998, *MNRAS*, (in press).

KAHABKA, P. & VAN DEN HEUVEL, E.P.J. 1997, *Ann.Rev.Astron.Ap.*, **35**, 69.

KING, A.R. 1988, *QJRAS*, **29**, 1.

KING, A.R. & RITTER, H. 1998, *MNRAS*, **293**, L42.

KING, A.R., KOLB, U. & BURDERI, L. 1996, *ApJ*, **464**, L127.

KOLB, U., ET AL. 1997, *ApJ*, **485**, L33.

KUULKERS, E. 1998, *New Astron. Rev.*, in press (astro-ph/9805031).

LASOTA, J.P. 1996, in *Compact Stars in Binaries*, IAU Symp **165**, 43.

MARSH, T.R., ROBINSON, E.L. & WOOD, J.H. 1994, *MNRAS*, **266**, 137.

MARTIN, A.C. ET AL 1995, *MNRAS*, **274**, L46.

MARTÍN, E.L. ET AL 1992, *Nature*, **358**, 129.

MARTÍN, E.L. ET AL 1996, *New Astron.*, **1**, 197.

MARTÍN, E.L., SPRUIT, H.C. & VAN PARADIJS, J. 1994, *A&A*, **291**, L43.

MARTÍN, E.L., CASARES, J., CHARLES, P.A. & REBOLO, R. 1995, *A&A*, **303**, 785.

McCLINTOCK, J.E. 1991, *Ann.NY Acad.Sci.*, **647**, 495.

McCLINTOCK, J.E. 1998, Proc. *8th Annual Astrophysics Conference in Maryland on "Accretion Processes in Astrophysical Systems"*.

McCLINTOCK, J.E. & REMILLARD, R.A. 1986, *ApJ*, **308**, 110.

McCLINTOCK, J.E. & REMILLARD, R.A. 1990, *ApJ*, **350**, 386.

MILLER, J.C., 1998, Proc. *12th Italian Conference on General Relativity and Gravitational Physics* World Scientific (in press).

MILLER, J.C., SHAHBAZ, T. & NOLAN, L.A. 1998, *MNRAS*, **294**, L25.

MIRABEL, I.F. ET AL. 1997, *ApJ*, **477**, L45.

MIRABEL, I.F. & RODRÍGUEZ, L.F. 1994, *Nature*, **371**, 46.

MORGAN, E. & REMILLARD, R. 1996, *ApJ*, **473**, L107.

NARAYAN, R., BARRET, D. & McCLINTOCK, J.E. 1997a, *ApJ*, **482**, 448.

NARAYAN, R., GARCIA, M.R. & MCCLINTOCK, J.E. 1997b, *ApJ*, **478**, L79.

O'DONOGHUE, D. & CHARLES, P.A. 1996, *MNRAS*, **282**, 191.

OROSZ, J.A. ET AL. 1994, *ApJ*, **436**, 848.

OROSZ, J.A., BAILYN, C.D., MCCLINTOCK, J.E. & REMILLARD, R.A. 1996, *ApJ*, **468**, 380.

OROSZ, J.A. & BAILYN, C.D. 1997, *ApJ*, **477**, 876 (and *ApJ*, **482**, 1086).

OROSZ, J.A., JAIN, R.K., BAILYN, C.D., MCCLINTOCK, J.E. & REMILLARD, R.A. 1998, *ApJ*, in press.

OROSZ, J.A., REMILLARD, R.A., BAILYN, C.D. & MCCLINTOCK, J.E. 1997, *ApJ*, .

PACZYNSKI, B. 1971, *Ann.Rev.Astron.Ap.*, **9**, 183.

PAVLENKO, E.P., MARTIN, A.C., CASARES, J., CHARLES, P.A. & KETSARIS, N.A. 1996, *MNRAS*, **281**, 1094.

REMILLARD, R.A., OROSZ, J.A., MCCLINTOCK, J.E. & BAILYN, C.D. 1996, *ApJ*, **459**, 226.

RHOADES, C.E. & RUFFINI, R. 1974, *Phys.Rev.Lett.*, **32**, 324.

SANWAL, D. ET AL. 1996, *ApJ*, **460**, 437.

SHAHBAZ, T., NAYLOR, T. & CHARLES, P.A. 1994a, *MNRAS*, **268**, 756.

SHAHBAZ, T., NAYLOR, T. & CHARLES, P.A. 1997, *MNRAS*, **285**, 607.

SHAHBAZ, T. ET AL. 1994b, *MNRAS*, **271**, L10.

SHAHBAZ, T., BANDYOPADHYAY, R. & CHARLES, P.A. 1998b, *MNRAS*, submitted.

SHAHBAZ, T., CHARLES, P.A. & KING, A.R. 1998a, *MNRAS*, submitted.

SHAHBAZ, T. & KUULKERS, E. 1998, *MNRAS*, **295**, L1.

SHAHBAZ, T., NAYLOR, T. & CHARLES, P.A. 1993, *MNRAS*, **265**, 655.

SHAHBAZ, T., ET AL. 1996, *MNRAS*, **282**, 977.

SUNYAEV, R.A. ET AL 1992, *ApJ*, **389**, L75.

TANAKA, Y. ET AL. 1995, *Nature*, **375**, 659.

TANAKA, Y. & LEWIN, W.H.G. 1995, in *X-ray Binaries*, 126: CUP.

TANAKA, Y. & SHIBAZAKI, N. 1996, *ARAA*, **34**, 607.

THORSETT, S.E. & CHAKRABARTY, D. 1998, *ApJ*, submitted.

THORSETT, S.E., ET AL. 1993, *ApJ*, **405**, L29.

VAN DER HOOFT, F. ET AL. 1997, *MNRAS*, **286**, L43.

VAN DER HOOFT, F. ET AL. 1998, *A&A*, **329**, 538.

VAN DER KLIS, M. 1995, in *X-Ray Binaries*, 252: CUP.

VAN PARADIJS, J. 1996, *ApJ*, **464**, L139.

VAN PARADIJS, J. 1998, in *The Many Faces of Neutron Stars*, eds. R. BUCCHERI, J. VAN PARADIJS, M.A. ALPAR, Kluwer Academic Publishers (astro-ph/9802177).

VAN PARADIJS, J. & MCCLINTOCK, J.E. 1994, *A&A*, **290**, 133.

VAN PARADIJS, J. & MCCLINTOCK, J.E. 1995, in *X-Ray Binaries*, 58: CUP.

VERBUNT, F. 1996, in *Compact Stars in Binaries*, IAU Symp **165**, 333.

VERBUNT, F., BELLONI, T., JOHNSTON, H.M., VAN DER KLIS, M. & LEWIN, W.H.G. 1994, *A&A*, **285**, 903.

WADE, R.A. & HORNE, K. 1988, *ApJ*, **324**, 411.

WAGNER, R.M., KREIDL, T.J., HOWELL, S.B. & STARRFIELD, S.G. 1992, *ApJ*, **401**, 97.

WARNER, B. 1995, in *Cataclysmic Variable Stars*, 117: CUP.

WHITE, N.E. & VAN PARADIJS, J. 1996, *ApJ*, **473**, L25.

ZHANG, S.N., CUI, W. & CHEN, W. 1997, *ApJ*, **482**, L155.

ŻYCKI, P.T., DONE, C. & SMITH, D.A. 1997, *ApJ*, **488**, L113.

Black holes in Active Galactic Nuclei: observations

By GREG M. MADEJSKI

Laboratory for High Energy Astrophysics, NASA/Goddard, Greenbelt, MD 20771

and Dept. of Astronomy, Univ. of Maryland, College Park, MD

This paper summarizes the observations which provide the best evidence for the presence of black holes in active galactic nuclei. This includes: X–ray variability; kinematical studies using optical emission lines as well as the distribution of megamaser spots; and the shape of the Fe Kα X–ray emission line. It also presents the current status of our understanding of jet-dominated active galaxies (blazars), and briefly reviews the currently popular AGN "Unification Schemes" based on orientation effects. Finally, it reviews the observations of the X–ray and γ–ray continuum, which, at least for the radio-quiet objects, is likely to be the primary form of their radiative output, and summarizes the best current models for the radiative processes responsible for the high-energy electromagnetic emission in radio-quiet AGN, as well as in jet-dominated blazars.

---

1. Introduction

Perhaps the most exciting astronomical observation leading to our current understanding of black holes has been the discovery of quasars. These celestial objects, originally found in the early sixties as point-like radio emitters, were identified with apparently stellar sources, possessing somewhat unusual spectra, with prominent emission lines. The identification by Schmidt (1963) of these lines as redshifted systems implied that quasars are distant and extremely luminous, commonly producing 10^{46} erg s^{-1}; this is a hundred times or more in excess of the total luminosity of all the stars in a galaxy. Sensitive imaging of the nebulosities which often surround them implied that quasars are nuclei of galaxies, and thus are higher-luminosity counterparts of the compact nuclei of Seyferts, studied some twenty years before the discovery of quasars as unusual emission line objects: hereafter, we assume that they are respectively the lower and higher luminosity end of the same population. A variety of scenarios were advanced to explain their nature, and this included multiple supernovae or massive spinning stars, but the proposal that quasars are powered by an accretion of surrounding matter onto a black hole, advanced in the mid-60s by Salpeter (1964) and Zeldovich & Novikov (1964), became the paradigm that we are developing and testing today. While this is a viable and very attractive paradigm, only the last few years brought a solid evidence for it, allowing also to measure the mass of the central object. It is important to note here that quasars are much more numerous at a redshift ~ 2 than they are locally, meaning that a substantial fraction of galaxies must have undergone the quasar phase. It is thus likely that many otherwise normal local galaxies harbor supermassive black holes, "dead quasars." In fact, as we discuss below, there is a number of relatively anonymous galaxies that show no signs of nuclear activity, but *do* show evidence for such black holes.

We present the observational evidence for black holes in AGN in Section 2; in Section 3, we discuss the effects of the orientation of the accretion disk surrounding the back hole on the appearance of the nucleus. In Section 4, we discuss the jet-dominated AGN known as blazars. In Section 5, we review the observations of AGN in X–rays and γ–rays, the bands that sample the regions closest to the black hole, and in Section 6, we review the radiation processes proposed to explain the emission in these bands.

2. Lines of Evidence for Presence of Black Holes in Active Galaxies

There are two general lines of argument that are used to "prove" the existence of black holes in AGN. The first attempts to measure the total mass within a volume, and argues that no other form besides a black hole can have these parameters. This is done either via estimation of the volume from variability data (via the light travel time arguments) and mass from the luminosity (via the Eddington limit); alternatively, this can be determined by a measurement of velocity of matter at a specified distance from the central object, essentially using Kepler's laws. The second method, discussed in more detail by Fabian (this volume), relies on the distortion of the emission line shapes caused by strong gravity resulting from the presence of a black hole, and we cover it here only briefly.

2.1. X–ray Variability

The variability of active galaxies generally shows the highest amplitude and the shortest time scales in the X–ray and γ–ray bands, which happen to be clearly separated from the optical/UV bands by the strong absorption of the interstellar medium in our own Galaxy. This rapid variability as well as other lines of argument indicate that the X–ray/γ–ray radiation arises closest to the central source, and in many cases, is the primary source of energy in active galaxies. While the total bolometric luminosity of quasars is often dominated by the optical and UV flux (see, e.g., Laor et al. 1997; for a recent review, see Ulrich, Maraschi, & Urry 1997), the bulk of this flux probably arises in more distant regions from the central source than the X–rays and γ–rays. The optical and UV radiation arising in the innermost regions of the nucleus, on the other hand, is most likely a result of reprocessing of X–ray/γ–ray photons. This – as well as the author's personal interest in X–rays – is the reason why this chapter focuses primarily on the high energy emission from quasars.

In general, the X–ray variability of quasars is aperiodic. While a measurement of periodic variability would give us a clue to the circumnuclear environment and thus the nature of the black hole, besides the ill-fated NGC 6814 (cf. Madejski et al. 1993), no strict periodicity was reported for any AGN. However, there were two reports of quasi-periodic variation of flux of active galaxies: NGC 5548 (Papadakis & Lawrence 1993) and NGC 4051 (Papadakis & Lawrence 1995) inferred from the EXOSAT data, but the quality of the data is only modest, and these still need to be confirmed. More statistically significant is the quasi-periodic variability of of IRAS 18325-5926 by Iwasawa et al. (1998), but this is inferred from only a few (< 10) cycles, and requires confirmation via further monitoring before drawing any detailed conclusions. Nonetheless, we understand relatively little about the details of variability of active galaxies, although the recent light curves are sufficiently good to discriminate if the time series are linear or non-linear; this is discussed later in this chapter.

In any case, this rapid variability implies a compact source size. This is of course the standard causality argument: no stationary source of isotropic radiation can vary faster than the time it takes for light to cross it. In the X–ray band, the power spectrum of variability generally is rather flat at long time scales, and above some characteristic frequency, it shows a power-law behavior, such that the variable power drops with decreasing time scale (see, e.g., McHardy 1989); the Fourier phases of these light curves show no coherence (cf. Krolik, Done, & Madejski 1993). As it was pointed by many authors, for this form of variability, the doubling time scale has no definitive meaning, but for the lack of better data, it suffices for the illustrative purposes: it is certainly valid as an order-of-magnitude relationship between the source radius r and the time scale for doubling of the source flux Δt such that $r < c\,\Delta t$. Again, for quasars, this is particularly true in the X–ray band, where the variability is most rapid: even the early

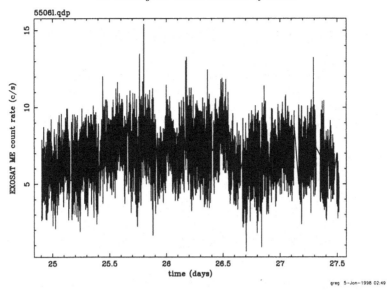

55061.qdp

greg 5-Jan-1998 02:49

FIGURE 1. X–ray light curve for the Seyfert galaxy NGC 5506, collected with the EXOSAT ME satellite (from Krolik, Done, & Madejski 1993).

X–ray data gave us a clue that quasars are very compact. For example, an X–ray light curve for the Seyfert galaxy NGC 5506 (by no means an extreme object) as observed by the ME detector onboard the EXOSAT satellite shown on Fig. 1 illustrates it well. This, and other observations of it, with redshift $z = 0.007$ and a 2 - 10 keV X–ray flux of $\sim 4 \times 10^{-11}$ erg cm^{-2} s^{-1}, imply a luminosity L_X of $\sim 10^{43}$ erg s^{-1}, corresponding to an Eddington mass of at least $\sim 10^5 \, M_o$. This corresponds to a Schwarzschild radius r_S of 3×10^{10} cm. The X–ray data show a doubling time of $\sim 10,000$ s, corresponding to $r < 3 \times 10^{14}$ cm, which, for a $10^5 \, M_o$ black hole, would imply that the X–ray emission arises from a region of radius $r_X \sim 10^4 \, r_S$. This example is by no means extreme – we made a number of assumptions that are probably even too conservative: more realistic assumptions imply a mass of $10^6 \, M_o$, and $r_X \sim 10^3 \, r_S$. A number of more extreme cases – including very luminous objects – were reported recently on the basis of the ROSAT data by Boller, Brandt, & Fink (1996). These are generally for the so-called "narrow-line Seyfert 1s," and we will return to those objects later. In brief, a doubling time scale of ~ 1000 s for a source with $L_X \sim 10^{44}$ erg s^{-1} is not uncommon, implying that the bulk of the X–ray emission arises around 10 - 100 r_S. However, the use of the observed variability time scale does not provide an "airtight" argument for the size of the emitting region, since the observed emission may well be anisotropic, yielding an underestimate of the emitting volume, as is almost certainly the case for blazars. We discuss this in more detail later on.

2.2. Kinematic Studies of Active and "Normal" Galaxies Using Optical Emission Lines

The other line of evidence for the presence of black holes in galaxies (both active and "normal") is the velocity field of the matter emitting closely to the nucleus. This kind of work has been recently reviewed by Kormendy & Richstone (1995), and it dates back to the ground-based observations made in the late 70s, when W. Sargent and collaborators showed that the stellar velocity dispersion in the radio galaxy M 87 increases to 350 km s^{-1} in the innermost 1.5″ from the nucleus. M 87 was in fact observed by the Planetary Camera by Ford et al. (1994) and the Faint Object Spectrograph

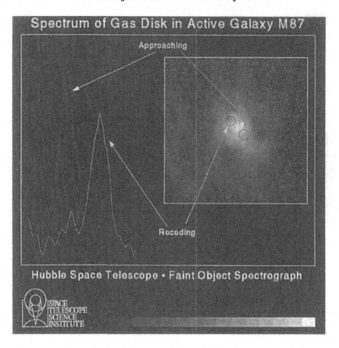

FIGURE 2. Illustration of the spectra of gas in the vicinity of the nucleus of the radio galaxy M 87 (STScI Public Archive).

by Harms et al. (1994) onboard the repaired Hubble Space Telescope, and the images showed the presence of a disk-like structure of ionized gas in the innermost few arc seconds. The spectroscopy provided a measure of the velocity of the gas at an angular distance from the nucleus of 0.25″ (corresponding to ~ 20 pc, or $\sim 6 \times 10^{19}$ cm), showing that in the reference frame of the object, it recedes from us on one side, and approaches us on the other, with a velocity difference of ~ 920 km s^{-1} (see Fig. 2). This implies a mass of the central object of $\sim 3 \times 10^9$ M_o, and besides a black hole, we know of no other form of mass concentration that can "fit" inside this region. As an aside, it is worth noting that M 87 is known to have a relativistic jet perpendicular to the disk structure mentioned above, expanding with the bulk Lorentz factor Γ_j of 4. As such, this object is probably just a blazar, with the jet oriented at an angle $\sim 40°$ to the line of sight, and thus is probably the closest to a hard evidence that blazars (which we discuss in more detail below) indeed *do* harbor black holes.

Besides M 87, similar spectroscopy observations done with Hubble Space Telescope revealed high stellar velocities in the central regions of a number of normal galaxies which otherwise show no evidence for an active nucleus. These were summarized recently by Ford et al. (1998), and include well-publicized observations (by L. Ferrarese, H. Ford, J. Kormendy and others) of NGC 6251, NGC 4261, NGC 4594, NGC 3115, but doubtless by now probably there are several new objects. Such high velocities in the innermost regions of these galaxies cannot be explained in any other way besides invoking the presence of massive $(10^8 - 10^9 \ M_o)$ black holes in their centers. (Interestingly, several of these galaxies also show weak radio jets!) Even for our own Milky Way galaxy, the infrared data and velocity measurements – readily performed from the ground (cf. Eckart & Genzel 1997), at much higher resolution than the HST data for external galaxies – reveal a "modest" nuclear black hole with a mass of $\sim 3 \times 10^6$ M_o. The evidence is building that supermassive black holes are quite common; recent estimates by Ford et al. (1998) as

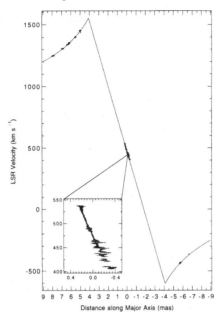

FIGURE 3. The distribution of megamaser spots in the Seyfert 2 galaxy NGC 4258 (from Miyoshi et al. 1995).

well as Ho et al. (1998) imply that they inhabit perhaps as many as half of all galaxies, and may well be the "dead quasars" of the past. Active galactic nuclei are thus probably only the "tip of the iceberg" of their population.

2.3. Megamasers in Active Galaxies

Perhaps the most elegant observation showing the presence of a Keplerian disk around a black hole – and thus capable of measuring the mass of the hole independently of the otherwise uncertain estimates of its distance – was the Very Large Baseline Interferometry observation of megamasers in the vicinity of the nucleus of Seyfert 2 galaxy NGC 4258 reported by Miyoshi et al. (1995). The masing activity can only be observed along a line of sight where the velocity gradient is zero, meaning that these can be seen at locations with masers either between us and their source of energy, or locations at $\sim 90°$ to the line of sight. Indeed, this is the spatial distribution of maser spots around the nucleus of NGC 4258, illustrated in Fig. 3. Specifically, these observations reveal individual masing spots revolving at distances ranging from ~ 0.13 pc to about twice that around the central object – presumably, again, a black hole – with a mass of $\sim 3.6 \times 10^7 \, M_\odot$. What is truly remarkable about these data is the near-perfect Keplerian velocity distribution, in a slightly warped disk-like formation; this implies that almost all the mass is located well within the inner radius where the megamasers reside. It is not possible to have a cluster of distinct, dark, massive objects responsible for such gravitational potential; at least some of the objects would escape on a relatively short time scale, and form a potential well with a different shape, which would now force a departure of the megamaser–emitting material from pure Keplerian motion (cf. Maoz 1995).

Since we know the central mass quite precisely, NGC 4258 has been a terrific laboratory to study the details of the accretion disk. In particular, this is a relatively low luminosity ($\sim 10^{42}$ erg s^{-1}) object, making it quite sub-Eddington, with $L/L_E \sim 3 \times 10^{-4}$. Such sub-Eddington sources are likely to obey unique solutions of accretion disk structure (see, e.g., Ichimaru 1977; Narayan & Yi 1994; Abramowicz et al. 1995), where the accreting gas is optically thin and radiates inefficiently, and the accretion energy that is dissipated viscously, is advected with the accretion flow. With this, as was argued by Lasota et al. (1996), the accretion disk in NGC 4258 can well be advection-dominated (but this does not *have* to be the case; see, e.g., Neufeld & Maloney 1995; Papaloizou et al. this volume). However, it is important to note that such low Eddington rate cannot be universal among quasars; if most of them radiated at such low L/L_E, the black hole masses of the most luminous sources would be much larger than expected on other grounds. However, an intriguing possibility (cf. Yi 1996) is that quasars in their "youth," when the black hole masses were more modest, had standard, "cold" (Shakura - Sunyaev) accretion disks, and this is why they were so luminous in the past. The masses of the black holes grew with time, and even if the mass rate supplied for accretion remained constant, L/L_E actually decreased, and thus the the inner accretion disks switched from "cold" (bright, Shakura - Sunyaev) phase to "hot" (fainter, advection-dominated) phase even if the rate of mass supply *did not* decrease. This is perhaps why some of the yesterday's bright quasars are today's dormant, "dark" black holes, revealed only via the kinematical studies mentioned above.

Nonetheless, NGC 4258 is probably *not* a unique object. We know of a large class of "low activity" active galaxies, known collectively as "Low Ionization Nuclear Emission Region" objects, or LINERs; they generally have low luminosity, coupled with the absence of the luminous inner disk as evidenced by the emission line ratios. However, unlike the *bona fide* Seyferts with low luminosity, which seem to vary relatively rapidly in X–rays, implying they are "scaled down" quasars with relatively low mass black holes – LINERs are known *not* to vary rapidly in any band, suggesting that the low activity is not due to a low black hole mass, but rather due to a low accretion rate (cf. Ptak 1997). As the nuclei of these objects are not very bright, the data are sparse, and thus the details of the radiative processes are poorly known; while workable models exist (see, e.g., Lasota et al. 1996), they still require more work on the details of the transition between the "standard" and advection-dominated regions of the disk.

2.4. Profile of the Fe K Emission Line

Perhaps the most convincing evidence that a strong gravitational field is present in active galactic nuclei comes from the recent measurements of the shape of the Fe Kα fluorescence line, arising in a geometrically thin, but optically thick accretion disk. This is discussed in more detail by Fabian (this volume), so what follows is a brief summary. The inner part of the disk is illuminated by X–rays. Because of the relative cosmic abundances and the fluorescence yields of various elements, the strongest discrete spectral feature predicted from the disk is the 6.4 keV fluorescent Fe Kα line; the strength (equivalent width) of the line of ~ 150 eV, as measured by Pounds et al. (1990), is in fact roughly consistent with predictions of George & Fabian (1991). Since this line arises from matter in motion, its profile is a tracer of the velocity field of the accreting matter.

The Asca observations of the X–ray bright Seyfert galaxy MCG-6-30-15 by Tanaka et al. (1995) indeed showed that the line (see Fig. 4) has a characteristic two-pronged shape expected to arise from matter flowing in a disk-like structure. The matter approaching us is responsible for the blue wing of the line, while that receding produces the red wing, with an additional redshift, since the photons are emitted in a strong gravitational field; the exact shape also depends on the inclination of the disk. A detailed spectral fitting of the line shape indicates that the emitted energy of the line is indeed 6.4 keV, while the bulk of its flux arises at < 10 r_S, implying in turn the presence of nearly neutral material very closely to the black hole. The analysis of a number of Seyfert spectra from the Asca archives by Nandra et al. (1997a) suggests that many Seyferts indeed show the Fe Kα

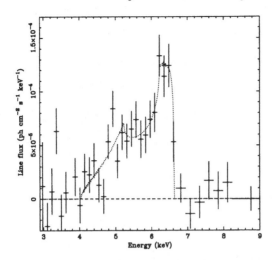

FIGURE 4. The shape of the Fe Kα fluorescence line, observed in the Seyfert galaxy MCG-6-30-15, showing the characteristic two-pronged shape expected when the emitting matter is in a disk-like structure orbiting closely to a black hole (from Tanaka et al. 1995).

line profiles that require an emission close to the black hole, but the quality of data is only modest. Fortunately, a number of more sensitive observatories – such as AXAF, Astro-E, XMM, and Constellation-X, will be launched in the next few years, providing ample opportunities for X–ray observations of effects of strong gravity.

3. Unifying Seyfert 1s and Seyfert 2s: the Orientation Effects

The megamaser source NGC 4258 is only the first of three active galaxies showing a spatial distribution of masing spots from which it is possible to measure the mass of the central object. The other two are the well-known NGC 1068 (Greenhill 1998), and NGC 4945 (Greenhill, Moran, & Herrnstein 1997). Both are also classified as Seyfert 2s, which from the observational side means that they show narrow emission lines, implying velocities on the order of 1,000 km s^{-1}; these lines show no variability, and thus it is generally accepted that they originate in a relatively large regions, on the order of 100 pc or more. Seyfert 1s as well as luminous quasars, on the other hand, exhibit generally very different spectra, with permitted emission lines, and these lines are usually broad, with velocities upwards of 1,500 km s^{-1}, often reaching 30,000 km s^{-1}. This, together with the variability of the lines that is commonly observed on time scales of weeks or months, implies that the broad line region is located much closer to the nucleus than the narrow line region.

The likely relationship between the two classes of active galaxies was revealed by the seminal observation of the well-known Seyfert 2 galaxy NGC 1068 by Antonucci & Miller (1985). The spectropolarimetric study of the Hβ line revealed that when observed in polarized light, the line is broad. They interpreted the polarization as due to electron scattering by material that is distributed preferentially along the symmetry axis of the system, and advanced the widely accepted scenario explaining the differences between the two types of Seyferts as an orientation effect. This is illustrated in Fig. 5; all Seyfert galaxies are surrounded by a geometrically and optically thick torus, with an inner radius of a fraction of a parsec. Such a torus can be, for instance, the outer regions of a severely warped disk, as recently suggested by Maloney et al. (1996). When an object is viewed along the axis of the torus, it is a Seyfert 1, revealing all the ingredients of the nucleus:

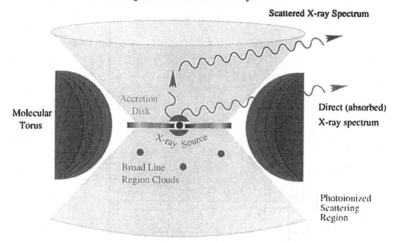

Scattered X-ray Spectrum

Molecular
Torus

Accretion
Disk

Direct (absorbed)
X-ray spectrum

X-ray Source

Broad Line
Region Clouds

Photoionized
Scattering
Region

Schematic model of the nucleus of an Active Galaxy

FIGURE 5. Schematic illustration of the "Unified Picture" of Active Galactic Nuclei: when viewed along or near the axis of the torus/disk, it is a Seyfert 1, while viewed along the plane, it is a Seyfert 2 (courtesy of Dr. C. Done, Univ. of Durham).

the broad line region and the unobscured X-ray source, both commonly varying on a short time scale. When viewed in or close to the plane of the torus, the broad line region is completely obscured, and the soft end of the X-ray spectrum is absorbed due to the photoelectric absorption by the material in the torus. The opening of the torus contains a "mirror" of ionized gas with free electrons, and these are responsible for the scattering of the broad line light into the line of sight, and hence the broad lines are seen only in polarized light.

In the context of this "unification" model, it is thus not surprising that all three megamasers with measured rotation curves are Seyfert 2 objects: to even observe a megamaser, we have to be located in the plane of material which has large column density and presumably is roughly co-axial with the torus. Likewise, the X-rays traveling closely to the plane of the torus encounter a larger column density, resulting in the greater photoelectric absorption. This is in fact, observed; even early X-ray spectra of Seyferts showed preferentially larger absorption in sources showing only narrow lines (Seyfert 2s), with the broad line region obscured by the same material that absorbs soft X-rays. Note that this is different than the case of only narrow emission lines present in some low-luminosity radio galaxies, such as M 87 (collectively known as the FR-I objects): there, the broad lines cannot be obscured, as we see no evidence of X-ray absorption. Furthermore, while Seyfert 2s sometimes do show highly ionized, narrow permitted lines (in addition to the commonly detected forbidden lines), FR-I objects *do not*, implying an absence of the very strong isotropic ionizing UV continuum. We will return to this point below.

4. Anisotropy and Doppler Effects: the Case of Blazars

The use of the observed variability time scale does not, however, provide an "airtight" argument for the size of the emitting region, since the observed emission may well be anisotropic, yielding an underestimate of the emitting volume. As pointed out by Rees (1967), if the radiating matter moves at a relativistic speed towards the observer, the variability time scale appears shortened by the Doppler effect. Furthermore, the

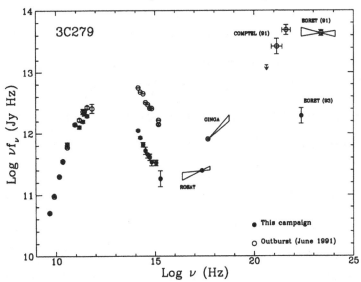

FIGURE 6. Broad-band spectrum of the GeV-emitting blazar 3C279, illustrating the dominance of the GeV γ–ray emission over other bands (from Maraschi et al. 1994).

flux of radiation is boosted in the direction of motion of the emitting matter, and thus the total luminosity inferred under an assumption of isotropy is an overestimate. We are nearly certain that this is the case for blazars. This sub-class of quasars consists of sources that generally show strong radio emission and are highly variable and polarized in all observable wavelengths, and also includes members with very weak or absent emission lines, known as BL Lacertae – type objects. High angular resolution radio observations, performed since the late 60s – using Very Long Baseline Interferometry – implied compact radio emission regions, often associated with a rapid change of their structure. With the assumption that the redshifts are indeed cosmological, these structures appeared to expand at transverse speeds exceeding c, and in many cases, they had shapes of jets. If the speed of the emitting matter is very relativistic, pointing closely to our line of sight, the apparent superluminal motion is just a projection effect, providing further support for the Doppler-boosting scenario for blazars.

In a number of aspects, blazars are the most extremely active galactic nuclei. Observations of many blazars by the EGRET instrument onboard the Compton Gamma-ray Observatory (CGRO) indicated that usually they are strong emitters in the GeV γ–ray band; we know of at least 50 such objects. The GeV emission can dominate the overall observed electromagnetic output of these sources, and this is illustrated in Fig. 6 for 3C 279, the first of the ERGET-discovered GeV blazars. In a few cases, the γ–ray emission has been observed with Cerenkov radiation telescopes to extend as far as the TeV range. This is particularly exciting, as TeV radiation can be observed from the ground, allowing a study of quite exotic phenomena occurring in quasars without the expense of space-borne platforms. Interestingly, we know of no "radio-quiet" MeV or GeV γ–ray emitting quasars, implying an association of the γ–rays with the compact radio source and thus a jet.

In many cases, the γ–ray emission from blazars is variable on a time scale of a day, and recent, simultaneous multi-wavelength monitoring observations indicate that the flux swings are reasonably well correlated between various bands, with the amplitude of variability increasing with the increase of the energy of the observing band. If the emission in the lower energy bands arises from a region co-spatial with that where the γ–rays originate – and the tracking of light curves in various bands supports this – then the opacity to the γ-γ pair production would imply large optical thickness. Similarly, even the early calculations indicated that if the radio sources are as luminous and as compact as the variability data imply, the cooling should be predominantly via the Compton process, with the radiating electrons losing all their energy via upscattering of the just–produced–photons, regardless of the radiation mechanism. Even the large γ–ray fluxes detected by EGRET are often orders of magnitude lower than the predictions of this self-Compton process. Both these difficulties, however, go away if the radio as well as the GeV emission are both Doppler-boosted, meaning that the true luminosity is lower, and source sizes are greater than inferred under an assumption of isotropy. In fact, we now believe that the entire broad-band continuum is produced in the relativistic jet (cf. Königl 1981; Ulrich et al. 1997), and statistical considerations imply that the bulk Lorentz factors Γ_j of these jets are on the order of 10 (cf. Vermeulen & Cohen 1994).

For the case of BL Lac – type blazars, the Doppler enhancement (going roughly as a square of Γ_j; see, e.g., Appendix A of Sikora et al. 1997)) is so strong that the continuum completely outshines the emission lines; in some cases, however, it is possible that the emission lines are absent to begin with. In any case, the mechanism for an acceleration and collimation of these powerful jets to such high relativistic speeds is not known, but general considerations of jet energetics are strongly suggestive that this energy is ultimately tapped from an accretion onto a compact object: the best picture today involves the so-called Blandford-Znajek mechanism (Blandford & Znajek 1977), where the power of the jet derives from the spin of the black hole.

4.1. Emission Processes in Blazars

The electromagnetic emission from jets in blazars – recently reviewed by Sikora (1997) and Ulrich et al. (1997) – is interesting in its own right, and the comparison of the broad-band spectra of blazars and the "radio-quiet" quasars shows it clearly. The strong GeV component which clearly dominates the spectra of the former (cf. Fig. 6) is entirely absent in the latter, and the relative strength of the GeV emission as compared to that in keV range is often different by a factor of ~ 100. The radio and optical emission shows polarization, and spectra are always non-thermal. All these facts argue that the radio-through-UV emission is produced by the synchrotron process, with ultrarelativistic electrons (with $\gamma_{el} \sim 1000$ or more) radiating in magnetic field on the order of a Gauss. The Compton upscattering of lower energy photons mentioned above is indeed present - and most current models suggest that it is responsible for the radiation observed above a few keV. In the lineless BL Lac type objects, we have no evidence for any strong external radiation field, and the dominant "seed" photons for Compton upscattering are likely to be produced by the synchrotron process internally to the jet. As was shown by Dermer, Schlikheiser, & Mastichiadis (1992) as well as by Sikora, Begelman, & Rees (1994) and Blandford & Levinson (1995), in blazars with emission lines, the external, diffuse radiation dominates; since this radiation is isotropic in the stationary frame, it appears Doppler-boosted in the frame of the jet.

This scenario predicts that there should be many objects with jets pointing farther away, up to the right angle to our line of sight. We now believe that for the quasar-type blazars, with strong emission lines, these are the giant, powerful radio galaxies. The difference between the quasar-type and BL Lac type blazars may well be related to the presence of the inner, extremely luminous accretion disk in the former. As suggested by M. Begelman, and, more recently Reynolds et al. (1996), some radio galaxies (as, for instance, M 87) have massive black holes, yet relatively modest luminosity, and essentially no broad emission lines; this might argue for a very sub-Eddington accretion rate, with a likelihood of an advection-dominated inner accretion disk. An intriguing possibility is that these low-luminosity radio galaxies are the "misdirected" BL-Lac - type blazars: the absence of the inner, luminous disk would then be responsible for the absence of the strong ionizing radiation and thus broad emission lines. This in turn would mean that in BL Lac - type blazars, there are no external photons to act as "seeds" to be Comptonized by the relativistic electrons in the jet, and the only "seed" photons are internal, produced by the synchrotron process.

4.2. Origin of the Difference Between "Radio-Quiet" Active Galaxies and Blazars

We are nearly certain that there is a clear *intrinsic* distinction between the jet-dominated blazars, and the more common, radio-quiet quasars. However, the reason for the different behavior of these two subclasses is far from certain, and this is primarily because we do not have a good understanding of the formation and acceleration of relativistic jets. Even though we cannot use the causality arguments to infer the presence of black holes in blazars solely from their variability – since we do not know for sure as to what extent the true variability time scales are shortened by Doppler boosting – more indirect evidence implies that ultimately, accretion onto a black hole powers blazars as well. As to the difference between the two categories, perhaps the most appealing scenario, advanced by Wilson & Colbert (1995), as well as by Moderski, Sikora, & Lasota (1998), is where the two classes differ by strong or weak spin of the black hole; again, the jet is formed and powered by tapping the rotational energy of the hole. This is somewhat similar to the distinction between the Galactic binary "microquasars" suggested by Zhang, Cui, & Chen (1997), and discussed by Charles (this volume). Nonetheless, this is an area of very active research where no clear conclusions have been reached. In particular, if the shape of the fluorescence Fe K line seen in the Seyfert galaxy MCG-6-30-15 (see above) and possibly in other Seyferts (Nandra et al. 1997a) indeed requires that the cold material powering the nucleus flows via a disk-like structure, this implies that the fluorescence occurs at at a distance $r < 3\ r_S$. Since the last stable orbit in a non-rotating (Schwarzschild) black hole is at $3\ r_S$ and beyond this, the matter is in a free fall, this would imply that the black hole is spinning (Kerr), where the last stable orbit can be substantially closer to the black hole, depending on its spin. However, we note that an alternative scenario, advanced by Reynolds & Begelman (1998) does *not* require a disk at $r < 3\ r_S$; nonetheless, this still requires X–ray emission at least at or just beyond $3\ r_S$, merely eliminating the requirement of *spinning* black hole, but it still implies that at least a non-rotating black hole is present.

5. Zooming in on a Supermassive Black Hole: High Energy Spectra of Radio-Quiet Active Galaxies

As we mentioned above, active galaxies are generally variable, and the most rapid variability is observed in the X–ray and γ–ray bands; since we are interested in understanding the regions closest to the black hole, these bands deserve the most detailed study. A recent, excellent article by Mushotzky, Done, & Pounds (1993) reviewed the X–ray spectra of AGN; however, this covered the status of observations before the results from CGRO became available, while we do include these results here. The most conclusive results are gleaned by a study of the non-blazar active galaxies, to avoid any potential contamination by the jet. Since no radio-quiet (= jet-less) active galaxy was detected above several hundred keV, the most relevant spectral region is the X–ray and soft γ–ray bands. This covers nearly four decades in energy, from ~ 0.2 keV to ~ 1 MeV, and often requires observations with multiple satellites, which must be made simultaneously, since active galaxies are variable in all bands.

The early X–ray observations of active galaxies by Mushotzky (1980), Halpern (1982), and Rothschild et al. (1983) implied that, to the first order, X–ray spectra of active galaxies are power laws with the energy index α (defined such that the flux density $S_\nu \propto \nu^{-\alpha}$) of about 0.7, modified by photoelectric absorption at the low energy end. A major advance came from sensitive observations with the ROSAT and Ginga satellites (covering respectively the bands of 0.1 to 2, and 2 to ~ 30 keV), revealing that the spectra are more complex, with a somewhat softer underlying continuum, with $\alpha \sim 0.9$. This continuum is modified by photoelectric absorption in the host galaxy of the quasar; the absorbing material can be either cold, or partially ionized, and this manifests itself as isolated edges, most notably of oxygen, seen in the ROSAT (and more recently, also in Asca) data. The Ginga spectra, reported by Pounds et al. (1990), showed a strong emission line at the rest energy of ~ 6.4 keV, presumably due to fluorescence of the K shell of iron, and a hardening above ~ 8 keV. These last two features were interpreted as signatures of reprocessing ("reflection") of the primary continuum in the cold material that is accreting onto the nucleus. The Fe K line is indeed the strongest expected, via the combination of the relatively high abundance of iron (decreasing with the atomic number Z), and fluorescence yield (increasing with Z); as pointed out by Makishima (1986), the line equivalent width as measured by Ginga is too high to be produced in absorbing material of any column.

Reprocessing by the matter that accretes onto the black hole is thus the most viable alternative. The hardening, predicted earlier in a seminal paper by Lightman & White (1988), arises as an additional spectral component, due to Compton reflection from cold matter. This is produced roughly at one Thomson depth, or when the equivalent hydrogen column density is $\sim 1.6 \times 10^{24}$ cm^{-2}. The cosmic abundances of elements are such that at low energies, the photoelectric absorption dominates, and unless the accreting matter is substantially ionized, this component emerges only above the last significant absorption edge, again from iron. The strengths of the reflection component and the line are in fact in agreement with theoretical predictions by George & Fabian (1991). With the intensity of the incident X–rays greatest closely to the central source (as inferred from the variability of the continuum), its kinematic and gravitational Doppler shifts are powerful diagnostics of the immediate circumnuclear region.

5.1. A Working Template: Spectrum of Seyfert 1 IC 4329a

Perhaps the brightest *bona fide* Seyfert 1 on the sky is the luminous object IC 4329a, and it has been observed simultaneously with the ROSAT (0.2 - 2 keV) and CGRO OSSE (50 - 1000 keV) (Fabian et al. 1993; Madejski et al. 1995). This left a gap between ~ 2 and ~ 50 keV, so the observations were supplemented by non-simultaneous data obtained by Ginga a few years earlier; the Ginga data were renormalized to match the ROSAT flux at 2 keV, so this combined data set provided a good representation of a broad band high energy spectrum of a Seyfert 1, and is illustrated in Fig. 7. The data indeed showed a modest photoelectric absorption due to neutral material (most likely the ISM of the host galaxy), plus an additional modest column of ionized absorber. Beyond ~ 2 keV, the spectrum is the primary continuum, with $\alpha \sim 0.9$. At ~ 6.4 keV, there is a strong Fe K line, which in the case of IC 4329a and many other Seyfert galaxies is broad, with σ of at least 200 eV. The detailed studies of this line in another object, MCG-6-30-15 Tanaka et al. (1995), show the characteristic two-pronged shape, implies relativistic motion in an accretion disk inclined to the line of sight as discussed above.

The region beyond the Fe K line shows general hardening of the spectrum, accompanied by the pseudo-edge due to Fe K; this is a signature of Compton reflection, as discussed above. The intensity of this component is consistent with the reflector being a semi-

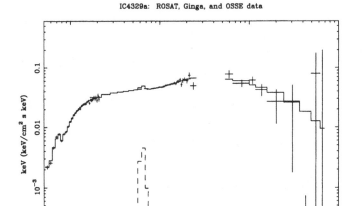

FIGURE 7. Broad-band high energy spectrum of bright Seyfert 1 galaxy IC 4329a. The ROSAT (0.1 - 2 keV) and OSSE (50 - 500 keV) data are simultaneous; the non-simultaneous Ginga (2 - 20 keV) data were scaled to match the ROSAT flux at 2 keV. The solid line is the best fit, including the underlying power law with an index $\alpha = 0.9$, absorbed at the low energies by a combination of neutral and ionized material and exponentially cut off at $E_c \sim 200$ keV, plus the two ingredients of reprocessing measured in the X–ray band: Compton reflection component, plus a Fe K line (from Madejski et al. 1995).

infinite plane; it peaks at ~ 30 keV, and beyond this, it steepens gradually, most likely due to the Compton recoil as well as the Klein-Nishina effects. Of course the underlying power law also can steepen there – we only observe a sum of the primary and reflected spectrum. The detailed fits (spectral decomposition) of the data imply that the primary spectrum is consistent with the power law having an exponential cutoff at an e-folding energy of ~ 200 keV. The recent simultaneous RXTE and OSSE observations of this object generally confirm this picture. A very important constraint on any theoretical models for radiation processes in active galaxies is the *absence* of strong annihilation line at 511 keV.

Is this high energy spectrum of IC 4329a unique? Given the variability of active galaxies, simultaneous observations in X–rays and soft γ–rays are required, but these are sparse; an alternative approach by Zdziarski et al. (1995) is to co-add many non-simultaneous observations. This in fact produced an average spectrum that is remarkably similar to the above picture, and thus any theoretical interpretation of the data for IC 4329a is probably valid for Seyfert nuclei in general.

5.2. The "Ultra-soft" Seyfert 1s: the Question of the "Soft Excess"

It is important to note that luminous objects similar to IC 4329a are a majority of Seyfert 1s, but there is one important sub-class of Seyferts that shows decidedly distinct X–ray spectra, differing from the above description in soft X–rays, below ~ 2 keV. These are the so-called "ultra-soft" Seyferts. First observed in the HEAO data by S. Pravdo and collaborators, they were suggested to be a possibly distinct class of active galaxies by F. Cordova on the basis of the Einstein Observatory Imaging Proportional Counter data. The EXOSAT data analyzed by Arnaud et al. (1985) as well as by Turner & Pounds (1989) showed that in a number of active galaxies, the extrapolation of the hard power law towards low energies (taking properly into consideration the absorp-

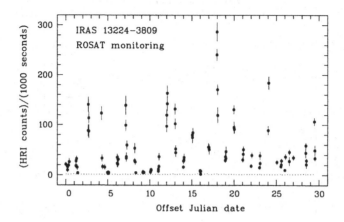

FIGURE 8. X–ray light curve from the Narrow-Line Seyfert 1 IRAS 13224-3809, collected with the ROSAT HRI, showing non-linear variability behavior (from Boller et al. 1997). Another non-blazar object where non-linearity is detected is 3C 390.3 (cf. Leighly & O'Brien 1997)

tion from our own Galaxy) *underpredicted* the soft X–ray flux, indicating that there is an additional component of X–ray emission. This, the so-called "soft excess," meant that in a given object, the ultra-soft component can co-exist with the hard power law, implying that active galaxies generally have two-component power law spectra, where either or both components are visible. It is important to note that this component appears in both low- and high-luminosity sources; for instance, it was observed in the Einstein Observatory data in the quasar PG 1211+143 by Elvis, Wilkes, & Tananbaum (1985). For the sources where both components are present, they intersect at $\sim 1 - 2$ keV (but this may be an observational artifact); the spectrum of the soft component is very soft (steep), and can be described as a power law with an index $\alpha \sim 2$ or even steeper (but a power law is usually not a unique model), while the hard component has just the "canonical" hard Seyfert 1 spectrum with $\alpha \sim 1$.

In the very few cases that the variability of both components has been measured, it appears that the two vary independently, showing no correlation between them (see, e.g., the case of Mkn 335 in Turner 1988). On the other hand, such a lack of correlated variability may be the result of a complex spectral deconvolution procedure, since only a tail of the soft excess component is observable due to intervening absorption in the ISM of the host, or our own Galaxy. In fact, a correlation of UV and EUV fluxes was clearly observed in the "soft excess" Seyfert NGC 5548 (Marshall et al. 1997). This implies some coupling of the soft excess component to the UV/X–ray reprocessing cycle discussed in more detail in Section 6 below. Similar correlation has been also detected in that source between the UV flux and soft X–ray residuals observed simultaneously in Ginga spectra (cf. Magdziarz et al. 1998). Since the EUV observations suggest much higher variability amplitude than the UV data, the lack of apparent correlation in fainter sources may be related to either variation of the cut off energy of the tail of the soft excess, or confusion with spectral index variations in the hard component.

"Ultra-soft" Seyferts appear to be quite common in the ROSAT all-sky survey, which is not surprising, since ROSAT is a very sensitive instrument below 2 keV. Detailed follow-up studies by Boller, Brandt, & Fink (1996) showed a very interesting result: the optical/UV emission lines in these objects are generally quite narrow, with the widths on the order of 1,000 - 3,000 km s^{-1}, as compared to $> 5,000$ km s^{-1} for "normal" Seyfert 1s. These are *not* similar to Seyfert 2s at all; these are widths of permitted lines, while Seyfert 2s only rarely show permitted lines. Important clues to the nature of these soft components may be in the variability patterns, although only very few well-sampled light curves exist. The EXOSAT data were consistent with the hard X–ray variability that is aperiodic but linear, meaning that the time series can be described as uncorrelated noise (cf. Czerny & Lehto 1997); an example of a light curve for this is shown in Fig. 1. The variability of soft X–rays, on the other hand, is often "episodic," with large flares (up to a factor of 100!) (see Fig. 8 for an example), and decidedly non-linear (cf. Boller et al. 1997). (In this case a "non-linear" behavior manifests itself qualitatively in an episodic, flare-like behavior such as that illustrated in Fig. 8; see, e.g., Vio et al. 1992. Quantitatively, a "non-linear" time series is a positive, definite one, which has the ratio of its standard deviation to its mean which is larger than unity; see, e.g., Green 1993.) This difference may imply different emission mechanisms in the hard and soft X–ray components; however two observational effects have to be considered before drawing any conclusions. First, the observational noise or/and presence of additional higher frequency variability in the hard component may effectively dissolve apparent signatures of non-linearity (cf. Leighly & O'Brien 1997). Second, if the variability related to the energy reprocessing does indeed originate from variations of the soft excess component (e.g., Magdziarz et al. 1998), then the signatures of non-linearity should be suppressed in the UV and the hard continuum.

Early modeling attempted to describe the "soft excess" as the tail end of the thermal, multi-blackbody emission from an accretion disk; however, this seems *not* to be the case, at least for the bright and well-studied NGC 5548. Magdziarz et al. (1998) have shown that the UV component in that object may be associated with rather cold disk continuum, with a temperature on the order of a few eV, while the soft excess requires a separate spectral component. We will return to this below.

5.3. High Energy Spectra of Seyfert 2s vs. Seyfert 1s

As it was mentioned above, the popular "unification" picture explains the differences between the spectra of Seyfert 1s vs. Seyfert 2s as due to the orientation effects. To the first order, the only difference that should be seen in the X–ray spectra of the two classes is the amount of photoelectric absorption, while the underlying continuum should be the same. Just as in the case of Seyfert 1s, this requires simultaneous observations by multiple satellites, only the problem is more acute here, as the large amount of absorption leaves generally fewer soft X–ray photons to allow for a sensitive measurement of the continuum. Nonetheless, the observations with the Ginga satellite by Awaki et al. (1991) revealed that X–ray spectra of Seyfert 2s are in fact equivalent to spectra of Seyfert 1s, absorbed by various column densities of cold gas. More detailed studies by Smith & Done (1996) implied that the continua of Seyfert 2s may be somewhat harder, but only marginally so; a more conclusive results should be obtained from observations by the Rossi X–ray Timing Explorer, which features a broader bandpass, extending to 50 keV (or, for brighter sources, even to 100 keV).

In any case, the inferred column densities in Seyfert 2s are $\sim 10^{22}$ cm^{-2} or more. In a few cases – as, for instance, the well-studied NGC 1068 – we only know that no primary X–ray continuum is seen, so the absorber must be quite Thomson-thick, with the absorbing column greater than $\sim 10^{25}$ cm^{-2}. A good, illustrative example of a nearly-extreme Seyfert 2 – but, with the primary continuum still barely penetrating the absorber – is NGC 4945, a Seyfert 2 which also shows megamaser emission, and thus is most likely observed in the plane of the putative torus. In the Asca and Ginga ranges (below ~ 10 keV), the source is relatively faint, but above ~ 10 keV, the spectrum rises sharply (Iwasawa et al. 1993). OSSE observations of it by Done, Madejski, & Smith (1996) re-

NGC 4945: Asca, Ginga, and OSSE data

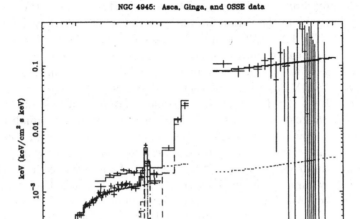

FIGURE 9. Broad-band high energy spectrum of hard X–ray - bright but heavily absorbed Seyfert 2 galaxy NGC 4945. The Asca (0.5 - 10 keV), Ginga (2 - 20 keV) and OSSE (50 - 500 keV) data are not simultaneous, and thus the overall spectrum may be inaccurate as a result of possible source variability. Nonetheless, the heavily absorbed primary continuum above ~ 10 keV is clearly discernible from the scattered / diffuse component below ~ 10 keV (from Done, Madejski, & Smith 1996).

vealed that at 50 - 100 keV, this is the second brightest radio-quiet active galaxy in the sky (see Fig. 9). The absorbing column is large, $\sim 4 \times 10^{24}$ cm^{-2}, and while the observations were not simultaneous, a comparison of Fig. 7 and Fig. 9 reveals that the underlying continuum, to the first order, is consistent with that of a Seyfert 1. Another interesting aspect of this source is the fact that the megamaser distribution implies a mass of the black hole of $\sim 10^6$ M$_\odot$ (Greenhill et al. 1997). With the bolometric luminosity of the nucleus of at least $\sim 10^{42}$ erg s^{-1} (Iwasawa et al. 1993; Done et al. 1996), this source radiates at a few percent of Eddington luminosity, and thus is unlikely to be advection-dominated, as may be the case for NGC 4258.

5.4. High Energy Spectra of High Luminosity Sources

So far, we discussed primarily the relatively low-luminosity quasars, and an obvious question to be asked is: does this general picture hold for the higher luminosity counterparts? In general, the answer is yes, but with some modifications. It is important here to compare "apples to apples," and in the case of quasars, this means selecting radio-quiet objects, as radio-loud quasars tend to have higher X–ray – to – optical flux ratios, which may be due to a contamination by a possible jet; unfortunately, this paucity of X–ray photons in the radio-quiet objects makes spectral studies somewhat more difficult. In general, the more luminous objects show a lower ratio of X–ray to optical luminosities (see, e.g., Kriss & Canizares 1985; Avni & Tananbaum 1986). The recent work by Laor et al. (1997) (using the ROSAT PSPC data) and Nandra et al. (1997b) (using Asca data) indicates that the continuum X–ray spectra of higher luminosity objects appear generally similar to those of the lower luminosity counterparts. Notable exception is an absence of the Fe K line and the Compton reflection component in quasars (see, e.g., Nandra et al. 1997b). With the more luminous central source, this may be the result of a nearly-complete ionization of the accreting material, such that the reflection component is present, but cannot be distinguished by its tell-tale Fe K line and spectral hardening above $\sim$ 8 keV: for an entirely ionized reflector, the incident or emerging photons encounter no photoelectric absorption, and the reflection is (below $\sim$ 30 keV) identical to the incident spectrum, with no Fe K line present. Unfortunately, those quasars are generally too faint to be studied in detail above $\sim$ 10 keV, where the only instruments currently available are non-imaging proportional counters such as the RXTE, dominated by uncertainties of the instrumental as well as the Cosmic X–ray Background. As a result, any detailed studies must await X–ray reflective optics sensitive beyond 10 keV, already under development; such telescopes are essential for studies of these faint objects, as they permit subtraction of background from the same image as the source, and thus will yield the best quality data for luminous quasars.

6. Radiation Processes in Radio-Quiet Active Galaxies

The availability of good quality high energy spectra permits us to constrain the possible emission mechanisms that can operate in active galaxies. We briefly discussed the case of blazars above; observationally, the continuum high energy spectra of radio-quiet objects are also decidedly non-thermal, but these mechanisms are probably somewhat different than the synchrotron + Compton model discussed for the blazar jets. The early work by Sunyaev & Titarchuk (1980), using the diffusion approximation, suggested that a power-law spectrum can be produced by a repeated Compton-upscattering of soft photons by a Compton-thick bath of hot electrons. A more general variant of this model is essentially what is used today to explain the primary, high energy spectrum in these objects as well as in the Galactic black hole candidates. In summary, a successful model has to explain a power law spectrum with an energy index α of $\sim$ 0.9 – 1, exponentially cutting off at $E_c \sim$ 200 keV: other spectral features are merely signatures of reprocessing.

There are essentially two flavors of the Comptonization model that can be applied to the isotropic emission in quasars: the original thermal Comptonization version, and a non-thermal variant. The difference between the two is related to the distribution of electron energies. The non-thermal version, developed by A. Zdziarski, A. Lightman, P. Coppi, as well as by C. Done, R. Svensson, G. Ghisellini, and A. Fabian (Zdziarski et al. 1990; Zdziarski & Coppi 1991) involves acceleration of particles to relativistic energies, and a subsequent pair cascade; these particles Comptonize UV photons, believed to be produced in abundance by the accretion disk. The attractive feature of this model is that the pairs thermalize to relatively low temperatures ($\sim$ a few keV), providing the medium which, again, upscatters the UV photons to form the "soft excess" discussed above.

One feature, however, predicted by this version of the model, is the annihilation line that should be present at $\sim$ 511 keV. No spectrum of any active galaxy collected so far with the CGRO OSSE detector showed such a feature (Johnson et al. 1997), and thus the model is somewhat out of favor, despite its natural ability to provide the "soft excess." Refinements to the thermal Comptonization model, primarily by Poutanen & Svensson (1996), allowed the quasi-analytical calculation in the optically thin - to intermediate regime. This is the most viable current model for radio-quiet active galaxies, and is discussed in more detail by Poutanen (this volume). Roughly, for regime relevant to the hard X–ray spectra of Seyferts, the energy of the exponential cutoff E_c determines the temperature of the Comptonizing plasma, such that $kT_{\mathrm{plasma}} \simeq E_c/1.6$, while the index of the power law, together with the cutoff, determine its optical depth τ_{Th}, such that $\tau_{\mathrm{Th}} \simeq 0.16 \ / \ (\alpha \times (kT_{\mathrm{plasma}}/m_e c^2))$ (Pietrini & Krolik 1995; Poutanen, Krolik, & Ryde 1997). For Seyferts, the Comptonizing plasma has to have a temperature of $\sim$ 100 keV, and optical depth $\tau \sim$ 1; the soft "seed" photons are available in abundance from the inner accretion disk, as evidenced by the presence of the reflection component (Zdziarski et al. 1997). However, a simple "sandwich" type structure (with a cold disk covered by a uniform corona) cannot work, as this would produce too many "seed" photons for Comptonization. Instead, an example of a good phenomenological model is a "patchy corona" above a surface of the disk, proposed by Haardt, Maraschi, & Ghisellini (1994); however, none of these models address the processes responsible for the particle acceleration.

The issue of the "soft excess" remains unresolved in the context of the above models. Spectral fitting to the data for the well-studied NGC 5548 imply that the soft excess can be produced by a relatively cold ($kT \sim 200$ eV) but optically thick ($\tau > 10$) plasma, while the hard continuum requires $kT \sim 50$ keV, but $\tau \sim 2$ (Magdziarz et al. 1998). Coexistence of such two phases may be related to the disk structure and dynamics of possible multi-phase transition regions (see, e.g., Magdziarz & Blaes 1998). In fact, such multi-phase medium appears naturally in local solutions of the disk corona transition layer (see, e. g., Różańska 1998). However, without high quality observational data on the spectral shape and temporal correlation between both components, so far, we lack clear clues as to the time evolution of the plasma energetics. The author's prejudice (based partially on the "episodic" nature of the soft light curves) is that we probably witness some form of a limit cycle operating in the inner disk, and thus the best avenue for this is a development of a more detailed theory for the structure of the inner accretion region, and in particular, the issue of stability of the transition region between the *bona fide* disk and the matter free-falling onto the black hole. However, any tests of theories require sensitive observations: especially needed are well-sampled light curves obtained simultaneously over a broad energy range, from the softest energies accessible (~ 0.1 keV) up to the end of the observable spectrum, in the MeV range. The prospects for such observations are very good: with the impending launch of AXAF, XMM, Astro-E, Integral, and, eventually, Constellation-X, we should have the data for the more definitive modeling.

Acknowledgements: The author wishes to acknowledge helpful comments from Drs. J. Krolik, P. Magdziarz, C. Done, E. Boldt, and M. Sikora, and figures from Drs. W. Brandt, K. Nandra, and L. Greenhill.

REFERENCES

ABRAMOWICZ, M., CHEN, X., KATO, S., LASOTA, J.-P., & REGEV, O. 1995, ApJ, 438, L37

ANTONUCCI, R.R.J, & MILLER, J.S. 1985, ApJ, 297, 621

ARNAUD, K.A., ET AL. 1985, MNRAS, 217, 105

AVNI, Y., & TANANBAUM, H. 1986, ApJ, 305, 83

AWAKI, H., KOYAMA, K, INOUE, H., & HALPERN, J. 1991, PASJ, 43, 195

BLANDFORD, R.D., & LEVINSON, A. 1995, ApJ, 441, 79

BLANDFORD, R.D., & ZNAJEK, R.L. 1977, MNRAS, 179, 433

BOLLER, T., BRANDT, W.N., & FINK, H. 1996, A&A, 305, 53

BOLLER, T., BRANDT, W.N., FABIAN, A.C., & FINK, H. 1997, MNRAS, 289, 393

CZERNY, B., & LEHTO, H. 1997, MNRAS, 285, 365

DERMER, C., SCHLIKHEISER, R., & MASTICHIADIS, A. 1992, A&A, 256, L27

DONE, C., MADEJSKI, G.M., & SMITH, D. 1996, ApJ, 463, L63

ECKART, A., & GENZEL, R. 1997, MNRAS, 284, 576

ELVIS, M., WILKES, B., & TANANBAUM, H. 1985, ApJ, 292, 357

FABIAN, A., NANDRA, K., CELOTTI, A., REES, M., GROVE, E., & JOHNSON, W. 1993, ApJ, 416, L57

FORD, H.C., ET AL. 1994, ApJ, 435, L27

FORD, H.C., TSVETANOV, Z.I., FERRARESE, L., & JAFFE, W. 1998, in *The Central Regions of the Galaxy and Galaxies*, proc. IAU Symp. 184, in press

GEORGE, I.M., & FABIAN, A.C. 1991, MNRAS, 249, 352

GREEN, A. 1993, PhD Thesis, University of Southampton, UK

GREENHILL, L.J. 1998, in *Radio Emission from Galactic and Extragalactic Compact Sources*, proc. IAU Coll. 164, eds. J. Zensus et al., ASP Conference Series, in press.

GREENHILL, L.J., MORAN, J.M., & HERRNSTEIN, J.R. 1997, ApJ, 481, L23

HAARDT, F., MARASCHI, L., & GHISELLINI, G. 1994, ApJ, 432, L95

HALPERN, J. 1982, PhD Thesis, Harvard University

HARMS, R., ET AL. 1994, ApJ, 435, L35

HO, L.C. 1998, in *The Central Regions of the Galaxy and Galaxies*, proc. IAU Symp. 184, in press

ICHIMARU, S. 1987, ApJ, 214, 840

IWASAWA, K., ET AL. 1993, ApJ, 409, 155

IWASAWA, K., FABIAN A.C., BRANDT W.N., KUNIEDA H., MISAKI K., REYNOLDS C.S., & TERASHIMA Y. 1998, MNRAS, submitted

JOHNSON, W.N., ZDZIARSKI, A.A., MADEJSKI, G.M., PACIESAS, W.S., STEINLE, H., & LIN, Y.-C. 1997, in *Proceedings of the Fourth Compton Symposium*, eds. C. D. Dermer, M. S. Strickman, & J. D. Kurfess (AIP: New York), AIP Conference Proceedings 410, p. 283

KÖNIGL, A. 1981, ApJ, 243, 700

KORMENDY, J., & RICHSTONE, D. 1995, ARA&A, 33, 581

KRISS, G.A., & CANIZARES, C. 1985, ApJ, 297, 177

KROLIK, J., DONE, C., & MADEJSKI, G. 1993, ApJ, 402, 432

LAOR, A., FIORE, F., ELVIS, M., WILKES, B., & McDOWELL, J. 1997, ApJ, 477, 93

LASOTA, J.-P., ABRAMOWICZ, M., CHEN, X., KROLIK, J., NARAYAN, R., & YI, I. 1996, ApJ, 462, 142

LEIGHLY, K., & O'BRIEN, P. 1997, ApJ, 481, L15

LIGHTMAN, A.P., & WHITE, T. 1988, ApJ, 335, 57

MADEJSKI, G.M., ET AL. 1995, ApJ, 438, 672

MADEJSKI, G.M., ET AL. 1993, Nature, 365, 626

MAGDZIARZ, P., BLAES, O., ZDZIARSKI, A., JOHNSON, W., & SMITH, D. 1998, MNRAS, in press

MAGDZIARZ, P., & BLAES, O. 1998, in Proc. IAU Symp. 188, Kyoto, Japan, in press

MAKISHIMA, K. 1986, in *The Physics of Accretion onto Compact Objects*, ed. K. Mason, M. Watson, & N. White (Springer-Verlag: Berlin), p. 249

MALONEY, P.R., BEGELMAN, M.C. & PRINGLE, J.E. 1996, ApJ, 472, 582

MAOZ, E. 1995, ApJ, 447, L91

MARASCHI, L., ET AL. 1994, ApJ, 435, L91

MARSHALL, H.L., ET AL. 1997, ApJ, 479, 222

McHARDY, I. 1989, in *Two Topics in X-ray Astronomy*, Proc. 23rd ESLAB Symp., eds. N. White, J. Hunt & B. Battrick, (ESA Publications: Paris), vol. SP-296, p. 1111

MIYOSHI, M., ET AL. 1995, Nature, 373, 127

MODERSKI, R., SIKORA, M., & LASOTA, J.-P. 1997, in *Relativistic Jets in AGNs*, eds. M. Ostrowski et al. (Astronomical Observatiory of the Jagiellonian University: Krakow) p. 110

MUSHOTZKY, R.F. 1980, Adv. Sp. Res., 3, 10

MUSHOTZKY, R.F., DONE, C., & POUNDS, K.A. 1993, ARA&A, 31, 717

NANDRA, K., GEORGE, I., MUSHOTZKY, R.F., TURNER, T.J., & YAQOOB, T. 1997a, ApJ, 477, 602

NANDRA, K., MUSHOTZKY, R.F., GEORGE, I., TURNER, T.J., & YAQOOB, T. 1997b, ApJ, 488, L91

NARAYAN, R., & YI, I. 1994, ApJ, 428, L13

NEUFELD, D.A., & MALONEY, P.R. 1995, ApJ, 447, L17

PAPADAKIS, I.E., & LAWRENCE, A. 1995, MNRAS, 272, 161

PAPADAKIS, I.E., & LAWRENCE, A. 1993, Nature, 361, 250

PIETRINI, P., & KROLIK, J. 1995, ApJ, 447, 526

POUNDS, K., NANDRA, K., STEWART, G., GEORGE, I., & FABIAN, A. 1990, Nature, 344, 132

POUTANEN, J., & SVENSSON, R. 1996, ApJ, 470, 249

POUTANEN, J., KROLIK, J., & RYDE, F. 1997, in *Proceedings of the Fourth Compton Symposium*, eds. C. D. Dermer, M. S. Strickman, & J. D. Kurfess (AIP: New York), AIP Conference Proceedings 410, p. 972

PTAK, A. 1997, PhD thesis, University of Maryland, College Park, MD

REES, M.J. 1967, MNRAS, 135, 345

REYNOLDS, C., DI MATTEO, T., FABIAN, A., HWANG, U., & CANIZARES, C. 1996, MNRAS, 283, L111

REYNOLDS, C., & BEGELMAN, M.C. 1998, ApJ, in press

ROTHSCHILD, R.E., MUSHOTZKY, R.F., BAITY, W.A., GRUBER, D.E., MATTESON, J.L., & PETERSON, L.E. 1983, ApJ, 269, 423

RÓŻAŃSKA, A. 1998, MNRAS submitted

SALPETER, E.E. 1964, ApJ, 140, 796

SCHMIDT, M. 1963, Nature, 197, 1040

SIKORA, M. 1997, in *Proceedings of the Fourth Compton Symposium*, eds. C. D. Dermer, M. S. Strickman, & J. D. Kurfess (AIP: New York), AIP Conference Proceedings 410, p. 494

SIKORA, M., BEGELMAN, M.C., & REES, M. 1994, ApJ, 421, 153

SIKORA, M., MADEJSKI, G.M., MODERSKI, R., & POUTANEN, J. 1997, ApJ, 484, 108

SMITH, D., & DONE, C. 1996, MNRAS, 280, 355

SUNYAEV, R., & TITARCHUK, L. 1980, A&A, 86, 121

TANAKA, Y., ET AL. 1995, Nature, 375, 659

TURNER, T.J. 1988, PhD Thesis, University of Leicester, UK

TURNER, T.J., & POUNDS, K. 1989, MNRAS, 240, 833

ULRICH, M.-H., MARASCHI, L., & URRY, C.M. 1997, ARA&A, 35, 445

VERMEULEN, R.C., & COHEN, M.H. 1994, ApJ, 430, 467

VIO, R., CRISTIANI, S., LESSI, O., & PROVENZALE, A. 1992, ApJ, 391, 518

WILSON, A., & COLBERT, E. 1995, ApJ, 438, 62

YI, I. 1996, ApJ, 473, 645

ZDZIARSKI, A.A., GHISELLINI, G., GEORGE, I.M., SVENSSON, R., FABIAN, A.C., & DONE, C. 1990, ApJ, 363, L1

ZDZIARSKI, A.A., & COPPI, P. 1991, ApJ, 376, 480

ZDZIARSKI, A.A., JOHNSON, W.N., DONE, C., SMITH, D., & McNARON-BROWN, K. 1995, ApJ, 438, L63

ZDZIARSKI, A.A., JOHNSON, W.N., POUTANEN, J., MAGDZIARZ, P., & GIERLINSKI, M. 1997, in The Transparent Universe, eds. C. Winkler at al. (ESA: Paris), SP-382, 373

ZELDOVICH, Y., & NOVIKOV, I. 1964, Sov. Phys. Dokl., 158, 811

ZHANG, S.N., CUI, W., & CHEN, W. 1997, ApJ, 482, L55

Physics of black holes

By IGOR NOVIKOV

Theoretical Astrophysics Center, Juliane Maries Vej 30, DK-2100 Copenhagen Ø, Denmark;

University Observatory, Juliane Maries Vej 30, DK-2100 Copenhagen Ø, Denmark;

Astro Space Centre of P.N. Lebedev Physical Institute,

Profsoyuznaya 84/32, Moscow, 117810, Russia;

Nordita, Blegdamsvej 17, DK-2100 Copenhagen Ø, Denmark

1. Introduction

Why are the problems of black holes so important for modern physics and astrophysics? The answer is obvious: black holes are absolutely unusual objects. What is involved is not just the investigation of yet another, even if extremely remarkable celestial body, but a test of the correctness of our understanding of the properties of space and time in extremely strong gravitational fields.

Of all conceptions of the human mind perhaps the most fantastic is the black hole. Black holes are neither bodies nor radiation. They are clots of gravity.

Black holes are very simple objects. Their properties are completely independent of the properties of the collapsed matter, of all the complexities of material structure and physical fields in it.

Nothing, however, can be more complex than a black hole: indeed, human imagination is unable to comprehend the degree to which space is curved and time flow is warped when a black hole is formed. The study of black hole physics extends our knowledge of the fundamental properties of space and time. Quantum processes occur in the neighborhood of black holes, so that the most intricate structure of the physical vacuum is revealed. Even more powerful (catastrophically powerful) quantum processes take place inside black holes (in the vicinity of the singularity).

It may be said that black holes are a door to a new, very wide field of study of the physical world.

In this paper I will give a brief review of some problems of the physics of black holes and discuss important new points. For systematic discussion of the problems see the books: Thorne et al (1986), Novikov and Frolov (1989), and Frolov and Novikov (1998).

At the beginning I will remind you of a few well known elementary facts.

A black hole is region in space-time from which no signal can escape to an external observer. A black hole's boundary is an event horizon. It is a globally defined null surface in four-dimensional spacetime.

After the gravitational collapse of a celestial body and formation of a black hole the external gravitational field of it asymptotically approaches a standard equilibrium configuration known as the Kerr-Newman field and characterized by just three numbers: mass, angular momentum and charge.

2. Physics outside the event horizon

Black holes reside in curved space. If a black hole has nonzero angular momentum then anything near a black hole will be dragged along by the vortex gravitational field. In this section I will consider a black hole without electric charge (Kerr black hole). The horizon's surface area can be written in terms of its mass M and angular momentum

$J = aM$, where a is an angular momentum per unit mass ($c = 1$, $G = 1$):

$$A = 4\pi(r_H^2 + a^2),$$ (2.1)

$$r_H = M + \sqrt{M^2 - a^2}.$$ (2.2)

The rotational energy of a Kerr black hole is the following

$$M_{rot} = M - \left[\frac{1}{2}M\left(M + \sqrt{M^2 - a^2}\right)\right]^{1/2}.$$ (2.3)

This rotational energy (energy of the vortex gravitational field) can be extracted (in principle) from a black hole.

The black hole is a clot of gravity, there is not any real matter on the horizon. In spite of this fact the horizon looks for an external observer (outside the black hole) and behaves as a physical membrane which is made from a two-dimensional viscous fluid with definite mechanical, electrical, and thermodynamic properties.

This remarkable viewpoint is known as the membrane paradigm (see Thorne et al. 1986, for a review). According to this paradigm the interaction of the horizon with the external universe is described in terms of familiar laws for the horizon fluid, e.g. the Navier-Stokes equation, Maxwell's equations, a tidal force equation, and the equations of thermodynamics. It is very important to emphasize that the membrane paradigm is not an approximation method or some analogy. It is an exact formalism which gives exactly the same results as the standard formalism of the General Relativity. Because the laws governing the horizon's behavior have familiar forms, they are powerful for understanding intuitively and computing quantitatively the interaction of black holes with complex environments.

In subsequent parts of this section we will consider some manifestations of the physical properties of the black hole's membrane, that resided in the three dimensional space.

2.1. *Mechanical properties of the horizon's membrane*

According to the membrane formalism, from the point of view of an external observer the hole's membrane has definite surface mass density and the surface stress tensor.

The formula for the mass density is

$$\Sigma = -\frac{1}{8\pi}\theta^H, \qquad \theta^H \equiv \frac{d(\Delta A)}{\Delta A dt},$$ (2.4)

where θ^H is a fractional change of area of a surface element per unit time of an observer at infinity. One can see that for the case of a black hole in equilibrium (for example, a Schwarzschild or a Kerr hole in the empty space) $\Sigma = 0$. The value of θ^H is always non-negative, consequently Σ is always non-positive.

There is surface pressure p^H in the membrane. For a Schwarzschild hole it is:

$$p^H = \frac{1}{32\pi M} \approx \left(10^{42}\frac{\text{dyne}}{\text{cm}}\right)\left(\frac{M_\odot}{M}\right).$$ (2.5)

From the point of view of the membrane formalism the gravity of a black hole in equilibrium is produced by p^H.

The horizon's shear viscosity η^H and the horizon's bulk viscosity ζ^H are correspondingly:

$$\eta^H = \frac{1}{16\pi} \approx 10^{37}\frac{\text{g}}{\text{sec}},$$ (2.6)

$$\zeta^H = -\frac{1}{16\pi} = -10^{37}\frac{\text{g}}{\text{sec}}.$$ (2.7)

Because the membrane paradigm regards a black hole as a two dimensional membrane with familiar mechanical properties it is rather easier to understand intuitively and compute quantitatively what happens with a black hole under some definite conditions. Let us consider a few examples.

If a black hole occurs as a result of gravitational collapse of an asymmetric celestial body (without rotation), then a nonspherical hole arises at the first moment. The hole's membrane is deformed and there is no balance between the surface pressure of the membrane and its gravity. So the membrane vibrates and radiates gravitational waves. The waves carry away the energy of the membrane deformation. This together with the membrane viscosity makes the horizon settle down into an absolutely spherical equilibrium shape.

Another example is a shape of the membrane of a rotating black hole. Centrifugal forces make a hole's membrane bulge out at its equator. The balance between the surface pressure, gravity and centrifugal forces determines the shape of the horizon's membrane.

Let us consider one very unusual property of the horizon's membrane. We emphasized above that the differential equations which describe the interaction of the horizon with the external universe are familiar physical laws (e.g. the Navier-Stokes equation and so on). But the solutions of the equations are determined also by the boundary conditions. In the case of standard physics the boundary conditions must be imposed at some initial moment or in the infinite past. That is not so for the hole's horizon! The point is that the horizon is a *globally* defined surface. It is a boundary between light-speed signals that can and those that cannot ever escape to spatial infinity. But this fact depends on the processes in the future, not in the past.

Whether a signal can escape depends on the region of spacetime to the future of the signal source. It means that the motion of the horizon at any moment of time depends not on what has happened to the horizon in the past but what will happen to the horizon in the future.

This property can be illustrated by the problem of a free fall of a thin spherical shell of a matter of mass ΔM into a Schwarzschild hole with mass M. The spacetime geometry is that of Schwarzschild both interior to the shell and outside it. In the interior the Schwarzschild mass is M and outside it is $M + \Delta M$. Now the light-speed signals with world lines at $r = 2M$ cannot be the boundary of the non-escape region because these signals and outgoing signals just outside $r = 2M$ will get caught and pulled into the hole by the added gravity of the shell when in the future the shell passes through them. The real boundary is generated by light-speed signals world lines which are just outside of the surface $r = 2M$. In the past, long before the shell arrives at the horizon this surface practically coincides with $r = 2M$. Then, as the shell nears it the surface (which is the real boundary, meaning the real horizon) starts to expand. This is because the world lines of its generators go farther an farther from $r = 2M$. This is their property in the Schwarzschild spacetime, and it does not depend on the approaching shell. When the shell finally passes through it, the added shell's gravity starts influence the motions of the generators of the surface, the horizon suddenly stops expanding and freezes at $r = 2(M + \Delta M)$. These behaviors of the horizon are dictated by the properties of propagation of the light-speed signals which generate the horizon and which have the property to propagate at $r = 2(M + \Delta M)$ *after* crossing with the shell. Thus, this behavior of the horizon *before* crossing with the shell (its expansion) depends on the events in the future (the crossing with the shell).

One refers to this dependence of future events as the "teleological" nature of the horizon (see Thorne et al 1986). I would like to emphasize that these behaviors looks as if the hole's membrane lives in time which flows in the opposite direction: from the future into the past. Indeed in this case the change of the size of the horizon looks very natural and causal. If we accept this point of view, we should consider the extraction of the shell from the hole, and just after this extraction of the shell from the membrane at $r = 2(M + \Delta M)$, the horizon starts contracting and settles down to $r = 2M$. We will see in Section 3 that this unusual property, namely, "feeling" information from the infinite future of the external observer, is a characteristic property not only of the horizon but also of the interior of a black hole.

2.2. *Black-Hole Electrodynamics*

A black hole horizon behaves as an electrically conducting sphere. To understand this let us ask what could be the external manifestation of the electric conductivity of a body in a flat spacetime. The simplest manifestation is the following. If one brings a positive electric charge close to a metal sphere then free electrons on the sphere's metal surface will be displaced with respect to the ions by the Coulomb electric forces. It polarizes the sphere. As a result, the electric field lines form a characteristic configuration in the space around the sphere. Now if one moves the charge parallel to the surface of the sphere from one position to another one, the characteristic configuration of the electric lines comes to a new place with some delay. This delay is determined by the resistivity of the sphere's metal surface. It turns out that if one brings a charge close to a non-spinning black hole, there is a similarity between the picture of the field lines in the vicinity of the black hole and the analogous picture in the vicinity of a metal sphere in a flat spacetime. Now the curvature of spacetime distorts the field line rather than displacement of real charges on the horizon. Nevertheless, it looks like the field of the charge polarizes the horizon.

If one moves the charge parallel to the hole's horizon to another position, then the configuration of the electric lines will settle down at the new place with some delay. Now it is determined by the finite time of propagation of electromagnetic signals. Nevertheless one can interpret it as an effective resistivity of the horizon.

In general one can say that a horizon's membrane behaves as a metal sphere with a surface resistivity equal to $R_H = 4\pi \approx 377 \text{ohms}$.

The membrane paradigm gives insight into possible behaviors of rotating black holes in interaction with magnetized plasma. We will draw an analogy with a dynamo. In its rotor the motion of wire coils in a magnetic field produces an electromotive force compelling the charges to flow through the conductor. A black hole is also a special dynamo of great size. If a spinning black hole is immersed in an external magnetic field, a powerful electric field will also develop in its vicinity. The magnetic field is created by the interstellar gas flowing into a black hole. The magnetic field lines will tend to rotate along with the spinning black hole. The motion of any magnetic field generates an electric field. In the case of a rapidly rotating, magnetized black hole, the electric field generated near its edges can produce an enormous voltage difference between the poles of the hole and its equatorial region:

$$\Delta V \approx (10^{20} \text{volts}) \left(\frac{a}{M} \right) \left(\frac{M}{10^9 M_\odot} \right) \left(\frac{B}{10^4 G} \right), \qquad (2.8)$$

where B is the magnetic field in the vicinity of the black hole. It is as though the spinning black hole was a huge battery. The electric field is responsible for accelerating the charged particles of the plasma and causing them to move along the magnetic lines of force. The total power output is

$$P \approx \left(10^{45} \frac{\text{erg}}{\text{sec}} \right) \left(\frac{a}{M} \right) \left(\frac{M}{10^9 M_\odot} \right) \left(\frac{B}{10^4 G} \right), \qquad (2.9)$$

Probably this mechanism is the main "engine" of the active galactic nuclei.

2.3. *Thermodynamics of black holes*

From many aspects of the thermodynamics of black holes, I will discuss the problem of the black hole's thermal quantum radiation and the related problem of the thermal atmosphere of a black hole.

S.Hawking (1974) claimed that a black hole should emit thermal radiation with temperature

$$T_H = \frac{\hbar}{8\pi k} M^{-1} \approx (10^{-7} K) \left(\frac{M_\odot}{M} \right). \tag{2.10}$$

How, in simple physical terms, could one understand that a black hole behaves like an ordinary body with temperature T_H. A key insight into thermal emission from a hole came from theoretical discoveries in the mid-1970s (see Unruh, 1976). The crucial point is the existence of the event horizon for some definite classes of observers. For example, an accelerated observer in an empty spacetime has a horizon. This observer cannot receive information from the region beyond the horizon. The virtual particles' vacuum fluctuation waves are not confined solely to the region above the horizon; part of each fluctuation wave is beyond the horizon and part is within the region which the observer can see. According to quantum mechanics this principle lack of information about vacuum fluctuation waves leads to the conclusion (for an accelerated observer) that they are real waves. As a result, this observer is bathed in a perfect bath of thermal radiation with temperature $T = \hbar a/(2\pi k)$, where a is the observer's acceleration. Since a static observer just above a Schwarzschild horizon can be viewed as analogous to an accelerated observer in flat spacetime with acceleration $a = c^2/z$, where z is the distance from the horizon, such an observer should feel himself bathed in thermal radiation with local temperature $T = \hbar/(2\pi k z)$. This thermal radiation forms a thermal atmosphere of the hole. The radiation, climbing up through the hole's gravitational field, would be redshifted by a factor $(1 - 2M/r)^{1/2}$. It will emerge with temperature T_H. Most of the photons and other particles fly upward a short distance and are then pulled back down by the hole's enormous gravity. A few of the particles manage to escape the hole's gravitational grip and evaporate into space. These particles form the Hawking radiation.

Note that a free falling observer does not feel this thermal atmosphere. He "sees" only vacuum fluctuations to consist of pairs of virtual particles.

The process of the Hawking quantum evaporation is very slow. The total lifetime is proportional to the cube of M. For a 20 solar mass black hole it is 10^{70} years.

In principle, the interactions of a black hole with the external Universe can change the process of extraction of the thermal energy from a black hole atmosphere drastically (see the review in referenced books).

3. Physics inside the event horizon

What can one know about the interior of a black hole? Does the internal geometry also approach the Kerr-Newman form soon after the collapse?

This problem was the subject of a very active investigation last decades. There is a great progress in these researches. We know some important properties of the realistic black hole's interior, but some details and crucial problems are still the subject of much debate.

A very important point for understanding the problem of the black hole's interior is the fact that the path into the gravitational abyss of the interior of a black hole is a progression in time. We recall that inside a spherical hole, for example, the radial coordinate is timelike. It means that the problem of the black hole interior is an *evolutionary problem*. In this sense it is completely different from a problem of an internal structure of other celestial bodies, stars for example.

In principle, if we know the conditions on the border of a black hole, we can integrate the Einstein equations in time and learn the structure of the progressively more and more deep layers inside the black hole. Conceptually it looks simple, but there are two types of principal difficulties which prevent realizing this idea consistently.

The first difficulty is the following. As we shall see in the subsequent discussion, the internal structure of a general black hole even soon after its creation depends crucially on the conditions on the event horizon at tremendous future of the external observer (formally at the infinite future). It is because the light-like signal can propagate inside a black hole from the tremendous future to those regions if the regions are deep enough in the hole.

This means that the structure of these regions depends crucially on the fate of the black hole at infinite future of an external observer. For example, it depends on the final state of the black hole evaporation, on possible collisions of the black hole with another black hole, and it depends on the fate of the Universe itself. It is clear that theoretitions feel themselves uncomfortable under such circumstances.

The second serious problem is related with existing of a singularity inside a black hole. Close to the singularity, where the curvatures of the spacetime approach Planck levels, the Classical General Relativity is not applicable. We have not the final version of the quantum theory of gravity yet, thus any extension of the discussion of physics in this region would be highly speculative. Fortunately, as we shall see, these singular regions are deep enough in the black hole interior and they are *in the future* with respect to overlying and *preceding* layers of the black hole where curvatures are not so high and which can be described by well-established theory.

The first attempts to investigate the interior of a Schwarzschild black hole have been made in the late 70's. It has been demonstrated that the parts of the interior of a black hole long after its formation and subsequent perturbations are virtually free of perturbations because the gravitational radiation from aspherical initial perturbations becomes infinitely diluted as it reaches these regions. But this result breaks down in the general case when the angular momentum or the electric charge is not equal to zero. The reason for that is related with the fact that the topology of the interior of the Kerr-Newman solution differs drastically from the Schwarzschild one. The key point is that the interior of the Kerr-Newman black hole possess a *Cauchy horizon*. This is a surface of infinite blueshift. Infalling gravitational radiation propagates inside the black hole along paths approaching the null generators of the Cauchy horizon, and the energy density of this radiation will suffer an infinite blueshift as it approaches the Cauchy horizon. As I will demonstrate, this infinitely blueshifted radiation together with the radiation scattered on the curvature of spacetime inside the black hole leads to formation of the curvature singularity along the Cauchy horizon.

3.1. *Formulation of the problem*

To specify the problem we will consider an isolated black hole (in an asymptotically flat spacetime) which was created as a result of a realistic collapse of a star without assumptions about special symmetries. At the beginning we will not consider the influence of the quantum evaporation on the internal structure of a black hole and leave this topic for the discussion at the end of the paper.

The initial data at the event horizon of an isolated black hole which determine the internal evolution for rather late period of time are known with precision because of the no hair property (see Novikov and Frolov, 1989). Near the event horizon we have a Kerr-

Newman geometry perturbed by a dying tail of gravitational waves. The fallout from this tail produces an inward energy flux decaying as an inverse power v^{-p} of advanced time, where $p = 4l + 4$ for a multipole of order l.

Now we should integrate the Einstein equations with the known boundary conditions to obtain the internal structure of the black hole. In general, the evolution with time into the black hole deeps looks like the following. The gravitational radiation penetrating the black hole and partly backscattered by the spacetime curvature can be considered roughly speaking as two intersecting radial streams of infalling and outgoing gravitational radiation fluxes, non-linear interaction of which leads to formation a nontrivial structure of the black hole interior. However in such a formulation it is a very difficult and still not solved completely mathematical problem, and it is necessary to make a few simplifying assumptions. In subsequent subsections we will investigate the problem step by step starting from very simple models and little by little will reveal main properties of this evolutionary process of the formation of the internal structure of a black hole.

A detail discussion of different steps in the analysis of the problem one can find in the following works: Poisson and Israel (1990), Ori (1992), Bonanno et al. (1995), Israel (1997) and references therein.

3.2. *Simplification*

Our first simplification is that we will consider a spherical charged black hole initially described by the Reissner-Nordstrom metric. A motive for this simplification is that the global structures of the rotating and the charged black holes are very similar, and as we shall see the essential physics is the same in both models. Of course, the assumption about spherical symmetry simplifies a lot the mathematical analysis.

The second simplification is that we imitate the ingoing and outgoing gravitational radiation by two intersecting radial streams of ingoing and outgoing lightlike particles following null geodesics. We assume that streams do not interact with each other and do not scatter on the background curvature. We will see that this approximation also works very well. The main reason is that the ingoing flux is the subject of infinite blueshift near the Cauchy horizon (which is a crucial reason for the whole analysis) and any further scattering should not be important. On the other hand, it turns out that the nature of the outgoing flux is not important at all for the formation of the essential properties of the black hole interior.

3.3. *Mass inflation*

We will see that this simple model incorporates all essential physics of the formation of a black hole interior for the general case of collapse of an asymmetric rotating body. The crucial point here is a tremendous growth of the black hole internal mass parameter, which was dubbed "mass inflation" (Poisson and Israel, 1990).

We start from revealing the mechanism responsible for the mass inflation. Consider a concentric pair of thin spherical shells in an empty spacetime without a black hole. One shell of mass m_{con} contracts, another one of mass m_{exp} expands, both with the speed of light. The contracting shell, when it initially has a radius greater than expanding one, does not create any gravitational action inside of itself, and the expanding shell does not feel the existence of the external shell. On the other hand the contracting shell is in the gravitational field of the expanding one. The mutual potential of the gravitational energy of the shells acts as a debit (binding energy) on the gravitational mass energy of the external contracting shell. Before the crossing of the shells the total mass of both of them, measured by an observer outside both shells, is equal to $m_{con} + m_{exp}$ and is constant because the debit of the increase (by modulus) of the negative potential energy is exactly balanced by the increase of the positive energies of photons blueshifted in the gravitational field of the internal sphere.

When the shells cross, at radius r_0, the debit is transformed from the contracting shell to the expanding one, but the blueshift of photons of the contracting shell survives. As a result masses of both spheres change. The increase of m_{con} is called mass inflation. The exact calculation shows that new masses m'_{con} and m'_{exp} are

$$\left. \begin{array}{l} m'_{con} = m_{con} + \frac{2m_{con}m_{exp}}{\epsilon} \\ m'_{exp} = m_{exp} - \frac{2m_{con}m_{exp}}{\epsilon} \end{array} \right\} \tag{3.1}$$

where $\epsilon \equiv (r_0 - 2m_{exp})$. The total mass-energy is, of course, conserved $m'_{con} + m'_{exp} = m_{con} + m_{exp}$. If ϵ is small (the encounter is just outside the horizon of m_{exp}) the inflation of mass of m_{con} can become arbitrary large. In this case m'_{exp} becomes negative, which means that it is trapped inside a black hole with the mass equal to $m_{con} + m_{exp}$.

It is not difficult to extend this result to the crossing of the shells inside a charged black hole. When the ingoing shell is very close to the Cauchy horizon m_{con} becomes arbitrary large. It describes the mass inflation.

3.4. *Black hole's interior.*

During the last years many different models of the ingoing and outgoing fluxes in the interior of the charged and rotating black holes have been discussed. In all these models the curvature of the spacetime diverges at the Cauchy horizon as a result of mass inflation. This means that the Cauchy horizon becomes singular.

It was demonstrated that though the tidal force in the reference frame of a freely falling observer grows infinitely when he approaches the singularity, its integral over proper time of this observer remains finite. Such a singularity which is quite weak was called mild singularity.

There is one more effect caused by the outgoing flux. The contraction is continuous until the Cauchy horizon shrinks to $r = 0$, and a stronger singularity occurs. This is the picture according to the classical general relativity. Now we should take into account the quantum effects.

3.5. *Quantum effects*

In previous discussions we emphasized that the internal structure of black holes is a problem of the evolution in time starting from boundary conditions on the event horizon for all moment of time up to the infinite future of the external observer.

It is very essential to know boundary conditions up to infinity because we observed that the essential events: mass inflation and singularity formation, happened along the Cauchy horizon which brought information from the infinite future of the external spacetime. However, even an isolated black hole in the asymptotically flat spacetime cannot exist forever. It will be evaporated by Hawking quantum radiation. So far we discussed the problem without taking into account this ultimate fate of black holes. Even without going into details it is clear that quantum evaporation of the black holes is crucial for the whole problem.

What can we say about general the picture of the black hole interior as accounted for by quantum evaporation? We have to restrict ourselves to general remarks only, because the work on the problem is still in progress. To account for quantum evaporation we have to change the boundary conditions on the event horizon as compared to the boundary conditions discussed above. Now they should include the flux of negative energy across the horizon which is related with the quantum evaporation. If the black hole is charged or/and rotates, then a process of decreasing of the charge and the angular momentum is more fast than decay of the mass. The last stage of the quantum evaporation when the mass of the black hole becomes comparable to the Planck mass, $m_{pl} = (\hbar/6)^{1/2} \approx 2.2 \times 10^{-5}$g, is unknown. At this stage the spacetime curvature near the horizon reaches l_{pl}^{-2}, where l_{pl} is the Plankian length, $l_{pl} = (G\hbar/c^3)^{1/2} \approx 1.6 \times 10^{-33}$cm. It means that from the point of view of the semi-classical physics a singularity arises here. Probably at this stage the black hole has characteristics of an extreme black hole, when an external event horizon and internal Cauchy horizon coincide.

We should emphasize that in any case the solution cannot be continued beyond the true singularity. As for the processes inside a true singularity, they can be treated only in the framework of an unified quantum theory incorporating gravitation which is unknown.

4. Acknowledgments

I am grateful to many of my colleagues for discussions and collaboration on topics mentioned in this talk and to Nora Kotok for her careful preparation of the manuscript.

I would like to thank the organizers of the Symposium Non-linear Phenomena in Accretion Disks around Black Holes for an excellent meeting.

This work was supported in part by the Danish Natural Science Research Council through grant No. 9401635 and also in part by Danmarks Grundforskningsfond through its support for the establishment of the Theoretical Astrophysics Center.

REFERENCES

BONANNO, A., DROZ, S., ISRAEL, W. & MORSINK, S.1995 *Proc. Roy. Soc* **A450** 553

FROLOV, V.P. & NOVIKOV, I.D. 1998, *Black Holes Physics: Basic concept and New Developments*, Kluwer Acad. Publ., in preparation.

HAWKING, S.W. 1974 *Nature*, **248**, 30

ISRAEL, W. 1997 in *Relativistic Astrophysics* (eds. B. Jones & D. Markovic), p. 173. Cambridge University Press.

NOVIKOV, I.D. & FROLOV, V.P. 1989, *Physics of Black Holes*, Kluwer Acad. Publ. (Russian version: 1986, Nauka, Moscow)

ORI, A. 1992 *Phys. Rev. Lett.* **68**, 2117

POISSON, E., & ISRAEL, W. 1990 *Phys. Rev.* **D41**, 1796

THORNE, K.S., PRICE, R.H. & MACDONALD, D.A. 1986, *Black Holes: The Membrane Paradigm*, Yale Univ. Press, New Haven.

UNRUH, W.G. 1976 *Phys. Rev.*, **D14**, 870

Physics of black hole accretion

By MAREK A. ABRAMOWICZ

Institute of Theoretical Physics, Göteborg University,

and Chalmers University of Technology, 412 96 Göteborg, Sweden

In their innermost parts, accretion discs around black holes are profoundly influenced by the general relativistic effects characteristic of gravitational fields of black holes. In particular, close to a black hole the relativistic Roche lobe overflow induces a dynamical mass loss from the inner edge of accretion discs and makes radial advection of heat to be the dominant cooling mechanism. This is because, for the relativistic Roche lobe overflow, the timescales of radial advection and thermal processes are about equal. Equality of these timescales is sufficient for advection to be the dominant cooling process in the thermal equilibria of accretion discs. It happens also far away from the black hole: typically, for optically thick accretion discs advection dominates when accretion rate is high (slim accretion discs), and for optically thin discs it dominates when accretion rate is low (ADAFs).

1. Introduction

A substantial part of the recent progress in the theory of black hole accretion comes from an improved understanding of the thermal equilibria of accretion discs. A local thermal balance between viscous heating Q^+ and radiative cooling Q^-_{rad},

$$Q^+ = Q^-_{\text{rad}}, \tag{1.1}$$

is one of the basic assumptions in the "standard" accretion disc paradigm based on the Shakura-Sunyaev accretion disc model (SSD). The SSD model describes accretion discs that are geometrically very thin: the vertical thickness H is much smaller than the corresponding radial distance from the black hole R. Standard discs are optically thick and they emit locally black body radiation. Their temperatures are low, $T_{max} \approx 10^7 (M/M_\odot)^{-1/4}(\dot{M}/\dot{M}_{Edd})^{1/4}$ K. Here $\dot{M}$ is the accretion rate, and $\dot{M}_{Edd} = L_{Edd}/c^2$ is the Eddington accretion rate which in turn is defined by the Eddington luminosity, $L_{Edd} = 1.3 \times 10^{38}(M/M_\odot)$ erg/sec. The SSD models are thermally stable far from the central black hole, where the total pressure is dominated by the gas pressure, but closer in, where the pressure is dominated by the radiation pressure, the SSD models are violently unstable. Thorne and Price (1975) have proposed that the instability may swell the inner part into much hotter, gas-pressure dominated, optically thin state. This idea was further developed in the classic paper by Shapiro, Lightman and Eardley (1976; SLE) who constructed a hot, optically thin model of accretion disc. In the SLE model electrons and protons have different temperatures, with $T_e \ll T_p$. The maximal proton temperature could be as high as $T_p \approx 10^{12}$ K. Despite the fact that the SLE discs are still violently thermally unstable, many authors have been routinely using the SLE models to "explain" high-energy spectra of some astrophysical objects containing accreting black holes. This rather embarrassing situation lasted till the most recent advent of ADAFs — advectively dominated accretion flows. The dominant cooling process which determines thermal equilibria of ADAFs is provided by a nonlocal advection of heat, and *not* as in the SSD and SLE models by local radiation cooling. ADAFs could be both optically thick or thin and they always have rather low radiative efficiency. Optically thin ADAFs are as hot as SLE models. Because ADAFs are radiatively inefficient and at the same time very hot, they are underluminous and have very hard spectra. This fits observed properties of several types of astronomical sources containing accreting black holes. For this reason, models of optically thin ADAFs are now widely used to explain detailed observational data. The interest in optically thin ADAFs is rapidly expanding.

Although the standard SSD model is more than twenty years old, only relatively recently a serious progress has been achieved in understanding its very physical foundation — the physical origin of turbulent viscosity. A correspondingly profound progress was also made in understanding details of radiative cooling processes in super hot, optically thin plasma. These remarkable developments came, in a certain sense, too late for the SSD and SLE models. Indeed, during the last five years, experts working on the accretion theory have adopted the view that both SSD and SLE models are inadequate in describing some most important properties of black hole accretion discs, and that they should be replaced by ADAFs, or by a hybrid disc model containing an ADAF close to the black hole, and an SSD farther out.

ADAFs have been discovered by theoreticians, through careful studies of fundamental issues concerning black hole accretion. The discovery was made independently at least three times in the past. ADAFs have been known previously as starved black holes, Polish doughnuts, ion tori, and slim discs. However, they have been generally accepted only recently, when links to observations were supplemented by a clear and complete theoretical picture. The first paper that introduced explicitly the very idea of ADAFs (Ichimaru, 1977) has not been noticed by accretion disc community. Later, Rees, Begelman Blandford & Phinney (1982) suggested that at very sub-Eddington accretion rates, accretion flows should be hot and optically thin, thus having a low radiative efficiency. The low radiative efficiency is also characteristic of spherical accretion at very high, near Eddington, accretion rates as first noticed by Katz (1977) and worked out in more details by Rees (1978) and Begelman (1978, 1979). Application to optically thick accretion discs has been elaborated in a series of papers discussing the Roche lobe overflow that very efficiently cools the innermost parts of black hole accretion discs (see *e.g.* Jaroszyński, Abramowicz & Paczyński 1980; Abramowicz 1981).

The unification of the subject was provided for optically thick case in the slim accretion discs paper by Abramowicz, Czerny, Lasota and Szuszkiewicz (1988), and for optically thin case by most recent ADAF series started by Abramowicz, Chen, Kato, Lasota and Regev (1995) and Narayan and Yi (1995).

In this article I concentrate on fundamental theoretical issues and describe, in general terms, the line of theoretical arguments that point to importance and universality of ADAFs as a model for the black hole accretion.

2. Astrophysical black holes

In the first section of this book, P. Charles and G. Madejski discuss the direct observational evidence, based mostly on very accurate mass estimates, that black holes do exist. Thus, it is certain that in about dozen X-ray sources in our Galaxy stellar black holes with masses of about 10 $M_\odot$ are present. In other galaxies a few supermassive black holes with masses greater than about 10^6 $M_\odot$ have been identified. Taking into account their huge masses, one may argue that these astrophysical black holes should be treated as classical objects: Hawking radiation and other quantum effect play no role for them. Indeed, as was discussed in the preceding section by I. Novikov, due to quantum effects all black holes emit radiation as if they were black bodies with the temperature that is inversely proportional to their masses. This is the famous Hawking radiation. The flux of Hawking radiation is inversely proportional to the fourth power of the mass, and because the area of the black hole is proportional to the square of the mass, Hawking luminosity is inversely proportional to the square of the mass. Obviously, Hawking luminosity corresponds to the rate at which the mass of the black hole is being radiated away. Dividing the mass by this rate one obtains the characteristic time for quantum processes, or the Hawking life time of the black hole. It is proportional to the cube of the black hole mass. For a black hole with the mass of 10^{15} g the life time is of the order of the Hubble time. For stellar black holes (that have masses greater than $1\ M_\odot = 10^{33}$ g) the corresponding life time is greater than 10^{54} Hubble times. Life times of supermassive black holes are still longer at least by a factor 10^{18}.

Because characteristic times for quantum processes are so much longer than all characteristic times for astrophyscial processes relevant to accretion, only classical aspets of black hole physics (*i.e.* involving the space time curvature) have so far been discussed in the context of black hole accretion discs. In the last paragraph of this article, I present a rather speculative idea that some quantum effects may be relevant for black hole accretion discs. The issue is connected to that discussed by Narayan *et al.* in paragraph 4.3 of their article in this volume.

3. Keplerian orbits around black holes

Paczyński made an ingenious discovery that the influence of the spacetime curvature on hydrodynamical and thermal properties of black hole accretion discs may be very accurately described in a Newtonian model of a non-rotating black hole (Paczyński & Wiita, 1980). The Paczyński & Wiita model for the black hole gravitational potential has been widely used by many authors because of its simplicity.

Paczyński's black hole is a Newtonian object with the gravitational potential $\Phi(R)$ given by,

$$\Phi = \frac{GM}{R - R_G}, \quad R_G = \frac{2GM}{c^2}, \tag{3.2}$$

where M is the mass of the black hole. I use spherical coordinates R, θ, ϕ centered at the black hole.

Free particles orbiting the Paczyński black hole on circular orbits have 'Keplerian' angular momentum $\ell_K(R) = \sqrt{(d\Phi/dR)^3} = GMR^{3/2}/(R-R_G)$. The function $\ell_K(R)$ has a minimum at $R = 3R_G \equiv R_{MS}$. Keplerian orbits with $R > R_{MS}$ are stable, and those with $R < R_{MS}$ are unstable. The unstable orbit with $R = 2R_G \equiv R_{MB}$ has zero binding energy. Thus, unstable orbits with $R < R_{MB}$ are unbound, *i.e.* they have positive binding energy. The same general properties of Keplerian motion around non-rotating black holes, *and* numerical values of R_{MS} and R_{MB} follow from exact calculations in general relativity. The angular momentum corresponding to the marginally stable orbit at $R = R_{MS}$ is $\ell_K(R_{MS}) \equiv \ell_{MS} = (3/2)^{3/2}R_G c$.

In the next section I explain why the fact that the Keplerian angular momentum is not monotonic, but has a minimum, has a crucial significance for black hole accretion discs.

4. Two types of accretion flows

Theoreticians have been discussing hydrodynamical equilibrium of "accretion tori" in which fluid lines are perfectly circular and pressure, gravity and centrifugal forces are in a perfect balance. Accretion flows are never in this state. However, as we shall see, a discussion of equilibria of accretion tori illuminates some most important and very general properties of black hole accretion flows.

When gravity and centrifugal force balance each other in a particular place inside an accretion torus, then, in this place, the angular momentum of matter must be equal to the Keplerian angular momentum discussed in the previous section. In this place, there is no pressure force, which means that the pressure gradient is zero, or that the pressure has minimum, maximum, or saddle point there. Let us assume for a moment that the angular momentum is constant inside the torus, $\ell = \ell_0 = \text{const}$. There are two possibilities: either $\ell_0 < \ell_{MS}$, or $\ell_0 > \ell_{MS}$.

It was first recognized by Abramowicz & Zurek (1981) that these two possibilities correspond to two quite different types of accretion flows: 'Bondi-like' and 'disc-like' accretion. Abramowicz & Zurek found that there is no smooth transition between these two types, because 'intermediate' solutions have globally unacceptable topology of the sonic point (Lu, 1985; Lu & Abramowicz, 1988). Muchotrzeb (1983) was the first who pointed out that the question which of the two types is chosen, depends mostly on the value of the viscosity coefficient α. Above the critical value, $\alpha_{\text{crit}} \approx 0.1$, the Bondi type, and below α_{crit}, the disc type is chosen. The parameter space considered by Muchotrzeb corresponded to small accretion rates and fixed outer boundary conditions. Her results have been discussed in great details and expanded by Kato and his collaborators (Matsumoto, Kato, Fukue & Okazaki, 1984; Abramowicz & Kato, 1989; and Okazaki, Kato & Fukue, 1987). Later, several other authors discussed detailed differences between the two types of black hole accretion flows in specific astrophysical situations, and examined the question of critical α.

Abramowicz & Zurek suggested that when one of the flow parameters experiences a secular change that leads to the change of the flow type, a shock may develop. In particular, Fukue (1987) and Abramowicz & Chakrabarti (1990) have presented a toy model for the shock formation based on the Bondi-disc (and disc-disc) jumps. The model employed free parameters that assumed *ad hoc* how much energy and entropy was lost by radiation emerging vertically from the shock. In a realistic model these parameters should not be assumed, but they should be calculated from first principles. Instead, most of the recent literature on radial shocks follows a different path: numerous 'solutions' are constructed based on *ad hoc* choices of the free parameters. In all global models carefully constructed numerically by several independent researchers, no standing radial shock have been found in stationary thin accretion discs. This may indicate that the *ad hoc* assumptions adopted by those who found such shocks may be unphysical (see Narayan *et al.*, this volume for more details, and Chakrabarti 1996 for a different view).

4.1. *The Bondi-like type*

When $\ell_0 < \ell_{MS}$, the straight line $\ell(R) = \ell_0$ representing the angular momentum in the torus lies *below* the parabola $\ell(R) = \ell_K$ representing the Keplerian angular momentum. The flow is sub-Keplerian everywhere, $\ell_0 < \ell_K$. This implies $\ell_0 \neq \ell_K$ inside the torus, which means that the pressure cannot have a critical point — it monotonically decreases outwards. In this case, there is no place in which the flow could be in hydrodynamical equilibrium: centrifugal force is too week to balance gravity. Qualitatively, this type of accretion flow resembles that of spherical accretion, when the fluid has zero angular momentum. For this reason, Abramowicz & Zurek called it the Bondi-like type.

For the Bondi-like type of the flow, radial and vertical components of velocity are not negligibly small, and nonlocal dissipation is dominant: a substantial part of locally dissipated energy, entropy and angular momentum is carried along (advected) with the matter and lost in the central black hole. Therefore, the Bondi-like accretion has a low radiative efficiency. For *spherical* accretion with high (near Eddington) rates, this was first noticed by Katz (1977). He has pointed out that radiation diffuses outward at a speed $v_{\text{diff}} = c/\tau$, where τ is the optical depth. With v being the velocity of matter, $v_{\text{diff}}/v = \dot{M}_{\text{Edd}}/2\dot{M}$. Thus, as $\dot{M}$ approaches $\dot{M}_{\text{Edd}}$, radiation is trapped within the accreting matter and cannot escape: $\dot{M}$ may be very large without L exceeding L_{Edd}.

4.2. The disc-like type

When $\ell_0 > \ell_{MS}$, the straight line $\ell(R) = \ell_0$ representing the angular momentum in the torus *crosses twice* the parabola $\ell(R) = \ell_K$ representing the Keplerian angular momentum: the flow is super-Keplerian in the region bounded by two radii, R_{cusp} and $R_{cent} > R_{cusp}$ at which $\ell_0 = \ell_K$. At R_{cent} the pressure has a maximum, at R_{cusp} it has a saddle point.

From the condition for the balance of gravitational, centrifugal and pressure forces, it follows that in a stationary equilibrium of a torus, surfaces of constant pressure P (and density ρ) must coincide with the 'equipotential' surfaces, $W(R,\theta) = \Phi(R) - \int \ell^2(R\sin\theta)(R\sin\theta)^{-3}d(R\sin\theta)$. Obviously, any stationary and finite distribution of rotating matter around a black hole must be bounded by a closed and finite $P = 0$ surface. Thus, such a stationary distribution could exist if and only if there exist a closed $W(R,\theta) = $ const equipotential surface around the black hole.

In constant angular momentum torii, the equipotential surfaces are given by,

$$W(R,\theta) = -\frac{GM}{(R - R_G)} + \frac{1}{2}\frac{\ell_0^2}{R^2\sin^2\theta}. \tag{4.3}$$

Suppose that there exist a closed surface $W(R,\theta) = $ const. This surface must cross the equatorial plane $\theta = \pi/2$ at least twice, at two different radii R_{in} and R_{out}, say. Let us consider the function $W_0(R) = W(R, \pi/2)$. (Notabene, this function is identical with the effective potential for radial motion of a particle with the angular momentum ℓ_0.) Obviously, $W_0(R_{in}) = W_0(R_{out})$ which means that $(dW_0/dR)_C = 0$ for a particular $R = R_C$ located between R_{in} and R_{out}. It is easy to see that from $(dW_0/dR)_C = 0$ it follows that $\ell_0 = \ell_K(R_C)$. However, we know that the minimum value of $\ell_K(R)$ is ℓ_{MS}. Thus, in the simplest case $\ell = \ell_0 = $ const, the necessary and sufficient condition for the existence of a closed equipotential surface — and therefore the necessary condition for the existence of a stationary and finite distribution of rotating matter around a black hole is $\ell_0 > \ell_{MS}$, otherwise it would be not possible to have $\ell_0 = \ell_K$. The above condition captures the essential point here. Indeed, for an *arbitrary* angular momentum distribution the general condition is quite similar: the angular momentum of matter should be greater than the Keplerian one, $\ell(R) > \ell_K(R)$, for at least some range of R.

If the condition $\ell(R) > \ell_K(R)$ is obeyed, then a part of the flow could be in a state *very close* to dynamical equilibrium. In this case, pressure gradient, gravitational, and centrifugal forces are almost in balance, and radial and vertical components of velocity are very small, $v_R \ll v_\phi$, $v_\theta \ll v_\phi$. For the flow in nearly dynamical equilibrium, there could be a substantial local reprocessing (e.g. by radiative losses) of locally dissipated energy, entropy and angular momentum, because fluid elements stay at the same radius during a large number ($\sim v_\phi/v_R \gg 1$) of dynamical periods.

5. Relativistic Roche lobe overflow and advective cooling

For disc-like accretion, equation $dW_0/dR = 0$ has then two solutions in the equatorial plane $\theta = \pi/2$: the first one $R = R_{cusp} < R_{MS}$, which corresponds to a maximum, and the second one $R = R_{cent} > R_{MS}$, which corresponds to a minimum of the effective potential $W_0(R)$. Let us define $W_R = W_0(R_{cusp}) = $ const. The equipotential surface $W(R,\theta) = W_R$ is called the Roche lobe. It crosses itself along the circle $R = R_{cusp}$

which is called 'the cusp'. The existence of the self-crossing Roche lobe is a qualitative property that in a wide range of parameters does not depend on details of the flow.

Matter distribution which overflows the Roche lobe cannot be in the equilibrium because it would touch the surface of the black hole (horizon), but no matter in hydrostatic equilibrium could be present at the horizon. Close to the cusp, accretion flow is always fully determined by the Roche lobe overflow. Suppose that far away from the cusp, the surface of the flow ($P = 0$) coincides with the equipotential surface $W(R, \theta) = W_S =$ const $> W_R$, *i.e.* matter distribution in the disc overflows the Roche lobe. This induces a mass loss from the disc through the cusp. Between the cusp and the black hole the flow is not in dynamical equilibrium: it is always of the Bondi-like type there. The local accretion rate at the cusp depends very strongly on the amount of the Roche lobe overflow. This provides a self-regulatory stabilization mechanism against thermal and viscous perturbations (Abramowicz 1981). To see how the mechanism works, consider a perturbation of a steady state accretion disc due to a local overheating at the cusp. The disc which is locally overheated will locally expand. This will increase the Roche lobe overflow and therefore also the local accretion rate will increase. In turn, the higher accretion rate will induce stronger advective cooling Q_{adv}^- through the cusp, which will suppress the effect of initial overheating and bring both the overflow and the accretion rate back to their steady state values.

Therefore, in a steady state, the amount of the Roche lobe overflow self-regulates according to the steady accretion rate which is fixed by the outer boundary conditions. One particular consequence of this is that, at the location of the cusp, the condition of thermal balance has the form,

$$Q^+ = Q_{\mathrm{adv}}^-. \tag{5.4}$$

This means that the innermost parts of *all* black hole accretion discs are *always* of the ADAF type (Abramowicz, 1981).

6. Slim accretion discs and ADAFs

It was initially thought, *cf.* Wheeler (1981), Shapiro & Teukolsky (1983), that the effect of the Roche lobe overflow stabilization could be relevant only at the immediate vicinity of the cusp. Later, in the context of slim discs, Lasota and Abramowicz have realized that the reason for stabilization of thermal modes by the Roche lobe overflow was connected to the fact that the thermal timescale of vertical expansion $\tau_{\mathrm{th}} \sim 1/(\alpha\Omega)$ was of the same order as the timescale of inward advection $\tau_{\mathrm{adv}} \sim r/v_R$. When this happens, the instability is advected inwards before it has a chance to grow, in an analogous way to what stabilizes the spherical accretion (Moncrief, 1980). The Roche lobe overflow gives just a particular (but important) example of the stabilization mechanism provided by advection.

In standard thin accretion discs the timescale for radial advection is given by the viscous timescale $\tau_{\mathrm{adv}} \sim \tau_{\mathrm{vis}} \sim \tau_{\mathrm{th}}/(H/R)^2$. Thus, stabilization of the thermal modes by advective cooling should occur when $H \sim R$. This is consistent with the fact that in general

$$Q_{\mathrm{adv}}^- \sim \left(\frac{H}{R}\right)^2 Q^+, \tag{6.5}$$

and therefore when $H \sim R$, the cooling is dominated by advection: $Q_{\mathrm{adv}}^- \sim Q^+$, and radiative cooling is thus insignificant in the heat balance.

Approximate semi-analytic models of slim disc have been reported in a conference proceedings (Abramowicz, Lasota & Xu 1986, hereafter ALX). Detailed numerical models have been published by Abramowicz, Czerny, Lasota & Szuszkiewicz (1988, hereafter ACLS). In contrast to the standard discs, slim discs have significant radial gradients of pressure and entropy, non-Keplerian rotation, and a significant radial heat transport that induces advective cooling. ALX suggested that due to nonlinearities connected with transitions from gas to radiation pressure and from radiative to advective cooling, sequences of optically thick accretion disc equilibria should bend twice in the $\ln \dot{M}$ versus $\ln \Sigma$ plane and form an S-curve. This was fully confirmed by exact numerical models calculated later by ACLS. The lower branch of the S-curve corresponds to stable, gas pressure dominated standard disc models. The middle branch corresponds to unstable, radiation pressure dominated standard models. Both lower and middle branches exist for sub-Eddington accretion rates. The upper branch exists at moderately super-Eddington accretion rates and corresponds to stable slim discs. Slim discs are optically thick, radiation pressure dominated, and have H slightly smaller than R. They are thermally stable, because the timescale of thermal vertical expansion is shorter than the timescale of advection. Thus, perturbations are advected inward before they could grow. Calculations of slim discs models have also been performed by Abramowicz, Kato & Matsumoto (1989), Kato, Honma & Matsumoto (1988) and later by Chen & Taam (1993). They have confirmed the results of ALX and ACLS, in particular the existence of the S-shaped equilibrium sequences. Electromagnetic spectra of slim discs are discussed by Szuszkiewicz, Malkan & Abramowicz (1996).

Having in mind the analogy with the dwarf novae S-curved equilibrium sequences, ALX suggested that slim discs should experience a limit cycle behaviour. Numerical simulation of Matsumoto, Honma & Kato (1989), Honma, Matsumoto & Kato (1991), and Szuszkiewicz & Miller (1997) confirmed this suggestion. However, it would be premature to attempt detailed explanations of the observed variability of some real objects as a slim disc limit cycle. Two problems are still unsolved: (i) results of simulations depend strongly on unknown and thus arbitrarily assumed properties of viscosity, (ii) in the non-stationary state the flow experiences optically thick-thin transitions which in absence of a full solution of the radiative transfer in discs are still treated by a phenomenological approach (Lasota & Pelat, 1991 see also Taam & Lin 1984).

In deriving the condition $H \sim R$ for the importance of advective cooling, no particular regime of optical depth has been assumed. Thus, the condition should also be valid for *optically thin* discs. Indeed, the very hot, optically thin discs always have $H \sim R$ and thus their thermal equilibria must be dominated by advective cooling.

The existence of thermal equilibria of advectively dominated, thermally stable, hot, optically thin accretion discs was recognized by Narayan and his collaborators at Harvard and by Abramowicz, Lasota and Regev during their collaboration at Technion in 1994. These new solutions have been anticipated in the paper on boundary layers by Narayan & Popham (1993), and possibility of them mentioned in Narayan & Yi (1994). Explicit solutions have been published by Abramowicz, Chen, Kato, Lasota & Regev (1995), who assumed that radiative cooling is given by bremsstrahlung and pointed out that they exist only below a certain accretion rate (which is smaller than the Eddington one), and by Narayan & Yi (1995) who considered a more realistic radiative cooling with synchrotron radiation and Comptonization.

7. Unification

Four physically distinct sequences of accretion discs thermal equilibria exists locally (Chen, Abramowicz, Lasota, Narayan & Yi, 1995). Two of these sequences correspond

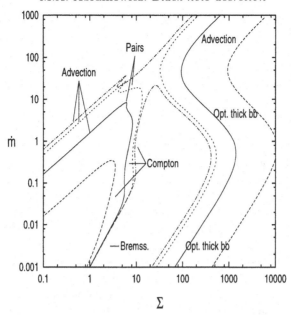

FIGURE 1. The four types of accretion flows. Labels indicate the most important cooling mechanisms. See text for detailed explanation (from Björnsson, Abramowicz, Chen & Lasota, 1996).

to a low viscosity, $\alpha < \alpha_{cr}(r)$, and the two others correspond to a high viscosity, $\alpha > \alpha_{cr}(r)$. The critical viscosity α_{cr} depends strongly on the radius, and weakly on the mass of the central accreting object. According to Chen, Abramowicz, Lasota, Narayan & Yi (1995), in the inner part of the disc $\alpha_{cr} \sim 0.1$. This value was changed by the more accurate treatment of microphysics and inner boundary condition by Björnsson, Abramowicz, Chen & Lasota (1996) who found $\alpha_{cr} > 1$. The two low viscosity sequences of models are further differentiated by whether the flow is optically thin or optically thick, and the two high viscosity flow sequences by whether advection is negligible or dominant. We shall now describe three of these four sequences in details. It would be convenient for the reader to consult Figure 1 and find the corresponding sequences and branches there. One should remember that this is a *local* classification at a fixed radius.

7.1. *Type (1): sequence SSD(gas)–SSD(rad)–slim ($\alpha < \alpha_{crit}$, $\tau > 1$, Q_{adv}/Q_{rad} varies)*

This is an "S" shaped sequence that consists of three branches: lower, middle and upper. The *lower* branch corresponds to gas a pressure dominated, radiatively cooled classical Shakura & Sunyaev solution. Disks here are cool, and optically thick. Radiation is of the black body or modified black body type. For low accretion rates, corresponding to those characteristic for dwarf novae, this branch itself bends into an S-shaped sequence of three sub-branches due to a strong opacity dependence on temperature. (There could be even more sub-branches depending on complications in heat transport, *e.g.* by a vertical convection.) For higher accretion rates the opacity is determined by electron scattering and the branch is both thermally and viscously stable. Most of what is firmly known about physics of accretion comes from a very detailed comparison of theory and observations for discs from the lower branch of this type. The *middle* branch is also a part of the classical Shakura & Sunyaev solution. It is optically thick, radiation pressure dominated, and cooled by radiation. It is thermally and viscously unstable. Thus, it has no astrophysical applications in modeling steady state discs. The *upper* branch was found by Abramowicz, Lasota & Xu. It is often called the *slim* accretion discs branch. It is optically thick, radiation pressure supported, and cooled by radial advection of heat. It is both thermally and viscously stable. It exists at moderately super-Eddington accretion rates and has radiative efficiency up to factor of 10 smaller than the classical Shakura & Sunyaev discs. The (modified) black body spectra calculated for slim discs explain the Big Blue Bump in the observed spectra of active galactic nuclei *far more* accurately than those calculated for the standard Shakura & Sunyaev ones.

7.2. *Type (2): sequence SSD(gas)–bridge–SLE ($\alpha > \alpha_{\rm crit}$, τ varies, $Q_{\rm adv}/Q_{\rm rad} \ll 1$)*

This sequence consists of three branches: right, upper, and left, that together form a shape of "⋂". The *right* branch is identical in properties (except that now the viscosity is higher) with the lower branch of the Type 1. The *upper* branch corresponds to discs that have $\tau \approx 1$ and $P_{\rm gas}/P_{\rm rad} \approx 1$. Because radiative transfer equation is very difficult to solve for plasma with $\tau \approx 1$, a phenomenological treatment ("bridging formulae") is used. It is firmly established, however, that the discs at this branch are thermally and viscously unstable. The *left* branch corresponds to the Shapiro, Lightman & Eardley models. It is optically thin, and very hot. Protons have much higher (approaching the virial regime) temperature than electrons. Radiative processes include bremsstrahlung, synchrotron radiation and Comptonization. Advective cooling is unimportant (but advective *heating* may be present). These discs are thermally unstable.

Wandel & Liang (1991) and Luo & Liang (1994) demonstrated that, one may construct global models that for large radii are Shakura & Sunyaev type, *i.e.* optically thick, and cool, while for small radii they are Shapiro, Lightman & Eardley type *i.e.* optically thin, and hot. In the 'bridging' region they have optical depth close to unity. Thus, they experience optically thick — optically thin transitions at some intermediate radii. Inner parts of these discs are predominantly cooled by radiation, and not by advection, and therefore they are violently thermally unstable.

7.3. *Type (3): sequence SLE–ADAF ($\alpha < \alpha_{\rm crit}$, $\tau < 1$, $Q_{\rm adv}/Q_{\rm rad}$ varies)*

This sequence consist of two branches, lower and upper, that together form a shape similar to asymmetric ">". The *lower* branch corresponds to SLE models, the *upper* branch corresponds to ADAFs. The ADAF discs are gas pressure dominated, very hot, and optically thin. *They are thermally and viscously stable.* Radiative efficiency is very low for them. The spectra of ADAFs are dominated by Comptonization of bremsstrahlung and synchrotron soft photons. Pairs are suppressed by strong advective cooling. Detail calculations show that these discs are excellent models for very hot, underluminous sources. Solutions of this type exist only *below* certain critical rate. It value depends strongly on detailed microphysics considered, but is typically sub-Eddington.

8. Conclusions

Advective cooling dominates thermal equilibria of the inner parts of all accretion discs around black holes. This is because the general relativistic effect of the Roche lobe overflow forces mass loss from the disc that induces a strong advective cooling.

Advective cooling is also important when $H \sim R$. This happens at high accretion rates for optically thick flows (slim accretion discs) and for small accretion rates for optically thin flows (ADAFs).

ADAFs are underluminous and very hot, with the proton temperatures reaching 10^{12} K. This resembles observational properties of several high energy sources, as explained in this book by R. Narayan *et al.* and J.-P. Lasota.

Theory predicts that ADAF and SSD models provide stable solutions for equations describing accretion flows with sub-Eddington accretion rates. Observations indicate that in some black hole accretion discs, the inner part is of the ADAF type, and the outer part of the SSD type. Observations also seem to indicate that in real objects every time when both ADAF and SSD solutions could be chosen, an ADAF is observed. R. Narayan called this observational indication "the strong ADAF principle". The question is, what physical process makes a transition from SSD to ADAF each time such a transition is mathematically possible.

In a work that is still in progress, Abramowicz & Björnsson (1998) made a conjecture: *the reason for the choice of ADAF rather than SSD solution by real objects is that ADAF produces entropy at a much lower rate than SSD.* The Prigogine (1961) principle states that a thermodynamical system, in which irreversible processes take place, seeks an equilibrium state through a configuration that minimizes the entropy production rate. Thus the "strong ADAF principle" would be seen as a particular example of a more general thermodynamical statement.

One could think that the fact that ADAF produces less entropy than SSD is trivial: dQ^+ is about the same for both solutions, and therefore the viscous and radiative entropy, $dS \approx dQ^+/T$, is smaller for ADAFs because the temperature is much higher for them. The question is, however, whether these entropy sources are the only ones that should be included. Indeed, the radial entropy flux through SSD is far smaller than that through ADAF, and thus the missing entropy may be supplied by the transition region. We calculated the entropy production by integrating viscous and radiative entropy production rates through the whole ADAF and SSD, and by including the entropy that is produced in the transition region, obtaining the same result: ADAF with the transition region produces far less entropy than the pure SSD. However, should the Bekenstein-Hawking entropy (see I. Novikov, this volume) be also included? We have strong arguments that it should not be, but this issue is connected to profound, unsolved problems in quantum theory of black holes. Then, could there be other entropy sources present in the system? For example, is entropy produced at the sound horizon, *i.e.* at the surface where the accretion speed equals the local speed of sound? Surely, some information must be lost there because the sound waves cannot propagate outwards in the supersonic flow (Visser, 1997).

These questions may establish a link between ADAFs and some problems in the quantum black hole theory.

REFERENCES

ABRAMOWICZ M.A. 1981 *Nature* **294** 235

ABRAMOWICZ M.A. & BJÖRNSSON G. 1998, work in progress

ABRAMOWICZ M.A. & CHAKRABARTI S. 1990 *Astrophys. J.* **350**, 281

ABRAMOWICZ M.A., CHEN X., KATO S., LASOTA J.-P., AND REGEV O. 1995 *Astrophys. J.* **438** L37

ABRAMOWICZ M.A., CZERNY B., LASOTA J.-P. & SZUSZKIEWICZ E. 1988 *Astrophys. J.* **332** 646

ABRAMOWICZ M.A. & KATO S. 1989 *Astrophys. J.* **336**, 304

ABRAMOWICZ M.A. KATO S. & MATSUMOTO R. 1989 *Publ. Astr. Soc. Japan* **41**, 1215

ABRAMOWICZ M.A., LASOTA J.-P. & XU C. 1986 in *Quasars*, (eds. G. Swarup & V.K. Kapachi), Reidel

ABRAMOWICZ M.A. & ZUREK W.H. 1981 *Astrophys. J.* **264** 314

BEGELMAN M.C. 1978 *Month. n. Royal Astr. Soc.* **184**, 53

BEGELMAN M.C. 1979 *Month. n. Royal Astr. Soc.* **187**, 237

BJÖRNSSON G., ABRAMOWICZ M.A., CHEN X., & LASOTA J.-P. 1996 *Astrophys. J* **467**, 99

CHAKRABARTI S.K. 1996 *Phys. rep.* **266**, 229

CHEN X., ABRAMOWICZ M.A., LASOTA J.-P., NARAYAN R., & YI I. 1995 *Astrophys. J.* **443** L61

CHEN X., & TAAM R.E. 1993 *Astrophys. J.* **412**, 254

FUKUE J. 1987 *Publ. Astr. Soc. Japan* **39**, 309

HONMA F., MATSUMOTO R., & KATO S. 1991 *Publ. Astr. Soc. Japan* **43**, 147

ICHIMARU S. 1977 *Astrophys. J.* **214**, 840

JAROSZYŃSKI M., ABRAMOWICZ M.A. & PACZYŃSKI B. 1980 *Acta Astr.* **30**, 1

KATO S., HONMA F., & MATSUMOTO R. 1988 *Publ. Astr. Soc. Japan* **40**, 709

KATZ J.I. 1977 *Astrophys. J.* **215**, 265

LASOTA J.-P., & PELAT D. 1991 Astron. Astrophys. **249**, 574

LU J.F. 1985 *Adiabatic accretion onto a black hole* Ph.d. thesis, (ISAS, SISSA)

LU J.F. & ABRAMOWICZ M.A. 1988 *Acta Astroph. Sinica* **8**, 1

LUO C. & LIANG E.P. 1994 *Month. not. Royal Astr. Soc* **266**, 386

MATSUMOTO R., HONMA F. & KATO S. 1989 in *Theory of Accretion Disks*, eds. F. Meyer & al. (Dordrecht: Kulver)

MATSUMOTO R., KATO S., FUKUE J. & OKAZAKI A.T. 1984 *Publ. Astr. Soc. Japan* **36**, 71

MONCRIEF V. 1980 *Astrophys. J.* **235**, 1038

MUCHOTRZEB B. 1983 *Acta Astron.* **79**, 1983

NARAYAN R. & POPHAM R. 1993 *Nature* **362**, 820

NARAYAN R. & YI I. 1994 *Astrophys. J.* **428** L13

NARAYAN R. & YI I. 1995 *Astrophys. J.* **444** 231

NARAYAN R. & YI I. 1995 *Astrophys. J.* **452** 710

OKAZAKI A.T., KATO S. & FUKUE J. 1987 *Pub. Astr. Soc. Japan* **39**, 457

PACZYŃSKI B. & WIITA P.J. 1980 *Astron. Astrophys.* **88** 23

PRIGOGINE I. 1961 *Introduction to Thermodynamics of Irreversible Processes*, Interscience Publishers, New York

REES M.J. 1978 *Phys. Script.* **17** 193

REES M.J., BEGELMAN M.C., BLANDFORD R.D., & PHINNEY E.S. 1982 *Nature* **295** 17

SHAPIRO S.L., LIGHTMAN A.P. & EARDLEY D.M. 1976 *Astrophys. J.* **204** 187

SHAPIRO S.L. & TEUKOLSKY S.A. 1983 *Black Holes, White Dwarfs and Neutron Stars*, Wiley

SZUSZKIEWICZ E., MALKAN M., & ABRAMOWICZ M.A. 1996 *Astrophys. J.* **458**, 474

SZUSZKIEWICZ E. & MILLER J.C. *Month. not. Royal Astr. Soc* **287**, 165

TAAM R.E., & LIN D.N.C. 1984 Astrophys. J. **287**, 761

THORNE K.S., & PRICE R.H. 1975 *Astrophys. J.* **195**, L101

VISSER M. 1997 gr-qc/9712010v2 15Dec1997

WANDEL A. & LIANG E.P. 1991 *Astrophys. J.* **380**, 84

WHEELER J.C. 1981 *Nature* **294** 230

Disc Turbulence and Viscosity

By AXEL BRANDENBURG

Department of Mathematics, University of Newcastle upon Tyne NE1 7RU, UK

Three-dimensional simulations of hydromagnetic flows in accretion discs provide strong evidence that the turbulence in discs is driven by a magnetic instability. Some basic results of those simulations are reviewed, current shortcomings discussed, and open questions and important issues are highlighted. The main motivation behind those simulations was simply to show that turbulence is self-sustained. However, an important quantitative outcome has been the determination of the magnitude of the Shakura-Sunyaev viscosity parameter α_{SS}. It is emphasized that α_{SS} cannot be considered a constant, as it does in fact depend on a number of factors: the magnetic field strength, the height above the midplane, and the magnitude of the velocity shear – to mention just a few. Given the availability of detailed simulations, it is now possible to address specific questions, for example what are the rates of Joule and viscous heating, where is the energy deposited, what are the values of turbulent Prandtl numbers, and how efficiently does the flow disperse and mix particles? Finally, the disc simulations have significantly affected and enhanced research in dynamo theory in different fields of astrophysics, because some of the ideas (dynamo-generated turbulence) may also apply to stars and galaxies.

1. Introduction

Accretion discs are a bit like waterfalls. Potential energy gets converted into kinetic and finally into thermal energy. The waterfall with the largest mass flux in Europe is Dettifoss in the northeast corner of Iceland with $\dot{M} = 1.5 \times 10^5$ kg/s. If Dettifoss were to be converted into a power plant, and if its efficiency was close to one hundred per cent, it would produce the equivalent of a luminosity of $L = 100$ MW. This is comparable with the power generated by an ordinary power plant, but less than the power produced by a nuclear power plant, which produces typically around 1000 MW. One is tempted to work out the change in water temperature per unit time due to viscous heating as the accretion stream splashes to the bottom. Equating the change of internal energy, $c_v \delta T$, with the potential energy difference suggests a temperature increase δT of only 0.1 Kelvin. This is of course consistent with common experience in that Icelandic rivers are known to be rather cold!

The mechanism by which potential energy is transferred via kinetic energy into heat is friction. However, in view of common day experience this must sound surprising, because we all know that a cup of tea would not get any warmer by stirring it. (On the contrary, but that is another matter.) Therefore, the way accretion discs produce emission is at first glance difficult to understand. However, at a second glance this is maybe not strange, because of the deep gravitational potential near the compact objects. (That does not apply to protostellar discs, but then those discs are not particularly hot either.) For Schwarzschild black holes almost 10% of the energy $E = mc^2$ of infalling material with mass m (c is the speed of light) can be extracted. On earth this is quite impossible, which is the reason why nuclear energy is here much more effective. Even on white dwarfs nuclear burning is still more effective than gravitational energy release.

There is however another much more important issue that is (or at least *was*) much more puzzling, namely the source of viscous dissipation and turbulence in discs. A related problem is the origin of enhanced (or turbulent) viscosity. With common day experience this does not appear to be a problem, because we are all used to the fact that turbulence is ubiquitous. However, this is not at all obvious in accretion discs. Convection, a strong driver of turbulence in stars, cannot be the ultimate driver of turbulence in discs, although convection may well be present in discs (e.g., Lin & Papaloizou 1980). For convection to develop, the inner layers close to the midplane have to be much hotter than the outer layers away from the midplane. However, viscous heating is the only mechanism that can possibly heat those layers. Convection transports heat away from those layers and thus cannot be responsible for setting up and maintaining a convectively unstable vertical entropy gradient. The other favourite source of turbulence is some nonlinear hydrodynamic shear instability (Dubrulle & Zahn 1991, Dubrulle 1993). However, attempts to identify such an instability in accretion discs have failed so far, and Balbus, Hawley, & Stone (1996) have shown, using simulations and analytical arguments, that purely hydrodynamic mechanisms cannot both draw energy from the shear and still transport angular momentum outwards. Although there are nonlinear shear instabilities in plane shear flow without rotation, rotation stabilizes such flows very efficiently. The simulations of Balbus, Hawley, & Stone (1996) have also shown that the onset of the nonlinear (finite amplitude) instability in the Rayleigh marginally stable case, $\partial(\varpi^2\Omega)/\partial\varpi = 0$, where ϖ is radius, can readily be simulated numerically even with relatively coarse resolution using numerical viscosities. This supports their conclusion that in the presence of rotation nonlinear hydrodynamical instabilities do not exist. There are examples of *unstable* Couette flow, but there is no example where the rotation profile is of any relevance to the case of accretion discs, i.e. where Ω decreases with radius like $\Omega \propto \varpi^{-q}$ and $0 < q < 2$.

With the rediscovery of the magnetic shearing (or magnetorotational) instability (Velikhov 1959, Chandrasekhar 1960, 1961) by Balbus & Hawley (1991, 1992) the situation has changed considerably. It is not just the fact that Balbus & Hawley have drawn attention to the importance of this instability for accretion discs, but more importantly the fact that they have shown persuasively that this, and nothing else, does actually work! In fact, Safronov (1972) did already draw attention to this instability. However, he was interested in protostellar discs and came to the conclusion that this instability would *not* work because of the low conductivity there. If his work had been in the context of accretion discs in general, maybe he would have come to another conclusion. Of course, a problem was that in the early seventies there were still many other possible sources of turbulence under consideration. One was hopeful that, if one mechanism turned out not work, some other would.

Anyway, one can say that the Balbus-Hawley instability, as it is now often called, is currently the only viable mechanism explaining turbulence in discs. Recent reviews of various aspects of the topic include those of Schramkowski & Torkelsson (1996), Brandenburg & Campbell (1997) and Balbus & Hawley (1998). The problem of cool, partially ionized protostellar discs has to be considered separately (§ 6.3). The Parker and other magnetic buoyancy instabilities must also play some role, once the Balbus-Hawley instability has developed, depending on the degree of stratification. What really matters for causing an instability similar to the Balbus-Hawley instability is simply the fact that magnetic fields couple different points in space, that would otherwise be independent. The coupling is elastic with an effective specific spring constant (per unit mass), $K = \omega_A^2 = k^2 v_A^2$, where v_A is the Alfvén speed, k some relevant wave number, and so ω_A is the Alfvén frequency. Now, if ω_A is smaller than some factor of order unity times the keplerian angular velocity Ω there is an instability. The instability appears to be much more general than that. In fact, there are now several examples where a harmonic oscillator in a keplerian orbit goes unstable if its frequency becomes comparable to or shorter than the angular frequency of the orbit. This similarity is developed further in a recent review by Brandenburg & Campbell (1997). Tidal disruption is a striking

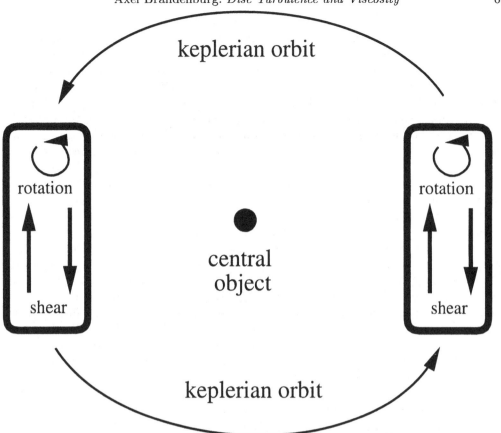

FIGURE 1. Sketch illustrating the symmetry between inward and outward directions. For the box on the right (a) the central object is to the left, so the inner parts move faster than the outer parts. For the box on the left (b) the central object is to the right. Again, the inner parts rotate faster. However, the shear flow in the two boxes (a) and (b) is exactly in the same direction. Curvature is needed to be able to tell in which direction the central object lies.

example. Here the harmonic oscillator corresponds to the radial p-mode oscillations of a star, which are of the order of 5 minutes for solar-type stars. Now if the orbital period becomes shorter or comparable to the 5 minutes oscillations the star disrupts. This can occur near black holes and is referred to as tidal disruption. This example seems to work also with g-modes. This theory has been applied to the generation of turbulence in clusters of galaxies (Balbus & Soker 1990); see also Lufkin, Balbus, & Hawley (1995) for nonlinear simulations.

The Balbus-Hawley instability is local in nature, i.e. it exists even in a local approximation. It also exists in global geometries (see Curry, Pudritz, & Sutherland 1994, Kumar, Coleman, & Kley 1994, Coleman, Kley, & Kumar 1995, Terquem & Papaloizou 1996, Ogilvie & Pringle 1996, Kitchatinov & Rüdiger 1997), but with some differences which do not affect the general conclusion. Ideally one would like to see turbulence simulations in global geometry, because the symmetry between inward and outward directions (see Fig. 1) is broken only then, and the system would 'know' whether the central object is to the left or to the right. There is work in progress by Drecker, Hollerbach, & Rüdiger (1998) trying to simulate dynamo-generated turbulence in a sphere. However, those simulations are incompressible and so the effect of gravity is only implicitly incorporated through curvature. Global turbulence simulations relevant to accretion discs have now begun to emerge. Matsumoto (private communication) has carried out global three-dimensional calculations in cylindrical geometry. However, the most detailed investigations to date have all been done using local simulations (Hawley, Gammie, & Balbus 1995, 1996, hereafter HGB95 and HGB96, Matsumoto & Tajima 1995, Brandenburg *et al.* 1995a, 1996a, hereafter BNST95 and BNST96, Stone *et al.* 1996, hereafter SHGB96). In one case (BNST96) an attempt has been made to break the symmetry between inward and outward directions by restoring terms of the order of H/R, where H is the vertical scale height and R the distance of the box from the central object.

We now discuss local calculations of accretion disc turbulence in more detail. Those simulations have in common the use of the shearing sheet approximation which, for the cross-stream (radial) direction, gives boundary conditions that are periodic with respect to fluid particles that follow the shearing motion in time. In the streamwise direction ordinary periodic boundary conditions are employed. As far as integral properties are concerned, for example the vertical magnetic flux or the total mass in the box, those quantities are conserved (assuming no mass loss in the vertical direction). This is an unfortunate restriction of all those models. The models by the various groups differ however in their vertical structure. HGB95 consider the uniform case with periodic boundary conditions in the vertical direction. SHGB96 have included vertical stratification, but they still use periodic boundary conditions in the vertical direction, although that is not a natural choice in that case. BNST95 also considered vertical stratification, but they used stress-free boundary conditions for the flow and assumed that the magnetic field is vertical on the upper and lower boundaries. The latter allows the horizontal (streamwise and cross-stream) magnetic flux to vary. This is an important property, because it enables the development of a net toroidal magnetic flux over the scale of the box. The self-consistent generation of magnetic fields, as opposed to the case of an imposed magnetic field, has been considered by BNST95, BNST96, HGB96, and SHGB96. A flow chart summarizing the relevant physical events is sketched in Fig. 2.

An important outcome of all simulations is the magnitude of the horizontal components of the Reynolds and Maxwell stresses. They are the terms that lead to angular momentum transport in the radial direction, as can be seen by inspecting the equation for angular momentum conservation in cylindrical polar coordinates, (ϖ, ϕ, z),

$$\frac{\partial}{\partial t}(\rho\varpi^2\Omega) + \boldsymbol{\nabla} \cdot [\varpi\,(\rho\boldsymbol{u}u_\phi - \boldsymbol{B}B_\phi/4\pi - \nu\rho\varpi\boldsymbol{\nabla}\Omega)] = 0. \tag{1.1}$$

When this equation is averaged we have

$$\frac{\partial}{\partial t}\langle\rho\varpi^2\Omega\rangle + \boldsymbol{\nabla} \cdot [\varpi\langle\rho\boldsymbol{u}u_\phi - \boldsymbol{B}B_\phi/4\pi - \nu\rho\varpi\boldsymbol{\nabla}\Omega\rangle] = 0. \tag{1.2}$$

In this equation the last term is small, because the microscopic viscosity ν is small. Much larger in comparison is the *turbulent* viscosity ν_t, which comes into play by *assuming* that averaged Reynolds and Maxwell stresses take a form similar to the viscous stress, but then obviously with ν being replaced by ν_t, i.e.

$$\frac{\partial}{\partial t}\langle\rho\varpi^2\Omega\rangle + \boldsymbol{\nabla} \cdot [\varpi\,(\langle\rho\boldsymbol{u}\rangle\langle u_\phi\rangle - \langle\boldsymbol{B}\rangle\langle B_\phi\rangle/4\pi - \nu_t\langle\rho\rangle\varpi\boldsymbol{\nabla}\Omega)] = 0, \tag{1.3}$$

where

$$-\nu_t\langle\rho\rangle\varpi\boldsymbol{\nabla}\Omega = \langle(\rho\boldsymbol{u})'u_\phi' - \boldsymbol{B}'B_\phi'/4\pi\rangle \equiv (\tau_{\phi\varpi}, \tau_{\phi\phi}, \tau_{\phi z}). \tag{1.4}$$

Here we have divided the various fields into mean and fluctuating parts denoted by a prime, so, for example, $\boldsymbol{u} = \langle\boldsymbol{u}\rangle + \boldsymbol{u}'$. We have also made use of the Reynolds rules (e.g.

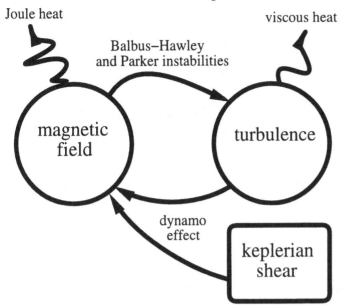

Joule heat

viscous heat

FIGURE 2. A flow chart of the different physical processes involved in accretion disc turbulence where no external magnetic field is imposed. Turbulence is generated from the magnetic field via Balbus-Hawley and Parker instabilities. The kinetic energy in the turbulence and the shear leads to the generation of magnetic fields via dynamo action. Both magnetic and kinetic energies are dissipated and produce Joule and viscous heating.

Krause & Rädler 1980) in order to write

$$\langle \rho u u_\phi - B B_\phi/4\pi \rangle = \langle \rho u \rangle \langle u_\phi \rangle - \langle B \rangle \langle B_\phi \rangle/4\pi + \langle (\rho u)' u'_\phi - B' B'_\phi/4\pi \rangle. \qquad (1.5)$$

In many applications it is assumed furthermore that $\nu_t = \frac{1}{3} u_t \ell$, where u_t is the turbulent rms velocity and ℓ some suitable correlation length. Neither of these two quantities are known, but if they are assumed to be some fraction of the sound speed and the vertical scale height, i.e. that

$$\nu_t = \alpha_{SS} c_s H, \qquad (1.6)$$

then the famous Shakura & Sunyaev (1973) prescription is obtained; hence the subscript SS on the nondimensional coefficient α_{SS}. Using keplerian rotation, i.e. $\partial\Omega/\partial \ln \varpi = -(3/2)\Omega$, and vertical hydrostatic equilibrium, $\langle c_s^2 \rangle = \Omega^2 H^2/2$, we have

$$\alpha_{SS} = \frac{\tau_{\varpi\phi}}{c_s H \langle \rho \rangle \frac{3}{2}\Omega} = \frac{\sqrt{2}}{3} \frac{\tau_{\varpi\phi}}{\langle \rho \rangle \langle c_s^2 \rangle} = 0.47 \times \frac{\tau_{\varpi\phi}}{\langle \rho \rangle \langle c_s^2 \rangle}. \qquad (1.7)$$

Thus, α_{SS} is 0.47 times the ratio of Reynolds and Maxwell stresses to the averaged gas pressure, $\langle p_{gas} \rangle = \langle \rho \rangle \langle c_s^2 \rangle$.

In the next section we summarize the results for the coefficient α_{SS} as found from the numerical simulations. We then discuss some aspects of the energetics and the large scale magnetic field generation. Finally we discuss a series of shortcomings of those models, some remaining questions and speculations, and then we have a look at neighbouring fields of research where cross-fertilization has occurred, or is bound to occur.

An important result for all applications is that α_{SS} is not a constant. We discuss this now in more detail, repeating some aspects raised already in the review by Brandenburg *et al.* (1996b).

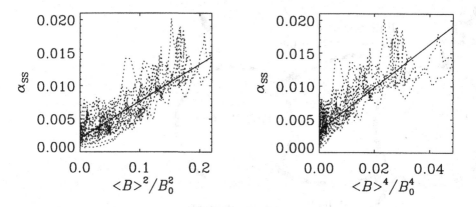

FIGURE 3. Dependence of the α_{SS} parameter on the magnetic field strength together with a least square fit against $\langle \mathrm{B} \rangle^2 / B_0^2$ (left hand panel) and $\langle \mathrm{B} \rangle^4 / B_0^4$ (right hand panel).

2. Alpha is not a constant

2.1. *Alpha depends on B*

The dependence of α_{SS} on the magnetic field strength B is probably the most important. In the models where an external magnetic field is imposed such a dependence is not surprising: the stronger the applied field, the larger the resulting stress and the larger the α_{SS} parameter. HGB95 have found that for their runs with an imposed toroidal field the stress scales with the field like $\tau_{\varpi\phi} = 0.51 \langle \boldsymbol{B}^2 \rangle / 8\pi \langle p_{\mathrm{gas}} \rangle$; see their Eq. (20). (Here and elsewhere we use a local cartesian coordinate system where x corresponds to the radial direction and y to the toroidal.) This gives

$$\alpha_{SS} = 0.47 \frac{\tau_{\varpi\phi}}{\langle p_{\mathrm{gas}} \rangle} \approx 0.12 \frac{\langle \boldsymbol{B}^2 \rangle}{B_0^2}, \qquad (2.8)$$

where $B_0^2 = 4\pi \langle \rho \rangle \langle c_s^2 \rangle$ is the square of the equipartition value with respect to the thermal energy.

It is remarkable that a similar dependence is found in the rather more general case where *no* external magnetic field is applied, but where an average field $\langle \boldsymbol{B} \rangle$ is generated self-consistently by dynamo action. In that case BNST96 suggested that the results for α_{SS} can be represented in the form

$$\alpha_{SS} \approx \alpha_{SS}^{(\mathrm{fit})} = \alpha_{SS}^{(0)} + \alpha_{SS}^{(B)} \frac{\langle \boldsymbol{B} \rangle^2}{B_0^2}. \qquad (2.9)$$

The parameters for the fit shown in the left hand panel of figure 3 are $\alpha_{SS}^{(0)} = 0.002$ and $\alpha_{SS}^{(B)} = 0.06$. Note that $\alpha_{SS} \neq 0$ even for vanishing mean field, $\langle \boldsymbol{B} \rangle \to 0$. In that case only the small scale magnetic field contributes to driving the turbulence (see HGB96). (The values of $\alpha_{SS}^{(B)}$ given in BNST96 should be divided by a factor $\langle \rho \rangle^{-2} \approx 5.2$ due to an error in their normalization of B_0^2.) However, the data points in figure 3 appear to deviate systematically from a straight line. Indeed, a fit of the form

$$\alpha_{SS} = \tilde{\alpha}_{SS}^{(0)} + \tilde{\alpha}_{SS}^{(B)} \frac{\langle \boldsymbol{B} \rangle^4}{B_0^4} \qquad (2.10)$$

appears to be somewhat better in that respect. In that case we find $\tilde{\alpha}_{SS}^{(0)} \approx 0.003$ and $\alpha_{SS}^{(B)} \approx 0.33$. However, a dependence on $\langle \boldsymbol{B} \rangle^4$ is theoretically less plausible and the

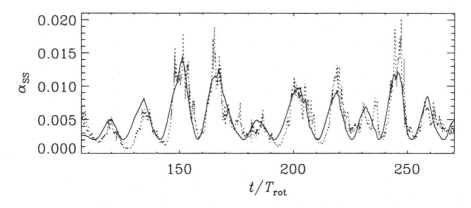

FIGURE 4. Comparison of the time dependence of $\alpha_{\rm SS}^{\rm (fit)}$ using the fit (2.9) (solid line) with $\alpha_{\rm SS}$ (dotted line). The orbital time is $T_{\rm rot} = 2\pi/\Omega$.

scatter in both cases is still large. The scatter is due to variations associated with the turbulence and will be discussed next.

2.2. Alpha fluctuates in time

The mean magnetic field depends on time and has both an average component that varies fairly gently and a fluctuating component that varies more vigorously. Therefore also $\alpha_{\rm SS}$ varies with time (figure 4). Even if the mean magnetic field dependence is removed using the fit (2.9) the fluctuations of $\alpha_{\rm SS}$ are still significant. The deviations from the fit, as measured by

$$\alpha_{\rm SS}^{\rm (fluct)}(t) = \alpha_{\rm SS}(t) - \alpha_{\rm SS}^{\rm (fit)}(B(t)), \qquad (2.11)$$

have no systematic time dependence, and the root-mean-square value of the fluctuations is about 0.002 for both fits, (2.9) and (2.10). Thus, we may conclude that $\alpha_{\rm SS}(t)$ can be written in the form

$$\alpha_{\rm SS} = \alpha_{\rm SS}^{\rm (fluct)} + \alpha_{\rm SS}^{(0)} + \alpha_{\rm SS}^{(B)} \frac{\langle \boldsymbol{B} \rangle^2}{B_0^2}, \qquad (2.12)$$

where we have used eq. (2.9) for the fit. A comparison of $\alpha_{\rm SS}(t)$ with $\alpha_{\rm SS}^{\rm (fluct)}(t)$ (figure 4) shows that the fit follows the evolution of the actual alpha parameter quite well, although the fit sometimes advances the evolution of $\alpha_{\rm SS}$.

2.3. Alpha depends on z or ρ

In the considerations above we have only looked at the vertically averaged stress. However, a disc shows significant variation in the vertical direction, so it is possible that $\alpha_{\rm SS}$ also depends on z, the height above the midplane. For example, all the models of outbursts of cataclysmic variables rely on S-shaped curves for the dependence of vertically integrated disc viscosity on vertically integrated density (Meyer & Meyer-Hofmeister 1982, Cannizzo et al. 1988). Those models assume that $\alpha_{\rm SS}$ is different on the upper and lower branches. The main reason for getting an S-shaped curve is a change in the ionization state, and thus the opacity. However, the detailed shape is model dependent. In order to calculate such a dependence from models it is assumed that $\alpha_{\rm SS}$ is independent of height. Brandenburg et al. (1996b) pointed out that this may not be justified. In the simulations of BNST95 the variation of the magnetic field with height is relatively small compared with the variation of the density. In other words, whilst the vertical scale height of the density is about half the vertical height of the box, i.e. $H \approx 0.5L_z$, the vertical scale height of the magnetic field is certainly larger than L_z. (This may seem surprising, but it is a fairly common situation in galactic discs, where the vertical scale height of the magnetic field may well be a few kpc, much larger than the scale height of the gas which is only about $100 \, \text{pc}$.)

The α-viscosity prescription was originally used in the context of vertically integrated models. If allowance for vertical dependence is made it is natural to continue using equations (1.4) with (1.6) and (1.7), so the vertical dependence of the horizontal component of the stress tensor, $\tau_{\varpi\phi}$, can be written in the form

$$\tau_{\varpi\phi} = -\nu_t \langle\rho\rangle_{\text{H}} \varpi \frac{\partial\Omega}{\partial\varpi} \approx \tfrac{3}{2}\sqrt{2}\alpha_{\text{SS}} \langle\rho\rangle_{\text{H}} \langle c_s^2\rangle_{\text{H}}, \tag{2.13}$$

for keplerian rotation. Here $\langle ...\rangle_{\text{H}}$ denotes horizontal (xy) averages. On the other hand, $\tau_{\varpi\phi}$ does not appear to decrease significantly with height like $\langle\rho\rangle_{\text{H}}$ does (see Brandenburg et al. 1996b). Therefore we are led to conclude that $\alpha_{\text{SS}} \propto \langle\rho\rangle_{\text{H}}^{-1}$ is a better approximation than just assuming α_{SS} to be independent of z. This does in fact follow directly from eq. (2.9), if $\alpha_{\text{SS}}^{(0)}$ is ignored and we assume that $\langle B\rangle_{\text{H}}$ is approximately constant with height, i.e.

$$\alpha_{\text{SS}} \approx \alpha_{\text{SS}}^{(\text{mag})} = \left(\alpha_{\text{SS}}^{(B)} \frac{\langle B\rangle_{\text{H}}^2}{4\pi\langle c_s^2\rangle_{\text{H}}}\right)\langle\rho\rangle_{\text{H}}^{-1}. \tag{2.14}$$

Since the vertical scale height of $\langle B\rangle_{\text{H}}^2$ is much larger than that of $\langle\rho\rangle_{\text{H}}$, and if $\langle c_s^2\rangle_{\text{H}}$ is approximately constant with height, it follows that $\alpha_{\text{SS}} \propto \langle\rho\rangle_{\text{H}}^{-1}$. Given that $\tau_{\varpi\phi}$ is approximately independent of height, eq. (2.13) could be replaced by a perhaps more direct representation of this fact, i.e.

$$\tau_{\varpi\phi} = -\hat{\alpha}_{\text{SS}}\Sigma c_s\varpi\frac{\partial\Omega}{\partial\varpi}, \tag{2.15}$$

where $\Sigma = \int_0^\infty \rho dz \approx H\langle\rho\rangle_{\text{H}}(0)$, with $\langle\rho\rangle_{\text{H}}(0)$ being the average density in the midplane. In our case $\hat{\alpha}_{\text{SS}} \approx 0.4\alpha_{\text{SS}}$. A vertical dependence of $\alpha_{\text{SS}} \propto \langle\rho\rangle_{\text{H}}^{-1}$, or, alternatively, the new dependence (2.15) in terms of $\hat{\alpha}_{\text{SS}}$, could significantly modify various properties including the S-curves obtained using eq. (2.13).

2.4. *Alpha depends on shear and vorticity*

There is, not surprisingly, a dependence of α_{SS} on the shear parameter $q = -\partial\ln\Omega/\partial\ln\varpi$, which measures the strength of the shear. If $q = 0$ the magnetic shearing instability shuts off, whereas in the case $q > 2$ the system is already hydrodynamically unstable (Rayleigh unstable). Abramowicz, Brandenburg, & Lasota (1996) found that α_{SS} increases monotonically with q, keeping all other input parameters unchanged. In particular, the values of c_s and Ω itself are unchanged. Abramowicz et al. have pointed out that, as $q \to 2$, the turbulence became more and more vigorous until they were unable to continue the simulation. They attempted a representation using only coordinate independent quantities such as the magnitudes of the shear and vorticity tensors, σ and ω, respectively. They found that α_{SS} is approximately proportional to the ratio of the magnitudes of the shear to viscosity tensors, i.e.

$$\langle\alpha_{\text{SS}}\rangle \propto \frac{\sigma}{\omega} = \frac{q}{2-q}. \tag{2.16}$$

The ratio σ/ω vanishes for $q \to 0$ and it tends to infinity for $q \to 2$. However, calculations of Drecker, Hollerbach, & Rüdiger (1998) of the Balbus-Hawley instability in a sphere with an imposed magnetic field do not show such a singularity near $q = 2$. The origin for this discrepancy may be related to different properties of the models.

Run	comment	resolution	L_y	β^{-1}	f^{-1}	$\beta^{-1}f^{-1}$	$\tan\phi$	$\langle\alpha_{SS}^{(mag)}\rangle$	$\alpha_{SS}^{(B)}$
O	no curv.	$31 \times 63 \times 32$	2π	0.03	2	0.06	0.07	0.004	0.07
A	no cooling	$31 \times 63 \times 32$	2π	0.01	2	0.02	0.09	0.002	0.09
B	short run	$31 \times 63 \times 32$	2π	0.01	4	0.02	0.08	0.002	0.15
C	high res.	$63 \times 127 \times 64$	2π	0.01	8	0.03	0.11	0.004	0.41
D	aspect rat.	$127 \times 63 \times 32$	4π	0.02	3	0.04	0.09	0.004	0.12
E	aspect rat.	$255 \times 63 \times 32$	8π	0.01	4	0.04	0.08	0.003	0.16

TABLE 1. Summary of parameters entering the equation for the value of α_{SS}. Note that in Run A cooling was turned off, so the temperature and hence the disc scale height increase with time. Therefore the temporal averages given for this run cannot readily be compared with those of the other runs. In all runs, except Run O, curvature terms of the form $1/R$ have been restored, so there is a nonvanishing mass accretion rate; see BNST96. However, within statistical errors, the values in the table are probably unaffected by this.

A dependence of the form (2.16) has several implications. First of all, it may provide a mechanism for limiting the disc thickness. A thick, pressure-supported disc (Abramowicz, Calvani, & Nobili 1980) is assumed to have constant angular momentum, i.e. $q \to 2$, so according to eq. (2.16) the viscosity will be very large, and one expects a large accretion velocity which would then rapidly lead to a state where the centrifugal force balances gravity as described by thin disc theory. The second example where eq. (2.16) could be applied is near black holes. Assuming that the dependence on σ/ω carries over to the relativistic regime near black holes, where σ/ω increases considerably, eq. (2.16) would predict a systematic increase of α_{SS} towards the inner parts of the disc, which tend to contribute strongest to the observed spectrum.

2.5. *Alpha depends on the numerical resolution*

Finally we mention the fact that our results for the averaged values of the mean magnetic field, $\langle B^2 \rangle$, and therefore also of α_{SS}, are not yet converged and tend to increase as the number of mesh points is increased. BNST96 found for the time averages $\langle\alpha_{SS}\rangle = 0.005$ for $31 \times 63 \times 32$ meshpoints and $\langle\alpha_{SS}\rangle = 0.007$ for $63 \times 126 \times 64$ meshpoints. The magnetic contribution to α_{SS}, see eq. (1.4), can be written as

$$\langle\alpha_{SS}^{(mag)}\rangle \equiv -0.47 \times \frac{\langle B'_\varpi B'_\phi \rangle}{4\pi\langle\rho\rangle\langle c_s^2\rangle} = -0.47 \times \frac{\langle B'_\varpi B'_\phi \rangle}{\langle B^2 \rangle} \frac{\langle B^2 \rangle}{\langle B \rangle^2} \frac{\langle B \rangle^2}{4\pi\langle\rho\rangle\langle c_s^2\rangle}; \qquad (2.17)$$

see (1.7). We now introduce the plasma-beta, $\beta = 2B_0^2/\langle B \rangle^2$, the tangent of the pitch angle, $\tan\phi \equiv -\langle B'_\varpi B'_\phi \rangle/\langle B^2 \rangle \approx -B'_\varpi/B'_\phi$, and the filling factor $f = \langle B \rangle^2/\langle B^2 \rangle$. With those definitions we can express $\langle\alpha_{SS}^{(mag)}\rangle$ in the form

$$\langle\alpha_{SS}^{(mag)}\rangle = 0.94\,\beta^{-1}f^{-1}\tan\phi. \qquad (2.18)$$

Except for the offset $\alpha_{SS}^{(0)}$ in eq. (2.9), eq. (2.17) is similar to eq. (2.9) if we identify $\alpha_{SS}^{(B)}$ with $f^{-1}\tan\phi$, i.e.

$$\alpha_{SS}^{(B)} = 0.47\,f^{-1}\tan\phi. \qquad (2.19)$$

This means that if eq. (2.9) is valid, the product of pitch and the inverse filling factor should be constant. Alternatively, if eq. (2.10) is valid, the product of pitch and inverse filling factor should increase with increasing field strength proportionally to β^{-1}. Table 2.5 summarizes some of the relevant parameters of the simulations of BNST96. However, the results do not confirm either of the two possibilities mentioned above. Instead, $\alpha_{SS}^{(B)}$ is roughly proportional f^{-1} and only weakly dependent on $\tan\phi$.

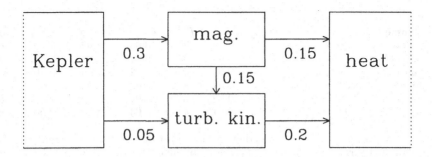

FIGURE 5. Diagram showing the energy fluxes between the various energy reservoirs. Most of the energy is being tapped from the keplerian shear via Maxwell stresses. However, the magnetic field drives turbulent motions which contribute significantly to heating the disc. The numbers denote energy fluxes in units of the average magnetic energy times Ω, as estimated from the simulation data. (Adapted from BNST95.)

3. Energetics and energy fluxes

Since we are dealing here with a magnetic instability it is not surprising that most of the energy that is tapped from the keplerian shear motion goes into magnetic energy. Subsequently, however, there is a conversion of magnetic energy into turbulent kinetic energy. This sink of magnetic energy lowers the amount of energy that has to go into Joule heating. The simulations of BNST95 have allowed us to estimate the relative magnitudes of the various energy fluxes. They found that, although most of the energy does indeed go first into magnetic energy, the energies that are eventually dissipated by Joule and viscous heating are more or less equal. The relative importance of the two heating processes may have consequences for the deposition of thermal energy into electrons and ions, which will be discussed next.

3.1. *Deposition of energy via Joule and viscous heating*

The relative importance of Joule and viscous heating is of particular significance for discs in active galactic nuclei, because there electrons are thought to be of lower temperature than the ions. On the other hand, if the Balbus-Hawley instability is really responsible for driving the turbulence one might expect that most of the heating takes place via Joule heating, which would deposit energy predominantly into the electron component.

However, there is typically a flux of energy from the magnetic field into the turbulent kinetic energy. This extra gain of kinetic energy due to the Balbus-Hawley and possibly other (secondary) instabilities enhances the viscous flux of energy and lowers the flux due to Joule heating; see figure 5. An important question is then whether the ratio between viscous and Joule heating depends on the microscopic magnetic Prandtl number $\mathrm{Pr}_M = \nu/\eta$, where ν is the kinematic viscosity and η the magnetic diffusivity. Note that these are not turbulent, but microscopic (i.e. 'molecular' or rather 'atomic') values for the viscosity and magnetic diffusivity. The expressions for viscous and Joule heating per unit volume are

$$Q_{\mathrm{visc}} = 2\nu\rho S^2, \quad Q_{\mathrm{Joule}} = \eta\mu_0 J^2, \tag{3.20}$$

where S is the strain tensor and J the electric current. For example, if Pr_M is large, i.e. $\nu \gg \eta$, one might expect $Q_{\mathrm{visc}} \gg Q_{\mathrm{Joule}}$. However, a small value of η does not necessarily imply that ηJ^2 is small, because J^2 increases when η decreases. This can be seen in simulations of coronal heating via nanoflares (Galsgaard & Nordlund 1996). The same is also true for viscous heating. Thus, one expects that the ratio $Q_{\rm visc}/Q_{\rm Joule}$ is roughly independent of Pr_M. Instead, $Q_{\rm visc}/Q_{\rm Joule}$ depends primarily on the rates at which kinetic and magnetic energies are tapped from the keplerian shear, and also on the rate at which magnetic energy can be converted into turbulent kinetic energy. If $F_{\rm kin} = \frac{3}{2}\Omega\langle\rho u_\varpi u_\phi\rangle$ and $F_{\rm mag} = -\frac{3}{2}\Omega\langle B_\varpi B_\phi\rangle/\mu_0$ are the rates at which kinetic and magnetic energy is being tapped from the shear, and if $W_{\rm Lor} = \int \boldsymbol{u}\cdot(\boldsymbol{J}\times\boldsymbol{B})\,dV$ is the work done by the Lorentz force ($W_{\rm Lor} > 0$ in the present case, where the turbulence is driven by the field), then we have on average in the statistically steady state

$$\tilde{Q}_{\rm visc} = F_{\rm kin} + W_{\rm Lor}, \tag{3.21}$$

$$Q_{\rm Joule} = F_{\rm mag} - W_{\rm Lor}. \tag{3.22}$$

where $\tilde{Q}_{\rm visc} = Q_{\rm visc} + Q_{\rm comp}$ is the heating from viscous dissipation and compressional heating, $Q_{\rm comp} = -\langle p\boldsymbol{\nabla}\cdot\boldsymbol{u}\rangle$; see BNST95. We should add here that due to discretization errors and finite time averages the actual numbers do not quite match this relation. This is also because certain integral relations that enter in the derivation of the relations above are numerically only approximately satisfied. Therefore the heating rates given in figure 5 have been adjusted in order to avoid confusion or misinterpretation.

It remains unclear how large $W_{\rm Lor}$ can be. If $W_{\rm Lor}/F_{\rm mag}$ approaches unity we have $Q_{\rm Joule} \ll Q_{\rm visc}$ and so most of the energy is dissipated in the 'conventional' way via viscous heating, which means that at first instance the ions will be heated. On the other hand, if $W_{\rm Lor}/F_{\rm mag}$ were small, most of the energy would go via Joule heating into the electrons. This may lead to a problem in that in active galactic nuclei the electron temperatures are observed to be smaller than the ion temperatures. However, as pointed out by Shapiro, Lightman, & Eardley (1976), the electrons cool faster, so even in that case the electron temperatures could well be below the ion temperatures. A recent discussion of this in connection with advection dominated accretion flows can be found in Bisnovatyi-Kogan & Lovelace (1997). Thus, the detailed properties of energy deposition unfortunately do not seem to lead to stringent observational constraints.

3.2. *The turbulent magnetic Prandtl number*

Although the microscopic Prandtl numbers may not be very important as far as macroscopic, or averaged, properties are concerned, the *turbulent* (ordinary and magnetic) Prandtl numbers may be be useful when trying to model large scale properties of the disc.

The value of the turbulent magnetic Prandtl number, $\mathrm{Pr}_M^{\rm (turb)} = \nu_t/\eta_t$, is important for the dragging of field lines from the interstellar medium into the disc. There is the common conception that the radial accretion flow will drag field lines to the inner parts of the disc. On the other hand, it is well known that this process has to compete against magnetic diffusion outwards (van Ballegooijen 1989, see also Pringle 1993 and Lubow, Papaloizou, & Pringle 1994). Efficient field line dragging may occur only for $\mathrm{Pr}_M^{\rm (turb)} \gg 1$. For $\mathrm{Pr}_M^{\rm (turb)} = 1$ the final result is not clear, the more so because the turbulent magnetic diffusion is really a tensor. More work is needed to determine the nature of field line slippage and viscous accretion flows.

3.3. *The turbulent Prandtl number*

The magnitude of the turbulent Prandtl number, $\mathrm{Pr}^{\rm (turb)} \equiv \chi_t/\nu_t$, where χ_t is the turbulent thermal diffusivity, could also play an important role. In many models of the vertical stratification of discs the effects of turbulent heat transport are neglected altogether and only the effects of radiation are taken into account.

Since the temperature in the boxes considered by BNST95 and BNST96 is almost independent of height, the specific entropy increases with height. According to mixing length theory (e.g. Rüdiger 1989) this must then lead to a turbulent enthalpy (or convective) flux, F_{conv}, towards the midplane, where

$$F_{conv} = -\chi_t \langle \rho \rangle \langle T \rangle \nabla \langle s \rangle. \tag{3.23}$$

The value of χ_t can be estimated, because all other quantities in eq. (3.23) are known. The result is that $Pr^{(turb)} \approx 0.1$. This value is relatively small, perhaps too small to explain significant modifications of the radial temperature dependence of this (Fröhlich & Schultz 1996). This value of χ_t should be compared with the radiative value, which cannot be done in the present simulations where radiation transport is not included. Furthermore, it is not clear that the results carry over to the case where the turbulence transports heat outwards and not inwards as in the present models.

3.4. *Compressive versus vortical motions*

For the Balbus-Hawley instability compressibility is not an essential ingredient (e.g. Ogilvie, & Pringle 1996). One may therefore expect compressibility to be weak in the simulations. On the other hand, the simulations are compressible and shocks may form, especially away from the midplane where the density is low and the Mach number high. It is therefore of interest to assess the relative importance of compressive and vortical motions. On the one hand, this is just another means of characterizing the type of motion taking place in the disc. On the other hand, the relative importance of compressive and vortical motions may have important implications for certain secondary effects, such as photon damping of compressive MHD waves (Agol, & Krolik 1998).

One way of quantifying the relative importance of compressive and vortical motions is by measuring the root-mean-square values of vorticity and velocity divergence. For example, in compressible convection the rms velocity divergence is about 10% of the rms vorticity (Brandenburg *et al.* 1996c). In accretion disc turbulence this ratio can easily be much larger, as can be seen from the first panel of figure 6 where we have plotted the ratio $(\nabla \cdot u)_{rms}/\omega_{rms}$ separately for each layer. Here, $(\nabla \cdot u)_{rms} = \langle (\nabla \cdot u)^2 \rangle^{1/2}$ and $\omega_{rms} = \langle \omega^2 \rangle^{1/2}$ are the rms values of velocity divergence and vorticity. One sees that this ratio is largest close to the midplane, where values between 0.4 and 0.7 are reached. Thus, compressive effects are actually important near the midplane, even though the Mach number is low there.

Another way of characterizing the flow is by separating it explicitly into vortical and compressive components,

$$u = u_{vort} + u_{comp}. \tag{3.24}$$

By expressing the two contributions in terms of vector and scalar potentials, $u_{comp} = \nabla \times \psi$ and $u_{vort} = \nabla \phi$, one can see that $\nabla \times u = \nabla \times u_{vort}$ and $\nabla \cdot u = \nabla \cdot u_{comp}$. In the second panel of figure 6 we show velocity powerspectra separately for vortical and compressive components of u. The ratio between the typical magnitudes of the spectral energies of vortical and compressive components of u is about ten, confirming that the ratio of the typical magnitudes of the velocity is around three. The spectra are too short to identify an inertial range, but even for the largest scales the slopes in the curves are typically k^{-2} or even steeper.

Finally, in figure 7 we look at the vertical profiles of rms velocity and Alfvén speed and the vortical and compressive Taylor microscales, u_{rms}/ω_{rms} and $u_{rms}/(\nabla \cdot u)_{rms}$,

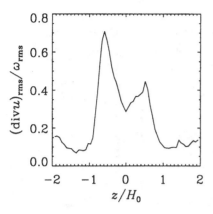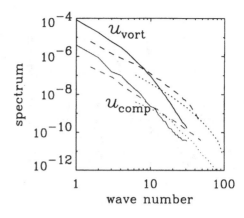

FIGURE 6. Left hand panel: ratio of rms velocity divergence to rms vorticity. Right hand panel: velocity power spectra of vortical and compressive components. Solid lines refer to power spectra taken in the y direction, whilst dotted and dashed lines refer to the x and z directions. The three spectra for the compressive motions are plotted as grey lines.

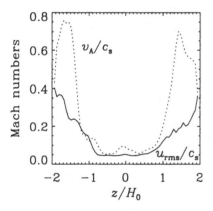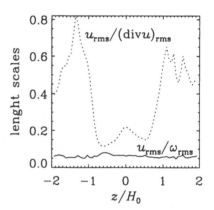

FIGURE 7. Left hand panel: the turbulent Mach number u_{rms}/c_s (solid line), and turbulent magnetic Mach number v_A/c_s (dotted line), as functions of height. Right hand panel: vortical and compressive Taylor microscales (see text).

respectively. We find that u_{rms}/ω_{rms} varies very little with height, whilst the scale $u_{rms}/(\nabla \cdot \boldsymbol{u})_{rms}$ is largest away from the midplane and smallest near the midplane.

4. The importance of modelling the large scale field

We have seen that the Shakura-Sunyaev alpha, α_{SS}, remains time dependent and varies with the large scale magnetic field; see equations (2.9) or (2.10). This type of variation is ignored in the standard accretion disc model – even if time dependence is essential, like in models for cataclysmic variables. In order to include this effect one would need to solve a set of model equations for the dynamo. We begin by discussing first the basic conclusions gained from the simulations and then turn to their phenomenological description.

It is remarkable that in the local simulations not only a small scale magnetic field is generated, but also a large scale field. However, the generated large scale magnetic field may be sensitive to the magnetic boundaries adopted, because the scales over which the

stretch - twist - fold   ...   - escape dynamo

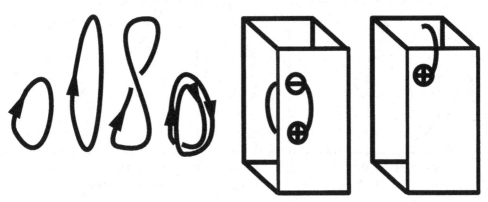

FIGURE 8. Sketch illustrating the enhancement of small scale flux from a stretch–twist–fold dynamo and the subsequent escape of flux contributing to a net horizontal field as long as the flux loop has escaped only partly through the boundary.

magnetic field varies is comparable with the size of the computational domain. In order to assess the sensitivity to changes in the magnetic boundary conditions we now compare two local simulations carried out with two rather 'orthogonal' conditions on $z = \pm L_z$,

$$B_x = B_y = \frac{\partial B_z}{\partial z} = 0 \quad \text{(vertical field condition)}, \tag{4.25}$$

$$\frac{\partial B_x}{\partial z} = \frac{\partial B_y}{\partial z} = B_z = 0 \quad \text{(perfect conductor condition)}. \tag{4.26}$$

The vertical field condition, eq. (4.25), imitates a vacuum boundary condition. However, for a proper vacuum boundary condition one would have to match the solution to a potential field solution, $\boldsymbol{B} = \boldsymbol{\nabla}\phi$ with $\nabla^2 \phi = 0$. This leads to a nonlocal condition in that the condition on one point depends on the field at all other points on the boundary. The condition becomes local in spectral space (e.g. Krause & Rädler 1980), and is usually implemented using Fourier transforms. However, the shearing sheet condition precludes the use of Fourier transforms in the cross-stream direction. A possible alternative would be to transform onto a nonorthogonal grid and to apply Fourier transform in the (inclined) direction in which the mesh is periodic. This approach has been adopted by Gammie (private communication) in order to solve the Poisson equation in simulations with self-gravity. Thus, in view of those difficulties, the vertical field condition eq. (4.25) is a sensible compromise – good enough for the purpose of a local model. In this connection it should be remembered that in a local model the potential field condition has its own unrealistic feature in that the field decays exponentially with height and not algebraically. Also, the medium in the disc corona is not insulating. A better, albeit much harder, approach is to solve for a force-free field outside. Anyway, the vertical field condition used so far can be physically motivated by saying that near the boundaries magnetic buoyancy tends to make the field emerge vertically from the boundary. This vertical field condition has been used extensively in simulations of magnetoconvection (e.g. Hurlburt & Toomre 1988).

Both, the vertical field condition as well as the vacuum condition have the property that toroidal flux is no longer conserved. This is a crucial property of a large scale dynamo. In figure 8 we illustrate why this is important. Suppose the dynamo in the disc generates only closed small-scale loops. A possible mechanism for this is the stretch–

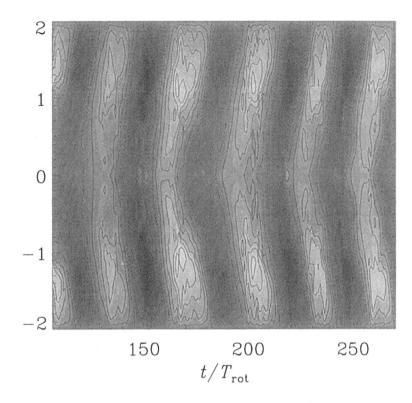

FIGURE 9. Space-time diagram of the averaged toroidal magnetic field in a simulation using the vertical magnetic field boundary condition. By imposing boundary conditions on $z = 0$ the field has been made strictly symmetric about the midplane. The field is oscillatory with a typical period of about 30 orbits.

twist–fold dynamo (Vainshtein, & Zeldovich 1972), which regenerates new field loops with twice the flux after each iteration. Since the loops are closed they contribute no net flux over the scale of the box, but when such a loop goes through the boundary there will be a nonvanishing net flux.

4.1. *The effect of boundary conditions in local simulations*

For most of the calculations carried out so far magnetic boundary conditions have been used that force the magnetic field to be vertical on the upper and lower boundaries. Technically this has the advantage that thus the horizontal components of the magnetic flux, i.e. the mean magnetic fields $\langle B_x \rangle$ and $\langle B_y \rangle$, are not restricted and can evolve freely (BNST95). In that case we find an oscillatory magnetic field, a space-time diagram of which is given in figure 9. Those results should be contrasted with the case of the perfect conductor boundary condition, of which a space-time diagram is shown in figure 10. We restarted this simulation from a previous snapshot obtained using the vertical field boundary condition. Therefore the initial condition was of even parity, and the field continued to show signs of oscillatory behaviour for the first ten orbits, but then the oscillations died away and the field began to settle into an antisymmetric configuration without cycles and yet finite amplitude.

In both cases a large scale field is generated. There are two main differences between

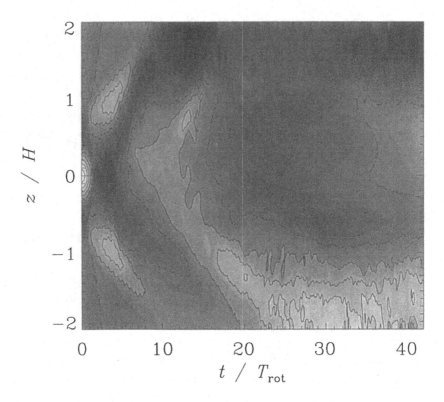

FIGURE 10. Space-time diagram of the averaged toroidal magnetic field in a simulation using the perfect conductor boundary condition for the magnetic field. The initial condition was taken from a snapshot of a simulation with the vertical field condition, where the field was oscillatory and of even parity. Until $t = 10\,T_{\text{rot}}$ the field continued to show signs of oscillatory behaviour, but it then turned to be of odd parity about $z = 0$ without being oscillatory.

the two cases with perfect conductor and with vertical field boundary conditions. Firstly, for the vertical field condition the toroidal field is of approximately even parity about the midplane, $z = 0$, whereas in the perfect conductor case the field is of approximately odd parity about $z = 0$. Secondly, the field is oscillatory if a vertical field condition is used, but non-oscillatory in the other case. Thus, obviously the behaviour of the magnetic field is quite dependent on the precise boundary conditions for the magnetic field. Thus, one might deduce that no sensible conclusions can be drawn from current local simulations. However, in the following we shall point out that this behaviour is quite consistent with what is expected from mean-field dynamo models in the same geometry. This may lend some support to the interpretation of those results in terms of mean-field models. However, it remains true that issues of whether or not the field is oscillatory require global modelling. We return to the results for global models of mean-field dynamos at the end of the next section.

4.2. *The effect of boundary conditions in mean-field models*

A way to understand large scale magnetic field generation in astrophysical bodies is the concept of an $\alpha\Omega$-dynamo. Here the dynamo alpha, α_{dyn} (with dimensions of velocity), is quite distinct from the nondimensional α_{SS} parameter. The relevant dynamo equation

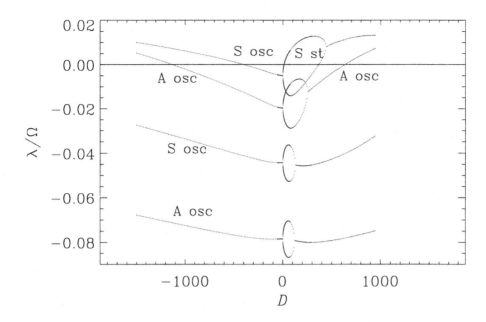

FIGURE 11. The first largest (normalized) eigenvalues, λ/Ω, of an $\alpha\Omega$ dynamo in a cartesian box with vertical magnetic field boundary conditions at top and bottom. For negative dynamo numbers, $D < 0$, the first excited mode has symmetric parity and is oscillatory (S osc) near $D = -400$, which is in qualitative agreement with the simulations.

is obtained by averaging the induction equation. In the resulting equations there are two important parameters, the α-effect and the turbulent magnetic diffusivity, η_t. For the particular simulations of BNST96 there is now quantitative information concerning the magnitudes of those two parameters. Recent work by Brandenburg & Donner (1997) and Brandenburg & Sokoloff (1998) suggests that $\alpha_{\mathrm{dyn}} \approx \mp 0.001\Omega H$ (in the upper and lower disc planes, respectively) and $\eta_t \approx (0.003 - 0.008)\Omega H^2$. Note that the sign of α_{dyn} changes at the midplane. This antisymmetry is connected with the fact that α_{dyn} is a pseudoscalar, i.e. it changes sign under a coordinate transformation with respect to reflection.

When solving the mean-field dynamo equations it is found that the lowest wave numbers of the magnetic field, corresponding to the largest possible scales, dominate the problem. Therefore boundary effects play an important role. For example, the nature of magnetic cycles is expected to depend on the geometry (local/global) and on boundary conditions. To address those issues we now compare our local simulations with mean-field dynamo models using the two boundary conditions (4.25) and (4.26).

We have calculated the first few largest eigenvalues, λ, of the mean-field dynamo equations using as boundary conditions either (4.25), see figure 11, or (4.26), see figure 12, and $\alpha_{\mathrm{dyn}}(z) = \alpha_0 \, (z/H)$. The relevant equations used in the calculation can be found in Brandenburg & Campbell (1997) and Brandenburg & Donner (1997), for example.

For oscillatory solutions there is a pair of complex conjugate eigenvalues, with frequencies $\pm \mathrm{Im}\lambda$. As the dynamo number, $D = q\alpha_0\Omega H^3/\eta_t^2$, changes those two modes with the same growth rate $\mathrm{Re}\lambda$ may split into two non-oscillatory modes. They are steady when the growth rate vanishes, so we refer to those modes as 'S st' and 'A st' for symmetric and antisymmetric fields, respectively. For negative values of α_0, i.e. for $D < 0$,

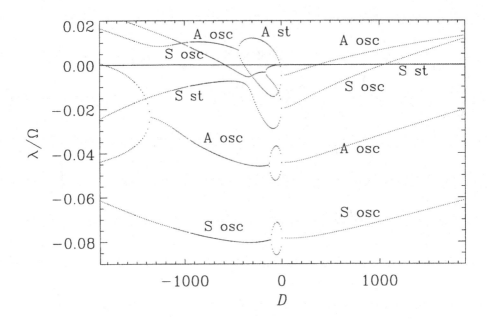

FIGURE 12. The first largest (normalized) eigenvalues, λ/Ω, of an $\alpha\Omega$ dynamo in a cartesian box with perfect conductor boundary conditions at top and bottom. For negative dynamo numbers, $D < 0$, the first excited mode is now antisymmetric about $z = 0$ and the field is nonoscillatory (A st) near $D = -10$, which is again in qualitative agreement with the simulations for this type of boundary condition.

we find that the first excited mode is symmetric and oscillatory (S osc) when (4.25) is used (see figure 11) and antisymmetric and non-oscillatory (A st) when (4.26) is used (see figure 12). Both of these properties agree with the results of the local turbulence simulations.

One could think of many reasons why the $\alpha\Omega$-dynamo may not provide an adequate description of the magnetic field behaviour in accretion discs. For example, there are indications that α_{dyn} and η_t may be wavenumber dependent (Brandenburg & Sokoloff 1998) and, of course, the effects of fluctuations are ignored. Nevertheless, it does allow some explicit modelling of the magnetic field, which otherwise would have been assumed to vanish altogether, and it does reproduce some gross properties of the magnetic field found in fully three-dimensional accretion disc simulations.

5. Deficiencies of current simulations

We have already mentioned briefly some main deficiencies of present models, namely the local nature and limited extent of the simulations in the vertical, radial and toroidal directions. Let us now discuss both of these restrictions in more detail.

5.1. *The limited vertical extent of the box*

Most of the published models extend to just a few density scale heights H. Whilst cooling keeps the value of H close to the initial value, any heating of the disc tends to increase the value of H and would therefore lower the number of scale heights within the box. At the same time, as we have already discussed, the magnetic field shows no signs of levelling off towards large heights. On the contrary, once the field has reached the upper and lower boundaries it starts to feel the boundaries and there are signs of an artifactual increase of the magnetic field near the boundaries (Nordlund, private communication).

Future work has to reveal, (i) what is the true vertical magnetic field distribution in discs and, (ii) how can that be modelled using local simulations. An important computational problem here is the fact that the density drops significantly towards the upper and lower boundaries, whilst the magnetic field has hardly changed. Consequently the Alfvén speed increases significantly towards the low plasma-beta corona of the disc, making the time step very short.

Recently, Matsumoto (private communication) has performed simulations of a box that covered about ten scale heights with a density contrast of about 1 : 1000 between the upper edge of the box and the midplane. He considers the case where z/ϖ reaches values of order unity and so he has to allow for deviations from the simple gravity law $g_z = -\Omega^2 z$. In his case the main objective was to study the effects resulting from the Parker instability. However at the moment his simulations cover only the first ten orbits.

5.2. *Geometry and radial boundary conditions*

An important restriction from the assumption of (sliding) periodic boundary conditions in the radial direction is the fact that the mass in the box cannot change, except in principle for vertical mass loss, which cannot occur in the present models either. Hence, with sliding periodic boundary conditions there can be no local accumulation of matter by having more input from the outside and less output towards the inner parts of the disc. Furthermore, the averaged vertical magnetic field is strictly conserved and since it is zero initially, it remains so for all times. Both of these problems can be removed by going to a model in cylindrical geometry with open boundary conditions in the radial direction. However, some experience needs to be gained to make sure that a model with open boundary conditions is stable.

5.3. *The toroidal extent of the box*

The boxes used by HGB95 and BNST95 had usually a toroidal extent of about six vertical scale heights. For shorter boxes the turbulence does not seem to have enough freedom to develop and the resulting values of α_{SS} are smaller than for larger boxes. Making the box much longer does not seem to have a strong effect on the value of α_{SS}; see BNST96 and figure 13, where we have reproduced images of vertical vorticity in a horizontal plane $z = H$ for runs with three different aspect ratios and, for the largest aspect ratio, two different values for the resolution in the longitudinal direction.

In the recent simulations of Matsumoto (private information) the toroidal extent has been increased to almost twenty vertical scale heights. Note that the Parker instability has wavelengths in the toroidal direction of ten or more density scale heights. It is therefore only now in the new simulations of Matsumoto that the Parker instability can fully develop. However, the toroidal extent alone was not sufficient, but it was probably the combination of increased toroidal and vertical extent together with an initially strong magnetic field (his plasma beta was initially unity). It will be important to see the full results of those simulations. The only problem is that there are at present difficulties with extending the simulations for much longer than about ten orbits. Part of this is difficulty is presumably the fact that the small size of the time step becomes very restrictive as the plasma beta becomes smaller than unity. Another difficulty is the fact that pressure gradients become ineffective in responding to fluid expansion or compression that are governed by the continuity equation. This may locally result in very low temperatures, which are difficult to handle if there are not enough mesh points.

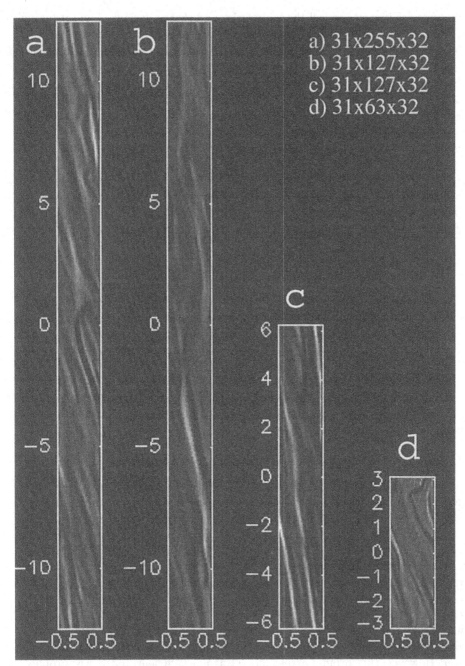

FIGURE 13. Images of vertical vorticity in horizontal planes for simulations with different toroidal extent. Note that the patterns do not seem to become longer, indicating that the most pronounced turbulence pattern is well accommodated in a box whose toroidal extent is just 2π.

5.4. *Lack of radiation transport*

The simulations of accretion disc turbulence in cartesian boxes all lack radiation transport. At best some kind of volume cooling is applied that works either in the entire computational box (BNST95) or only near the surface layers (Brandenburg *et al.* 1995b). In the former case the disc tends to be isothermal, which is clearly unrealistic. In the latter case the disc can continue to heat up in the inner parts, and only the outer parts are kept at a low temperature. This leads to a sudden drop in specific entropy near the location where the upper and lower cooling layer begin which, in turn, can lead to the development of convective layers. However, those layers do not seem to contribute significantly to the overall stress and the value of α_{SS}.

The time is ready now to incorporate realistic fully nonlocal (both optically thin and thick) radiation transport into the shearing box models. Work in this direction is currently in progress (Caunt 1998). The numerical techniques are similar to those used in solar and stellar granulation calculations (Nordlund 1982). Discs in cataclysmic variables are particularly well suited for this work, because here the temperatures and densities in the disc are comparable to those in the sun. It should in principle be possible to construct models along the upper and lower branches of the S-curve in the $(\Sigma, \nu\Sigma)$ parameter space. This would extend the one dimensional models of Meyer & Meyer-Hofmeister (1982) and Cannizzo et al. (1988) with prescribed α_{SS} profiles to the three-dimensional case with self-consistently calculated turbulence without making any assumptions about the nature and properties of turbulent dissipation.

6. Questions and speculations

In this section we discuss some issues that can be addressed either now, or that one would like to address in the near future using more realistic models.

6.1. *Vortices*

Rapidly rotating bodies, such as the earth and Jupiter, show large scale vortices. In the case of the earth those vortices are associated with cyclones (low pressure) and anticyclones (high pressure), whereas in the case of Jupiter the Great Red Spot is actually an anticyclone (high pressure). It is quite possible that vortices form also in accretion discs (Abramowicz et al. 1992; see Bracco et al, this volume). Local simulations, however, have so far not revealed the existence of vortices (BNST95), which could be due to the inhibiting effects of the magnetic field on the development of such coherent structures (Dubrulle & Valdettaro 1992).

It has recently been proposed that vortices are a global mode of the AKA-effect (von Rekowski, Kitchatinov, & Rüdiger 1998). The AKA-effect describes an instability of the mean-field hydrodynamic equations (Frisch, She, & Sulem 1987, Kitchatinov, Rüdiger, & Khomenko 1994), similar to the α-effect in the mean-field induction equation – therefore the name AKA, which means Anisotropic Kinetic Alpha effect. Again, however, local simulations of accretion disc turbulence have not yet indicated the existence of this effect (BNST95). One possible explanation for the absence of this effect may be related to the fact that the AKA-effect requires a non-galilean invariant forcing, i.e. some forcing that moves relative to the gas in the disc. It appears difficult to explain such a forcing, unless there is a companion star constantly perturbing the velocity field.

In the nonmagnetic case Hawley (1987) found long-lived vortices in thick accretion disc simulations, and Goodman, Narayan & Goldreich (1987) found an analytic solution of a long-lived eddy, which they refer to as a planet solution. Balbus & Ricotti (1998) have extended this solution to the magnetic case with uniform pressure. However, simulations are needed to see whether or not those solutions do indeed correspond to long-lived structures.

6.2. *Relative importance of the Parker instability*

It has been debated whether or not the Parker instability plays any role in producing turbulence and to reinforcing the magnetic field. Tout & Pringle (1992) suggest that most of the *vertical* magnetic field is generated by the Parker instability. On the other hand, HGB95 have pointed out that in their simulations, since there is no vertical gravity, there is no stratification and so the Parker instability is eliminated. Therefore it can be argued that the Balbus-Hawley instability alone is sufficient to produce turbulence and reinforce the magnetic field via dynamo action without invoking the Parker instability.

Real accretion discs are stratified in the vertical direction, and thus the Parker instability will always be present. In fact, it is quite possible that it might then even contribute to driving the turbulence. To estimate the relative importance of those two mechanisms, BNST95 compared the magnitudes of the magnetic tension term, $\boldsymbol{B} \cdot \nabla \boldsymbol{B}$, and the magnetic buoyancy term, $-\nabla(\boldsymbol{B}^2/2)$. Unlike the case of convective dynamos (Nordlund *et al.* 1992), where the magnetic pressure gradient is larger than the magnetic tension term, BNST95 found that in discs the two are approximately equally large. They used this to suggest that Parker and Balbus-Hawley instabilities were roughly equally important for driving turbulent motions in their simulations. However, more accurate comparisons have not been carried out so far. One way to do this is to project the flow field onto the eigenfunctions of the two instabilities. A similar procedure has been adopted in the context of stellar oscillations (Bogdan, Cattaneo & Malagoli 1993).

6.3. *Protostellar discs*

Whether or not magnetic fields are of crucial importance in protostellar discs is somewhat unclear. Stepinski *et al.* (1993) argues that the degree of ionisation is so low that the conductivity will be insufficient to retain a magnetic field. If that is the case the Balbus-Hawley instability could not be responsible for driving turbulence in such discs. On the other hand, the Balbus-Hawley instability is probably the only mechanism that is known to produce turbulence and viscosity. However, self-gravity may provide another mechanism producing turbulence (Gammie, private communication).

Whether or not the Balbus-Hawley instability works depends not only on the conductivity, but also on the ratio of the neutral-ion collision frequency to the orbital frequency (Blaes & Balbus 1994). Recently, Regös (1997) has estimated that the ratio of those frequencies is around 300, large enough for the Balbus-Hawley instability to operate and for dynamo-generated turbulence to work (BNST95). Nevertheless, if the conductivity is really too low or, more precisely, the magnetic Reynolds number much below one hundred there is no way MHD effects could be important for driving turbulence, enhanced viscosity, and thus angular momentum transport.

Of course, it is possible that the estimates for the conductivity are too pessimistic. It is also possible that turbulence is really absent near the midplane of the disc, and that only the outer parts of the disc are turbulent (Gammie 1996). Furthermore, there is no doubt that both the inner parts (near the protostar) as well as the outer parts (well beyond 1 AU) are sufficiently ionised (Stepinski *et al.* 1993). Hence, in those regions MHD turbulence will be possible. This might then lead to a situation where mass is transported everywhere, except in a strip near 1 AU (for the solar nebula). This could lead to an accumulation of matter near the outer edge of the nonmagnetic strip. It is conceivable that this might then lead to an unstable situation. This type of scenario needs to be studied in more detail before some sensible conclusions can be drawn.

In any case, most parts of a protostellar disc (i.e. everywhere except near the midplane around 1 AU for the solar nebula) will be in a state of hydromagnetic turbulence. Therefore existing simulations can be used to study turbulent processes in protostellar discs such as the formation of planetesimals from dust. Hodgson & Brandenburg (1998) have recently attempted such a study. They have addressed the question whether or not dust accumulates in cyclonic vortices, as was recently suggested by Barge & Sommeria (1994) and Tanga *et al.* (1996). Similar ideas have also been put forward recently by Klahr & Henning (1997). However, the answer seems to be no for several reasons. First of all, in the simulations of BNST95 vortices are short lived, which could be due to the presence of the magnetic field (Dubrulle & Valdettaro 1992). Secondly, the strong radial shear in discs opposes the tendency for particles to accumulate in anticyclones (Hodgson & Brandenburg 1998).

An important next step in this line of research is to include agglomeration processes in such particle calculations to see whether or not turbulence enhances the collision frequency of particles to the extent that particles would stick together more easily and form planetesimals more rapidly.

6.4. *Some comments regarding 2-D models with alpha viscosity*

Traditionally the alpha-viscosity prescription has been used to construct one-dimensional accretion disc models (Shakura-Sunyaev 1973, Novikov & Thorne 1973). This approach proved to be quite successful, which is partly due to the fact that the phenomenological parameter α_{SS} enters the final results only with low powers (e.g. Frank, King, & Raine 1992). An increase of α_{SS} by a factor of ten would lower the disc temperature only by a factor 1.6, for example.

The alpha-viscosity has also been used to produce one-dimensional models of the vertical disc structure. In § 2.3 we have already mentioned that simulations indicate an increase of the effective value of α_{SS} with height above the midplane. In the present section we want to discuss the use of the alpha-viscosity (or any type of turbulent viscosity) in models in more than one dimension.

It is not the dimension of the model as such that gives rise to concern, but rather the complexity of the resulting flow pattern. In one-dimensional models, and surely in some models in more than one dimension, the flow acts just in such a way as to bring the system into a relaxed state. However, in general a flow may result from some instability. For example large scale convection may be produced because of an unstable vertical or radial entropy gradient. Another obvious example of an instability is the Balbus-Hawley instability. In all those cases a curious situation may occur. Take the example of the Balbus-Hawley instability, which leads to fingering in the radial direction in a meridional plane in the presence of a vertical magnetic field (Hawley & Balbus 1991). The presence of (turbulent) viscosity may affect the onset of the instability. We refer to this instability as a macroscopic instability, so as to distinguish from the microscopic instability which was responsible for producing the turbulence, which in turn gives rise to an enhanced turbulent viscosity, ν_t.

The macroscopic instability will be suppressed when the turbulent decay rate $\nu_t k^2$ (for perturbations with wave number k) becomes comparable with the growth rate of the instability, which is of the order of Ω, i.e.

$$\nu_t k^2 \gtrsim \Omega \quad \text{for macro-stability.} \tag{6.27}$$

Assuming now $\nu_t = \alpha_{SS} \Omega H^2$ and $k \approx 2\pi/H$ we have

$$\alpha_{SS} \gtrsim (2\pi)^{-2} \approx 0.03 \quad \text{for macro-stability.} \tag{6.28}$$

This condition is normally not satisfied, because current simulations indicate $\alpha_{SS} = \mathcal{O}(0.01)$. However, we have to keep in mind that our estimates were quite crude and that α_{SS} may turn out to be much larger in more realistic simulations. Also, of course,

α_{SS} is time dependent, which complicates the issues further. More important at the moment seems to be the question of what is the meaning of an instability of a state that is already unstable. In the most optimistic interpretation it could mean that the turbulent state shows large scale flows, but that would mean that we have to rely on the detailed equations describing the mean flow using turbulent transport coefficients. In fact, this mean-field approach can only be, if anything, an approximation and is likely to be only of little use for any more sophisticated applications like the ones discussed here.

There is a similar example in the stellar context, where the outer convection zones are usually modelled using mixing length (or mean-field) theory. The turbulent thermal diffusivity and kinematic viscosity are given by the profiles of turbulent velocity and mixing length, that are obtained from mixing length models. One can then calculate a Rayleigh number for the convective shell and finds that it is usually supercritical. This has led to the speculation that some large scale convection might develop on top of a turbulent background and that the solar differential rotation might even be unstable (e.g. Rüdiger 1989 and references therein). In principle such large scale flows may even be physical and could resemble large scale flows that have been seen in laboratory convection (Krishnamurti & Howard 1981). Another possibility might be that, since in the sun the turbulent transport coefficients are due to convective turbulence, their values should really lead to a marginally stable state. The reason why the numbers do not quite yield a marginally stable state may be related to the fact that mean-field theories are inaccurate and do not include detailed physics such as rotation, magnetic fields, boundary effects, and the global geometry of the problem. This issue has been discussed by Tuominen *et al.* (1994), where many references to earlier work can also be found.

In conclusion, the study of flow patterns obtained by solving the mean-field hydrodynamic equations in two or even three dimensions using turbulent viscosities may be an interesting exercise, but it is at present unclear whether those flows occur in reality.

6.5. *The origin of the dynamo-alpha in discs*

We have mentioned in § 4.1 that the dynamo alpha, α_{dyn}, is negative in the upper disc plane (BNST95), which is in contrast to basic ideas in dynamo theory (e.g. Krause & Rädler 1980). The following calculation may shed some light on this question. It does reproduce the sign of α_{dyn} that is seen in the simulations. It also yields a natural relationship between the two rather different quantities α_{dyn} and α_{SS}.

We assume that the vertical motions are governed by magnetic buoyancy, so

$$\frac{\partial u_z'}{\partial t} = -\frac{\rho'}{\rho}g = \frac{(B^2)'}{8\pi p}g \approx \frac{\langle B_y \rangle B_y'}{4\pi p}g, \qquad (6.29)$$

where primes refer to deviations from some mean value, ρ is density, p is gas pressure, and g is gravity. We adopt a local cartesian coordinate system, where y corresponds to the azimuthal direction and x to the radial direction in cylindrical polar coordinates. The resulting electromotive force is then

$$\mathcal{E}_y = \langle u_z' B_x' - u_x' B_z' \rangle \approx \langle u_z' B_x' \rangle = +\langle B_y \rangle \frac{\langle B_x' B_y' \rangle}{4\pi p} g\tau, \qquad (6.30)$$

where τ is some relevant time scale. Now, because of shear ($\partial u_y / \partial x < 0$) we have $\langle B_x' B_y' \rangle < 0$. The dynamo alpha quantifies the magnitude of the component of the electromotive force in the direction of the mean field. Therefore,

$$\mathcal{E}_y = \alpha_{\text{dyn}} \langle B_y \rangle + ..., \qquad (6.31)$$

and so we have (ignoring higher order terms)

$$\alpha_{\mathrm{dyn}} = +\frac{\langle B'_x B'_y\rangle}{4\pi p}g\tau. \tag{6.32}$$

In accretion disk theory the *negative* ratio of the horizontal Maxwell stress and the gas pressure is about twice the Shakura-Sunyaev viscosity parameter α_{SS}; see eq. (1.7). Also, since $g = \Omega^2 z$, we can write

$$\alpha_{\mathrm{dyn}} \approx -2\alpha_{\mathrm{SS}}\Omega^2 z\tau \tag{6.33}$$

or, in terms of the inverse Rossby number $\mathrm{Ro}^{-1} = 2\Omega\tau$,

$$\frac{\alpha_{\mathrm{dyn}}}{\Omega H} \approx -\alpha_{\mathrm{SS}}\,\mathrm{Ro}^{-1}\,\frac{z}{H}. \tag{6.34}$$

The effects of rotation and shear are now hidden in the fact that the stress $\langle B'_x B'_y\rangle$ is negative, which is due to the negative shear. This estimate also assumes that the thermal expansion of buoyant tubes is small compared with the magnetic contraction due to the $\mathbf{B}\cdot\nabla\mathbf{B}$ term. Otherwise the sign may be the conventional one. In fact, the values of α_{dyn} obtained from the above estimate are far too optimistic compared with the values obtained in the simulations; see § 4.1. This suggests that α_{dyn} is governed by some more delicate balance with other effects that tend to cancel each other. Thus, a proper analysis is called for. However, at present there is no other calculation that explains even the sign of α_{dyn} that is seen in the simulations.

6.6. *Further applications, implications, and related developments*

We have mentioned in the beginning that the simulations of Balbus-Hawley turbulence and dynamo action (dynamo-generated turbulence) have had a tremendous impact on dynamo research in general. Simulations of accretion disc turbulence have produced strong large scale magnetic fields, even magnetic activity cycles, which all seem to be important properties of stellar dynamos. This has spawned related studies in at least two different directions. On the one hand, simulations of accretion disc turbulence resemble in some ways the conditions relevant to the interstellar turbulence of the stratified galactic disc on the scale of a few vertical density scale heights. On the other hand, the simulations have emphasized the importance of shear for causing large scale magnetic fields, which has led to simulations of convective stellar turbulence with imposed differential rotation with a profile similar to that suggested by helioseismology. In the following we briefly report on those two strands of ongoing investigations in a little more detail.

6.6.1. *Galaxies*

The main difference between turbulence in discs and galaxies is perhaps the presence of external drivers of turbulence in galactic discs, such as supernova explosions, stellar winds, etc. Korpi, Brandenburg, & Tuominen (1998) have recently started to investigate the effect of supernova explosions on stratified MHD shear flows with an initial toroidal magnetic field. The simulations have not yet been run for long enough, but it should be possible to assess in the near future whether the Balbus-Hawley instability leads to significant forcing of the turbulence, or whether the turbulence is mostly due to the supernova explosions or other external drivers. Longer calculations at lower resolution (Korpi *et al.* 1998) have indicated that the energy input from shear is surprisingly small compared with the energy input from supernova explosions. This is consistent with the fact that the inverse Rossby number, $\mathrm{Ro}^{-1} = 2\Omega H/\langle\mathbf{u}^2\rangle^{1/2}$, is much smaller in galaxies (about 1/2) than in accretion discs (20 in the simulations of BNST95).

6.6.2. *Stars*

Convective turbulence is clearly present in all late-type stars with outer convection zones. This was always thought to be the main driver of stellar dynamos. Recent simulations of Brandenburg, Nordlund, & Stein (1998) have suggested, however, that a substantial amount of energy can be tapped from the differential rotation in a similar fashion as in accretion discs. In those simulations a solar-like differential rotation profile has been added to convection simulations. Whether or not the Balbus-Hawley instability, or some other instability plays the key role is still unclear. Gilman & Fox (1997) have recently studied an instability in a sphere using a two-dimensional approach ignoring radial extent. However, this instability is unlikely to be important in the present case, because it predicts only the $m = 1$ mode to be unstable. Another proposal came from Schmitt (1985), who found that magnetostrophic waves, that are destabilized by a vertical gradient of the magnetic field, cause an α-effect which may explain the growth of large scale magnetic field seen in the simulations.

7. Conclusions

In this review we have highlighted some of the main results obtained recently using local simulations of hydromagnetic turbulence in accretion discs. The main objective for the future will be to construct global models of magnetized accretion discs. There will be two main strands of future work. On the one hand fully three-dimensional turbulence simulations will be produced that allow for realistic accretion flows and time dependence. On the other hand, local models will be improved (for example radiation transport included) and their average behaviour parameterized so that more realistic (and yet one-dimensional) accretion disc models can be constructed. Clearly, the main emphasis here will lie in combining Shakura-Sunyaev type models with dynamo models. Work in that direction has been pursued by Campbell (1992, 1997).

Already at this point some of the parameterizations suggested from simulation data have proven useful. For example the proposal that α_{SS} is proportional to the mean magnetic energy density has been used to put constraints on models of energy extraction from rotating black holes by the Blandford-Znajek (1978) process, see Ghosh & Abramowicz (1997). Curiously enough, whilst α_{SS} depends on the ill-known magnetic field strength, in the problem considered by Ghosh & Abramowicz the field strength drops out of the problem.

Another problem concerns the origin of turbulence in protostellar discs (i.e. is the magnetic Reynolds number large enough?) and the nature of outbursts in cataclysmic variables. Also, we would like to know the effect of the global magnetic field, which either connects with the wind of the disc or with other parts in the disc. This could significantly modify the torque acting on the disc. Global simulations should be able to address these questions.

It is a pleasure to thank my main collaborators in this field, Åke Nordlund, Bob Stein and Ulf Torkelsson for the many interesting discussions we had, especially during the Iceland meeting. I thank Eric Agol for asking me about the relative importance of compressive and vortical motions in our simulation. I am grateful to David Moss, Gordon Ogilvie, and Ulf Torkelsson for commenting on an earlier draft of this review. I am happy to acknowledge partial support from the Isaac Newton Institute, where I have written parts of this review.

REFERENCES

ABRAMOWICZ, M. A., CALVANI, M. & NOBILI, L. 1980 Thick accretion disks with super-eddington luminosities. *Astrophys. J.* **242**, 772-788.

ABRAMOWICZ, M. A., BRANDENBURG, A. & LASOTA, J.-P. 1996 The dependence of the viscosity in accretion discs on the shear/vorticity ratio. *Monthly Notices Roy. Astron. Soc.* **281**, L21-L24.

ABRAMOWICZ, M. A., LANZA, A., SPIEGEL, E. A. & SZUSZKIEWICZ, E. 1992 Vortices on accretion disks. *Nature* **356**, 41-43.

AGOL, E. & KROLIK, J. 1998 *Photon damping of waves in accretion disks.* (preprint).

BALBUS, S. A. & SOKER, N. 1990 Resonant excitation of internal gravity-waves in cluster cooling flows. *Astrophys. J. Letters* **357**, 353-366.

BALBUS, S. A. & HAWLEY, J. F. 1991 A powerful local shear instability in weakly magnetized disks. I. Linear analysis. *Astrophys. J.* **376**, 214-222.

BALBUS, S. A. & HAWLEY, J. F. 1992 A powerful local shear instability in weakly magnetized disks. IV. Nonaxisymmetric perturbations. *Astrophys. J.* **400**, 610-621.

BALBUS, S. A., HAWLEY, J. F. & STONE, J. M. 1996 Nonlinear stability, hydrodynamical turbulence, and transport in disks. *Astrophys. J.* **467**, 76-86.

BALBUS, S. A. & RICOTTI, M. 1998 "On nonshearing magnetic configurations in differentially rotating disks," *Astrophys. J.* (in press).

BALBUS, S. A. & HAWLEY, J. F. 1998 Instability, turbulence, and enhanced transport in accretion disks. *Rev. Mod. Phys.* **70**, 1-53.

BARGE, P. & SOMMERIA, J. 1995 Did planet formation begin inside persistent gaseous vortices? *Astron. Astrophys.* **295**, L1-L4.

BISNOVATYI-KOGAN, G. S. & LOVELACE, R. V. E. 1997 Influence of ohmic heating on advection-dominated accretion flows. *Astrophys. J.* **486**, L43-L46.

BLANDFORD, R. D. & ZNAJEK, R. L. 1977 Electromagnetic extraction of energy from Kerr black holes. *Monthly Notices Roy. Astron. Soc.* **179**, 433-456.

BLAES, O. M. & BALBUS, S. A. 1994 Local shear instabilities in weakly ionized, weakly magnetized disks. *Astrophys. J.* **421**, 163-177.

BOGDAN, T. J., CATTANEO, F. & MALAGOLI, A. 1993 On the generation of sound by turbulent convection. I. A numerical experiment. *Astrophys. J.* **407**, 316-329.

BRANDENBURG, A. & CAMPBELL, C. G. 1997 Modelling magnetised accretion discs. In *H. Spruit, & E. Meyer-Hofmeister* (ed. Accretion disks - New aspects), pp. 109-124. Springer-Verlag.

BRANDENBURG, A., NORDLUND, Å., STEIN, R. F. & TORKELSSON, U. 1995a Dynamo generated turbulence and large scale magnetic fields in a Keplerian shear flow. *Astrophys. J.* **446**, 741-754. (BNST95)

BRANDENBURG, A., NORDLUND, Å., STEIN, R. F. & TORKELSSON, U. 1995b Dynamo generated turbulence is discs. In *Small-scale structures in three-dimensional hydro and magnetohydrodynamic turbulence* (ed. M. Meneguzzi, A. Pouquet, & P. L. Sulem), pp. 385-390. Lecture Notes in Physics **462**, Springer-Verlag.

BRANDENBURG, A., NORDLUND, Å., STEIN, R. F. & TORKELSSON, U. 1996a The disk accretion rate for dynamo generated turbulence. *Astrophys. J. Letters* **458**, L45-L48. (BNST96)

BRANDENBURG, A., NORDLUND, Å., STEIN, R. F. & TORKELSSON, U. 1996b Dynamo generated turbulence is disks: value and variability of alpha. In *Physics of Accretion Disks* (ed. S. Kato *et al.*), pp. 285-290. Gordon and Breach Science Publishers.

BRANDENBURG, A., JENNINGS, R. L., NORDLUND, Å., RIEUTORD, M., STEIN, R. F., TUOMINEN, I. 1996c Magnetic structures in a dynamo simulation. *J. Fluid Mech.* **306**, 325-352.

BRANDENBURG, A. & DONNER, K. J. 1997 The dependence of the dynamo alpha on vorticity. *Monthly Notices Roy. Astron. Soc.* **288**, L29-L33.

BRANDENBURG, A. & SOKOLOFF, D. 1998 "Local and nonlocal magnetic diffusion and alpha-effect tensors in shear flow turbulence," *Geophys. Astrophys. Fluid Dyn.* (submitted).

BRANDENBURG, A., NORDLUND, Å., STEIN, R. F. 1998 "Simulation of a convective dynamo with imposed shear," *Astron. Astrophys.* (to be submitted).

CAMPBELL, C. G. 1992 Magnetically-controlled disc accretion. *Geophys. Astrophys. Fluid Dyn.* **63**, 197-213.

CAMPBELL, C. G. 1997 *Magnetohydrodynamics in Binary Stars.* Kluwer Academic Publishers, Dordrecht.

CANNIZZO, J. K., SHAFTER, A. W. & WHEELER, J. C. 1988 On the outburst recurrence time for the accretion disk limit cycle mechanism. *Astrophys. J.* **333**, 227-235.

CAUNT, S. E. 1998, PhD thesis (University of Newcastle)

CHANDRASEKHAR, S. 1960 The stability of non-dissipative Couette flow in hydromagnetics. *Proc. Natl. Acad. Sci.* **46**, 253-257.

CHANDRASEKHAR, S. 1961 *Hydrodynamic and Hydromagnetic Stability.* Dover Publications, New York., pp. 384

COLEMAN, C. S., KLEY, W. & KUMAR, S. 1995 The oscillations and stability of magnetized accretion discs. *Monthly Notices Roy. Astron. Soc.* **274**, 171-207.

CURRY, C., PUDRITZ, R. E. & SUTHERLAND, P. 1994 On the global stability of magnetized accretion disks: axisymmetric modes. *Astrophys. J.* **434**, 206-220.

DRECKER, A., HOLLERBACH, R. & RÜDIGER, G. 1998 "Viscosity alpha in rotating spherical shear flows with an external magnetic field," *Monthly Notices Roy. Astron. Soc.* (in press).

DUBRULLE, B. 1993 On the local stability of accretion disks. *Icarus* **106**, 59-678.

DUBRULLE, B. & VALDETTARO, L. 1992 Consequences of rotation in energetics of accretion disks. *Astron. Astrophys.* **263**, 387-400.

DUBRULLE, B. & ZAHN, J.-P. 1991 Nonlinear instability of viscous plane Couette flow Part 1. Analytical approach to a necessary condition. *J. Fluid Mech.* **231**, 561-573.

FRANK, J., KING, A. R. & RAINE, D. J. 1992 *Accretion power in astrophysics.* Cambridge University Press.

FRISCH, U., SHE, Z. S. & SULEM, P. L. 1987 Large-scale flow driven by the anisotropic kinetic alpha effect. *Physica* **28D**, 382-392.

FRÖHLICH, H.-E. & SCHULTZ, M. 1996 The vertical structure of the galactic gaseous disk and its relation to the dynamo problem. *Astron. Astrophys.* **311**, 451-455.

GALSGAARD, K. & NORDLUND, Å. 1996 Heating and activity of the solar corona: I. boundary shearing of an initially homogeneous magnetic-field. *J. Geophys. Res.***101**, 13445-13460.

GAMMIE, C. F. 1996 Layered accretion in T Tauri disks. *Astrophys. J.* **457**, 355-362.

GILMAN, P. A. & FOX, P. A. 1997 Joint instability of latitudinal differential rotation and toroidal magnetic fields below the solar convection zones. *Astrophys. J.* **484**, 439-454.

GHOSH, P. & ABRAMOWICZ, M. A. 1997 Electromagnetic extraction of rotational energy from disc-fed black holes: the strength of the Blandford-Znajek process. *Monthly Notices Roy. Astron. Soc.* **292**, 887-895.

GOODMAN, J., NARAYAN, R. & GOLDREICH, P. 1987 The stability of accretion tori – II. Nonlinear evolution to discrete planets. *Monthly Notices Roy. Astron. Soc.* **225**, 695-711.

HAWLEY, J. F. 1987 Non-linear evolution of a non-axisymmetric disk instability. *Monthly Notices Roy. Astron. Soc.* **225**, 677-694.

HAWLEY, J. F. & BALBUS, S. A. 1991 A powerful local shear instability in weakly magnetized disks. II. Nonlinear evolution. *Astrophys. J.* **376**, 223-233.

HAWLEY, J. F., GAMMIE, C. F. & BALBUS, S. A. 1995 Local three-dimensional magnetohydrodynamic simulations of accretion discs. *Astrophys. J.* **440**, 742-763. (HGB95)

HAWLEY, J. F., GAMMIE, C. F. & BALBUS, S. A. 1996 Local three dimensional simulations of an accretion disk hydromagnetic dynamo. *Astrophys. J.* **464**, 690-703. (HGB96)

HODGSON, L. S. & BRANDENBURG, A. 1998 Turbulence effects in planetesimal formation. *Astron. Astrophys.* **330**, 1169-1174.

HURLBURT, N.E. & TOOMRE, J. 1988 Magnetic fields interacting with nonlinear compressible convection. *Astrophys. J.* **327**, 920-932.

KITCHATINOV, L. L., RÜDIGER, G. & KHOMENKO, G. 1994 Large-scale vortices in rotating stratified disks. *Astron. Astrophys.* **287**, 320-324.

KITCHATINOV, L. L. & RÜDIGER, G. 1997 Global magnetic shear instability in spherical geometry. *Monthly Notices Roy. Astron. Soc.* **286**, 757-764.

KORPI, M. J., BRANDENBURG, A. & TUOMINEN, I. 1998 Driving interstellar turbulence by supernova explosions. *Studia Geophys. et Geod.* (in press).

KORPI, M. J., BRANDENBURG, A., SHUKUROV, A. & TUOMINEN, I. 1998 Vortical motions driven by supernova explosions. In *P. Franco & A. Carraminana* (ed. Interstellar Turbulence), Cambridge University Press.

KLAHR, H. H. & HENNING, T. 1997 Particle-trapping eddies in protoplanetary accretion disks. *Icarus* **128**, 213-229.

KRAUSE, F. & RÄDLER, K.-H. 1980 *Mean-Field Magnetohydrodynamics and Dynamo Theory.* Akademie-Verlag, Berlin; also Pergamon Press, Oxford.

KRISHNAMURTI, R. & HOWARD, L. N. 1981 Large-scale flow generation in turbulent convection. *Proc. Natl. Acad. Sci. USA* **78**, 1981-1985.

KUMAR, S., COLEMAN, C. S. & KLEY, W. 1994 The axisymmetric instability in weakly magnetized accretion discs. *Monthly Notices Roy. Astron. Soc.* **266**, 379-385.

LIN, D. N. C. & PAPALOIZOU, J. C. B. 1980 On the structure and evolution of the primordial solar nebula. *Monthly Notices Roy. Astron. Soc.* **191**, 37-48.

LUBOW, S. H., PAPALOIZOU, J. C. B. & PRINGLE, J. E. 1994 Magnetic field line dragging in accretion discs. *Monthly Notices Roy. Astron. Soc.* **267**, 235-240.

LUFKIN, E. A., BALBUS, S. A. & HAWLEY, J. F. 1995 Nonlinear evolution of internal gravity-waves in cluster cooling flows. *Astrophys. J.* **446**, 529-540.

MATSUMOTO, R. & TAJIMA, T. 1995 Magnetic viscosity by localized shear flow instability in magnetized accretion disks. *Astrophys. J.* **445**, 767-779.

MEYER, F. & MEYER-HOFMEISTER, E. 1982 Vertical structure of accretion disks. *Astron. Astrophys.* **106**, 34-42.

NOVIKOV, I. D. & THORNE, K. S. 1973 Astrophysics of black holes. In *C. DeWitt & B. S. DeWitt* (ed. Black holes - Les astres occlus), pp. 343-450. Gordon & Breach, New York.

NORDLUND, Å. 1982 Numerical Simulations of the Solar Granulation I. Basic Equations and Methods. *Astron. Astrophys.* **107**, 1-10.

NORDLUND, Å., BRANDENBURG, A., JENNINGS, R. L., RIEUTORD, M., RUOKOLAINEN, J., STEIN, R. F. & TUOMINEN, I. 1992 Dynamo action in stratified convection with overshoot. *Astrophys. J.* **392**, 647-652.

OGILVIE, G. I. & PRINGLE, J. E. 1996 The non-axisymmetric instability of a cylindrical shear flow containing an azimuthal magnetic field. *Monthly Notices Roy. Astron. Soc.* **279**, 152-164.

PRINGLE, J. E. 1993 Cosmogony of stellar and extragalactic jets. In *Astrophysical Jets* (ed. D. Burgarella, M. Livio & C. O'Dea), pp. 1-13. Cambridge University Press.

REGÖS, E. 1997 Magnetic viscosity in weakly ionized protostellar discs. *Monthly Notices Roy. Astron. Soc.* **286**, 97-103.

RÜDIGER, G. 1989 *Differential rotation and stellar convection: Sun and solar-type stars.* Gordon & Breach, New York.

SAFRONOV, V. S. 1972 *Evolution of the protoplanetary cloud and formation of the Earth and the planets.* Israel Program for Scientific Translation, Jerusalem.

SCHMITT, D. 1985 *Dynamowirkung magnetostrophischer Wellen.* PhD thesis, University of Göttingen.

SCHRAMKOWSKI, G. P. & TORKELSSON, U. 1996 Magnetohydrodynamic instabilities and turbulence in accretion disks. *Astron. Astrophys. Rev.* **7**, 55-96.

SHAKURA, N. I. & SUNYAEV, R. A. 1973 Black holes in binary systems. Observational appearance. *Astron. Astrophys.* **24**, 337-355.

SHAPIRO, S. L., LIGHTMAN, A. P. & EARDLEY, D. M. 1976 A two-temperature accretion disk model for Cygnus X-1: Structure and spectrum. *Astrophys. J.* **204**, 187-199.

STEPINSKI, T. F., REYES-RUIZ, M. & VANHALA, H. A. T. 1993 Solar nebula magnetohy-drodynamic dynamos - kinematic theory, dynamical constraints, and magnetic transport of angular-momentum. *Icarus* **106**, 77-91.

STONE, J. M., HAWLEY, J. F., GAMMIE, C. F. & BALBUS, S. A. 1996 Three dimensional magnetohydrodynamical simulations of vertically stratified accretion disks. *Astrophys. J.* **463**, 656-671. (SHGB96)

TANGA, P., BABIANO, A., DUBRULLE, B. & PROVENZALE, A. 1996 Forming planetesimals in vortices. *Icarus* **121**, 158-170.

TERQUEM, C. & PAPALOIZOU, J. C. B. 1996 On the stability of an accretion disc containing a toroidal magnetic field. *Monthly Notices Roy. Astron. Soc.* **279**, 767-784.

TOUT, C. A. & PRINGLE, J. E. 1992 Accretion disc viscosity: A simple model for a magnetic dynamo. *Monthly Notices Roy. Astron. Soc.* **259**, 604-612.

TUOMINEN, I., BRANDENBURG, A., MOSS, D. & RIEUTORD, M. 1994 Does solar differential rotation arise from a large scale instability? *Astron. Astrophys.* **284**, 259-264.

VAN BALLEGOOIJEN, A. A. 1989 Magnetic fields in the accretion disks of cataclysmic variables. In *Accretion disks and magnetic fields in astrophysics* (ed. G. Belvedere), pp. 99-106. Kluwer Academic Publishers, Dordrecht.

VON REKOWSKI, B., KITCHATINOV, L. L. & RÜDIGER, G. 1998 "Global vortex systems on standard-accretion disk surfaces," *Monthly Notices Roy. Astron. Soc.* (submitted).

VAINSHTEIN, S. I. & ZELDOVICH, YA. B. 1972 Origin of magnetic fields in astrophysics. *Sov. Phys. Usp.* **15**, 159-172.

VELIKHOV, E. P. 1959 Stability of an ideally conducting liquid flowing between cylinders rotating in a magnetic field. *Sov. Phys. JETP* **36**, 1398-1404. (Vol. 9, p. 995 in English translation)

The role of electron-positron pairs in hot accretion flows

By GUNNLAUGUR BJÖRNSSON

Science Institute, University of Iceland, Dunhagi 3, IS-107 Reykjavik, Iceland

Electron-positron pairs are copiously produced in a hot effectively optically thin plasma cloud when two conditions are met: i) The temperature of the plasma is sufficiently high, $kT/m_e c^2 \gtrsim$ 0.1 and ii) the optical depth to pair production by photon-photon interactions is sufficiently large. The latter condition is especially important, as only photon-photon interactions can produce pairs in any appreciable quantity. It can be translated into a condition on the luminosity to size ratio of the plasma cloud being sufficiently large. Inclusion of $e^{\pm}$-pairs in studies of accretion flows is still at a rather primitive stage, having been attempted only under rather strong assumptions. As Advectively Dominated Accretion Flows are generally very hot, but also very low luminosity phenomena, pair effects may appear to be unimportant.

1. Introduction

Accretion is a common process in astronomical environments and plays an important role in galactic black hole candidates and at the center of Active Galactic Nuclei. It is the most efficient way of converting gravitational potential energy into radiation. The general theory of accretion is reviewed e.g. by Pringle (1981) and by Frank *et al.* (1992).

The differential equations describing accretion onto compact objects are, in the standard accretion disc model, simplified to a set of algebraic equations by neglecting all radial gradients. The crucial assumption here being that the disc is geometrically thin, in the sense that $H/r \ll 1$, where H is the disc scale height and r the radial distance from the central object. Thus simplified, the equations were known to have two possible families of solutions, optically thick solutions (Shakura & Sunyaev 1973; Novikov & Thorne 1973), and optically thin solutions (Shapiro, Lightman & Eardley 1976). The optically thick solutions are never sufficiently hot to explain the X- and γ-rays observed from compact sources (see e.g. Krolik, this volume, for a discussion of optically thick flows). The optically thin solutions, while sufficiently hot, are thermally unstable (e.g. Piran 1978), and thus cannot represent a real description of accretion onto compact objects. For the lack of anything better, the optically thin solutions have been studied extensively, either directly or as a part of more complex models of high energy radiation from compact objects (for a review see e.g. Svensson 1997; Poutanen, this volume).

A third family of solutions was discovered by Ichimaru (1977). These solutions are characterized by the cooling timescale of the protons being longer than the inflow timescale. Only a small fraction of the dissipated energy is radiated from the flow, most of it gets advected into the black hole. The solutions are thermally stable and are in fact primitive versions of what is now called advectively dominated solutions. Variants of these solutions, the thick ion-tori, were discussed by Rees *et al.* (1982). Such flows are underluminous for their accretion rates (Fabian & Canizares, 1988; Fabian & Rees, 1995), and the gas becomes very hot, easily reaching virial temperatures.

The relationship between the families of solutions in general, and the properties of the third family in particular, has been intensely studied over the past few years (e.g. Narayan & Yi, 1995; Abramowicz *et al.* 1995; Chen 1995; Chen *et al.* 1995; Björnsson *et al.* 1996; see Narayan *et al.* this volume for a review).

It was early realized, both from observations and theory, that production of electron-positron pairs would be important in cosmic sources of X- and γ-rays, as the electron temperature could reach $10^9 - 10^{10}$ K. Considering pair production in a hot plasma cloud by particle-particle interactions only, Bisnovatyi-Kogan *et al.* (1971), established the existence of a maximum plasma temperature above which pair equilibria was not possible. Later studies included particle-photon and photon-photon interactions, as well as photon sources external to the plasma cloud, confirmed the existence of maximum temperature and outlined a region in parameter space where no pair equilibria are possible (Lightman 1982; Svensson 1982, 1984; Zdziarski 1985).

In these early studies, the hot cloud was assumed to be spherical and the pair equilibrium properties were shown to depend on two parameters, the *compactness parameter*, ℓ, and the *proton optical depth*, $\tau_{\rm p}$. The compactness parameter of the plasma is defined as $\ell = L\sigma_{\rm T}/Hm_ec^3$, where L is the luminosity of the cloud, $\sigma_{\rm T}$ is the Thomson cross section and H is the cloud radius. The cloud radius is here denoted by H to imply a direct interpretation as scale height within the accretion disc models. The proton optical depth is just the Thomson scattering depth of the ionized electrons, $\tau_{\rm p} = n_{\rm p}\sigma_{\rm T}H$. Here, $n_{\rm p}$, is the number density of protons in the plasma. Note that $\tau_{\rm p}$ also has a direct interpretation in accretion disc models as $\tau_{\rm p} = \sigma_{\rm T}\Sigma/2m_{\rm p}$, with Σ being the disc surface density.

One of the main results of these early studies was that pair production by particle-particle and particle-photon interactions could never produce more pairs than about 10% of the proton density by number. Higher pair densities require photon-photon interactions, and thus very compact radiation fields. Followup studies considered more general environments (e.g. Kusunose 1987; Begelman, Sikora & Rees 1987; Lightman, Zdziarski & Rees 1987), the stability properties of pair equilibria (Sikora & Zbyszewska 1988; Björnsson & Svensson 1991a), and non-thermal pair production (e.g. Guilbert, Fabian & Rees 1983; Svensson 1987; Zdziarski & Lightman 1985, 1987). See Svensson (1986, 1990) for a review of pair plasma properties and the history of pair plasma research.

2. Pairs in hot flows

The very first attempts to include $e^{\pm}$-pairs in any kind of an accretion flows were that of Zhang & Fang (1978) who considered optically thick models, and Liang (1979), who considered the two-temperature optically thin model of Shapiro, Lightman & Eardley (1976). Liang realized that the temperature was sufficiently high for pair production to be potentially important, but as the physics of pair plasmas had not been developed at that time, his treatment of the pair physics was rather approximate. Only pair production by photon-photon interactions was considered, and the radiation field was assumed to be a pure Wien spectrum. Despite these rather crude approximations, his results were remarkably complete, although these were somewhat hard to interpret with the then current pair plasma knowledge. In fact, with hindsight, many of the results obtained by others ten years later, can be extracted from Liang's paper (see e.g. Björnsson & Svensson, 1992).

2.1. *Geometrically thin models*

Kusunose & Takahara (1988) attempted to calculate the structure of thermal two-temperature optically thin accretion discs, including $e^{\pm}$-pairs, with a pure bremsstrahlung radiation field. Their approach was not entirely self-consistent, however, as Comptonization of the internally generated bremsstrahlung photons was neglected. This was corrected and extended by Kusunose & Takahara (1989, 1990). Their main result was that physically consistent solutions could only be obtained for relatively low accretion rates, $\dot{m}_{cr} = \dot{M}/\dot{M}_E \lesssim 0.3$, where $\dot{M}_E$ is the accretion rate corresponding to the Eddington limit. In addition, the equilibrium pair density of these solutions was always very low, $z = n_+/n_p \ll 1$, where n_+ denotes the positron density. High pair density solutions, $(z > 1)$, were also shown to exist, but these were unphysical as the proton temperature was greater than the virial temperature. Above the critical accretion rate, that increases with increasing viscosity parameter, α, no equilibrium solutions were possible in a radial annulus around $r = 16r_g/3$, where $r_g = 2GM/c^2$. This result was confirmed in a detailed and thorough study by White & Lightman (1989), who showed that $\dot{m}_{cr} \sim \alpha^{0.25}$.

Tritz & Tsuruta (1989) performed a detailed study of pairs in accretion discs, considering both thermal and non-thermal pair production. In their thermal models the pair density is always low, both in disc as well as in quasi-spherical accretion. The pairs therefore, do not have any appreciable effect on the disc structure as they never reach higher densities than a few per cent of the proton density by number. Non-thermal pair production, being more efficient, can easily give rise to pair number densities considerably larger than the proton number density, and thus can influence the disc structure rather strongly (see also Pi-bo *et al.* 1993).

The original interpretation of why no equilibrium solutions could be found for accretion rates above $\dot{m}_{cr}$, was that the pair annihilation rate was not able to keep up with pair production rate. Björnsson (1990), Björnsson & Svensson (1991b, 1992) and Kusunose & Mineshige (1992), showed that this interpretation was incorrect, and in fact equilibrium solutions could be found at all radii and for all accretion rates. The solutions obtained for high accretion rates (above $\dot{m}_{cr}$), are related to the high pair density solutions found for accretion rates below $\dot{m}_{cr}$. They were therefore, also unphysical as again the proton temperature was considerably higher than the virial temperature. In these unphysical solutions, typically $\tau_p \gtrsim 0.1$ (corresponding to $\Sigma \gtrsim 0.25$ g/cm^2), the compactness parameter, $\ell \gtrsim 10$, and the pair density could be very high, a factor of 10 to 100, higher than the proton density for large values of α.

Björnsson & Svensson (1992), also pointed out that the existence of the unphysical high pair density solutions implied that radial gradients could not be neglected in the disc structure equations. In particular, radial pressure gradients could be strong, implying a deviation of the rotation law from Keplerian. In fact, the radial solution profiles for the high pair density solutions showed infinite radial gradients at certain radii, clearly in contradiction with the assumptions of the geometrically thin disc models.

2.2. *Advective flows*

In recent years the term *advectively dominated accretion flow* has come to mean that advection of energy is very important, and ADAF's really refers to flows where advection of thermal energy dominates over radiative cooling processes (see Narayan *et al.* this volume for a thorough review).

It was early realized, however, that if $e^\pm$-pairs would be produced in a hot accretion flow, advection of the pairs them selfs could be important. Indeed, among the first attempts to include pairs in accreting flows considered pairs in quasi-spherical accretion, allowed for pair-advection, but neglected the advection of energy.

2.2.1. *Advection of pairs*

Considering the effects of thermal and non-thermal pair production on quasi-spherical accretion, but using rather approximate pair production rates, Begelman, Sikora & Rees (1987), concluded that strong cooling effects by the pairs could remove the pressure support in the innermost part of flow. This would happen inside a critical radius, r_{cr}, that mainly depends on the accretion rate, but also on the dominant cooling processes in the plasma. As a consequence, in the presence of angular momentum, the flow would then collapse to a central disc-like structure with a hot two-temperature quasi-spherical flow outside. They suggested that a time dependent limit-cycle behaviour might result. No direct calculations, however, were presented to substantiate this conclusion. Their outline of r_{cr} as a function of $\dot{m}$ was more accurately calculated by Tritz & Tsuruta (1989), but again without direct calculations of how the flow collapse would proceed. The pair densities at r_{cr} turned out to be negligible in models with thermal pair production, but could be substantial in models with non-thermal pair production (Tritz & Tsuruta, 1989). In a time dependent study of the disc structure equations, including $e^{\pm}$-pairs, White & Lightman (1990) kept the gradient terms in *both* the energy balance equations of the protons and the pairs. An advective term was added to the pair balance equation, effectively increasing the local annihilation rate. Considering only the very innermost disc region and starting the calculations from the (unphysical) pair dominated solution in this region, the disc was observed to indeed collapse to an optically thick configuration. The collapse was observed to take place on a time scale that is approximately the sum of the pair creation time scale and the hydrodynamical timescale. Although this result is indicative of what a fully time dependent calculation may reveal, it's validity is restricted by the fact that it is obtained over a very narrow radial range ($3 < r/r_g < 10$), and all variables were held fixed at the outer radius.

Assuming non-thermal pair production and pair advection in a quasi-spherical flow, Lightman, Zdziarski & Rees (1987), showed that pair effects would increase the photon trapping radius, defined as the distance from the center where the radiation diffusion speed equals the inflow speed of the plasma. Due to pair effects, the effective Eddington limit would also decrease (see also Svensson 1987).

2.2.2. *Advection of energy*

Early attempts to solve for the advectively dominated flow structure, approximated the entropy gradient term in the energy equation with a constant throughout the disk, thus avoiding the solution of the differential equation directly. With this approximation, the disk structure equations remained algebraic and could be solved at any radius as before. This *local* approach was used e.g. by Abramowicz *et al.* (1995), Narayan (1996) and Narayan *et al.* (1996). The *global* solutions of the problem, *i.e.* the proper solution of the differential equations with the appropriate boundary conditions, with varying degree of complexity of microphysical ingredients but all excluding pair processes, have been presented by e.g. (Chen 1995; Chen *et al.* 1997; Narayan *et al.* 1997).

Björnsson *et al.* (1996), calculated the structure of two temperature discs, assuming the radiation field to be Comptonized bremsstrahlung, and including thermal pair-production. They accounted for advection in the local approximation, and performed a detailed study of the $\dot{m} - \Sigma$-parameter space in the manner of Abramowicz *et al.* (1995). Figure 1 shows their map of parameter space. The importance of the dominant cooling process or radiation field is indicated next to each part of the curves. It is clear from the figure that the pair effects are restricted to a rather small region in parameter space as compared to Comptonization effects. In fact, Figure 2 shows the approximate region that Björnsson *et al.* (1996) outline as potentially important for pair production. In the local picture this is at the boundary between the optically thin and optically thick solutions, and is in fact a region in parameter space where almost all ingredients in the models can influence the solution. Rather high values of α and $\dot{m}$ are required and the region appears to be concentrated around $\Sigma \approx 10$. A detailed investigation of this region, using global solutions and including all the relevant microphysics is yet to be carried out.

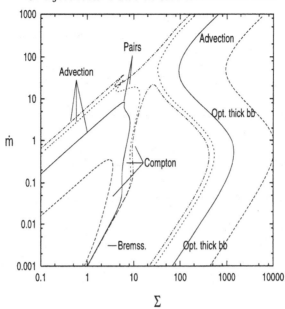

FIGURE 1. Thermal equilibria of accretion disks with advection and detailed microphysics. Note the upward bending of the optically thin bremsstrahlung branch on the lower left due to the Comptonization and the pair effects. Here, $r = 5r_g$ and a constant advective term is assumed. The long dashed, solid, short dashed and dot-dashed curves correspond to $\alpha = 0.1, 1.0, 3.0$ and 4.0, respectively. (From Björnsson et al. (1996)).

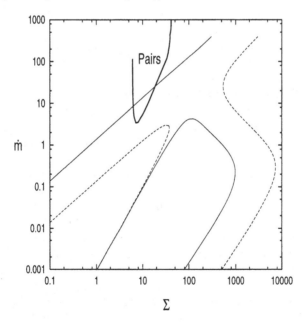

FIGURE 2. Thermal equilibria of accretion disks with no pairs (from Chen et al. 1995). Solid curves correspond to $\alpha = 1.0$ and the dashed ones to $\alpha = 0.1$. In the region bounded by the heavy solid curve and marked *Pairs*, pair processes may be important for the disk structure. (from Björnsson et al. (1996))

Simultaneously and independently, Kusunose & Mineshige (1996), investigated the importance of pairs in ADAFs, using much the same ingredients as Björnsson *et al.* (1996). They also used the local approach so their results are therefore very similar to that of Björnsson *et al.* (1996). As mentioned in the previous subsection, Kusunose & Mineshige (1996), also included in the pair balance equation a term describing the advection of leptons (see also Tritz & Tsuruta, 1989). This term is mathematically equivalent to the energy advection term in the proton energy equation, and was therefore treated in the same way, i.e. it was assumed to be a constant. Kusunose & Mineshige (1996) concluded that this term was not very important, modifying the solutions only very slightly.

Misra & Melia (1996), attempted to solve the global disc structure problem including electron-positron pairs, including the advection term in the proton energy equation, but solving the pair balance locally. They only included particle-particle interactions, and particle-photon interactions, and so excluded processes that could have given rise to high pairs densities. The pair number densities in their solutions are therefore always very low (less than 10% of the proton number densities), and are never very important for the structure of the disc. They also allowed for pair escape (see Misra & Melia, 1995), but as pair densities where never high, pair escape did not have any major effect on the disc structure.

In an extensive application of ADAFs to model spectral states of black hole X-ray binaries, Esin *et al.* (1997), assume that particle-particle pair production dominates and that advection of pairs can neglected, but allow for advection of pair energy. In all cases they consider, the resulting pair density is very low, $z \lesssim 10^{-5}$.

3. Conclusions

The result, that $e^{\pm}$-pairs do not seem to be important in ADAFs, is mostly obtained in a *local* treatment. It has already influenced a number of recent papers (e.g. Nakamura *et al.* 1996; Manmoto *et al.* 1997), as pair effects are routinely neglected by assumption. In these papers, the global two-temperature problem is solved, and the electron temperatures become as hot as 10^{10} K. Neglecting pair processes in a pure ADAF may be a valid assumption for most models, as their luminosity is generally very low, although their temperature can be very high. Thus efficient pair production may be unimportant. However, if the results obtained in the local analysis of ADAFs can be trusted, there is a region in disk parameter space $(\dot{m}, \alpha)$, where pair processes may be very important. It is important that this region be explored by solving the global problem, as the spectral shape and evolution may be dominated by pairs. In fact, Manmoto *et al.* (1997), show that in the innermost disc region, the electron radiative cooling is balanced by advective electron *heating*, with the electron-ion energy exchange being unimportant inside $\approx 10r_g$. Almost all the emission from their model comes from within $10r_g$.

The physics of pair plasmas is now well understood. Due to the strong non-linearity of the pair production processes, those have either been approximated in previous treatments, or the disc structure equations have been simplified by neglecting gradient terms, or assuming that they are constant throughout the disc. Several attempts have thus been made at including them in a a self-consistent solution of advectively dominated accretion flows, but only in the local approximation. A few fragmentary results indicate their importance even in the full global problem. Despite ample computational facilities a self-consistent global solution including pairs is yet to be attempted.

This work was supported in part by the Research Fund of the University of Iceland.

REFERENCES

ABRAMOWICZ, M.A., CHEN, X.M., KATO, S., LASOTA, J.P. & REGEV, O. 1995 Thermal equilibria of accretion disks *ApJ* **438**, L37

BEGELMAN, M.C., SIKORA, M., & REES, M.J. 1987 Thermal and dynamical effects of pair production on two-temperature accretion flows *ApJ* **313**, 689

BISNOVATYI-KOGAN, G.S., ZEL'DOVICH, YA.B., & SUNYAEV, R.A. 1971 Physical processes in a relativistic plasma of low density *Soviet Astr. – AJ*, **15**, 17

BJÖRNSSON, G. 1990 A study of electron-positron pair equilibria in models of compact X- and gamma-ray sources Ph.D. Thesis, University of Illinois.

BJÖRNSSON, G., & SVENSSON, R. 1991a On the stability of electron-positron pair equilibria and the X-ray variability of active galactic nuclei *MNRAS* **249**, 177

BJÖRNSSON, G., & SVENSSON, R. 1991b in *Structure and Emission Properties of Accretion Disks*, IAU Coll. 129, (eds. C. Bertout, S. Collin-Souffrin, J.-P. Lasota & J. Tran Thanh Van) 379, Editions Frontiers, Gif-sur Yvette, France

BJÖRNSSON, G., & SVENSSON, R. 1992 Hot pair-dominated accretion disks *ApJ* **394**, 500

BJÖRNSSON, G., ABRAMOWICZ, M.A., CHEN, X.M. & LASOTA, J.-P. 1996 Hot Accretion Disks Revisited *ApJ* **467**, 99

CHEN, X. 1995 Hot accretion discs with advection *MNRAS* **275**, 641

CHEN, X., ABRAMOWICZ, M.A., LASOTA, J.-P., NARAYAN, R. & YI, I. 1995 Unified description of accretion flows around black holes *ApJLett.* **443**, L61

CHEN, X., ABRAMOWICZ, M.A., LASOTA, J.-P. 1997 Advection-dominated Accretion: Global Transonic Solutions *ApJ* **476**, 61

ESIN, A.A., MCCLINTOCK, J.E. & NARAYAN, R. 1997 Advection-dominated Accretion and the Spectral States of Black Hole X-Ray Binaries: Application to Nova Muscae 1991 *ApJ* **489**, 865

FABIAN, A.C., CANIZARES, C.R. 1988 Do massive black holes reside in elliptical galaxies? *Nature* **333**, 829

FABIAN, A.C., REES, M.J. 1995 The accretion luminosity of a massive black hole in an elliptical galaxy *MNRAS* **277**, L55

FRANK, J., KING, A., & RAINE, D. 1992, *Accretion Power in Astrophysics*, Cambridge University Press.

GUILBERT, P.W., FABIAN, A.C., REES, M.J. 1983 Spectral and variability constraints on compact sources *MNRAS* **205**, 593

ICHIMARU, S. 1977 Bimodal behavior of accretion disks: theory and application to Cygnus X-1 transitions *ApJ* **214**, 840

KUSUNOSE, M. 1987 Relativistic thermal plasmas: time development of electron-positron pair concentration *ApJ* **321**, 186

KUSUNOSE, M. & TAKAHARA, F. 1988 Two-temperature accretion disks with electron-positron pair production *PASJ* **40**, 435

KUSUNOSE, M. & TAKAHARA, F. 1989 Two-temperature accretion disks in pair equilibrium - Effects of unsaturated Comptonization of soft photons *PASJ* **41**, 263

KUSUNOSE, M. & TAKAHARA, F. 1990 Two-temperature accretion disks with electron-positron pairs - Effects of Comptonized external soft photons *PASJ* **42**, 347

KUSUNOSE, M. & MINESHIGE, S. 1992 Geometrically thin, hot accretion disks - Topology of the thermal equilibrium curves *ApJ* **392**, 653

KUSUNOSE, M. & MINESHIGE, S. 1996 Effects of Electron-Positron Pairs in Advection-dominated Disks *ApJ* **468**, 330

MISRA, R. & MELIA, F. 1995 Hot Accretion Disks with Electron-Positron Pair Winds *ApJ* **449**, 813

MISRA, R. & MELIA, F. 1996 The Crucial Effects of Advection on the Structure of Hot Accretion Disks *ApJ* **465**, 869

LIANG, E.P.T. 1979 Electron-positron pair production in hot unsaturated Compton accretion models around black holes *ApJ* **234**, 1105

LIGHTMAN, A.P. 1982 Relativistic thermal plasmas - Pair processes and equilibria *ApJ* **253**, 842

LIGHTMAN, A.P., ZDZIARSKI, A.A, & REES, M.J. 1987 Effects of electron-positron pair opacity for spherical accretion onto black holes *ApJLett.* **315**, L113

MANMOTO, T., MINESHIGE, S. & KUSUNOSE, M. 1997 Spectrum of Optically Thin Advection-dominated Accretion Flow around a Black Hole: Application to Sagittarius A* *ApJ* **489**, 791

NARAYAN, R. & YI, I. 1995 Advection-dominated Accretion: Underfed Black Holes and Neutron Stars *ApJ* **452**, 710

NARAYAN, R. 1996 Advection-dominated Models of Luminous Accreting Black Holes *ApJ* **462**, 136

NARAYAN, R., MCCLINTOCK, J.E. & YI, I. 1996 A New Model for Black Hole Soft X-Ray Transients in Quiescence *ApJ* **452**, 821

NARAYAN, R., KATO, S. & HONMA, F. 1997 Global Structure and Dynamics of Advection-dominated Accretion Flows around Black Holes *ApJ* **476**, 49

NAKAMURA, K.E., MATSUMOTO, R., KUSUNOSE, M., & KATO, S. 1996 Global Structures of Advection-Dominated Two-Temperature Accretion Disks *PASJ* **48**, 761

NOVIKOV, I.D., & THORNE, K.S. 1973 Astrophysics of Black Holes in *Black Holes* (ed. C. De-Witt & B. DeWitt) 343, Gordon & Breach.

PRINGLE, J.E. 1981 Accretion discs in astrophysics *ARA&A* **19**, 137

PIRAN, T. 1978 The role of viscosity and cooling mechanisms in the stability of accretion disks *ApJ* **221**, 652

PI-BO, Y., JIA-LIN, G., & LAN-TIAN Y. 1993 Two-temperature accretion disks with electron-positron pairs *Chin. Astron. Astrophys.* **17**, 126

REES, M.J., BEGELMAN, M.C., BLANDFORD, R.D., & PHINNEY, E.S. 1982   Ion-supported tori and the origin of radio jets *Nature* **295**, 17

SHAKURA, N.I. & SUNYAEV, R.A. 1973 Black holes in binary systems. Observational appearance *A&A* **24**, 337

SHAPIRO, S.L., LIGHTMAN, A.P., & EARDLEY, D.M. 1976 A two-temperature accretion disk model for Cygnus X-1 - Structure and spectrum *ApJ* **204**, 187

SIKORA, M. & ZBYSZEWSKA, M. 1986 A stability analysis of electron-positron pair equilibria of a two-temperature plasma cloud *Acta Astronomica* **36**, 255

SVENSSON, R. 1982 Electron-positron pair equilibria in relativistic plasmas *ApJ* **258**, 355

SVENSSON, R. 1984 Steady mildly relativistic thermal plasmas - Processes and properties *MNRAS* **209**, 175

SVENSSON, R. 1986 Physical Processes in Active Galactic Nuclei in *Radiation Hydrodynamics in Stars and Compact Objects* (ed. Mihalas, D. & Winkler, K.-H.), 325, Springer

SVENSSON, R. 1987 Non-thermal pair production in compact X-ray sources - First-order Compton cascades in soft radiation fields *MNRAS* **227**, 403

SVENSSON, R. 1990 An introduction to relativistic plasmas in astrophysics in *Physical Processes in Hot Cosmic Plasma* (ed. Brinkmann, W., A.C. Fabian & Giovanelli, F.), 357, Kluwer

SVENSSON, R. 1997 X-rays and Gamma Rays from Active Galactic Nuclei in *Relativistic Astrophysics: A Conference in Honor of Professor I.D. Novikov's 60th Birthday* (ed. B.J.T. Jones & D. Markovic), 235, Cambridge University Press

TRITZ, B.G. & TSURUTA, S. 1989 Effects of electron-positron pairs on accretion flows *ApJ* **340**, 203

WHITE, T.R. & LIGHTMAN, A.P. 1989 Hot accretion disks with electron-positron pairs *ApJ* **340**, 1024

WHITE, T.R. & LIGHTMAN, A.P. 1990 Instabilities and time evolution of hot accretion disks with electron-positron pairs *ApJ* **352**, 495

ZDZIARSKI, A.A. & LIGHTMAN, A.P. 1985 Nonthermal electron-positron pair production and the 'universal' X-ray spectrum of active galactic nuclei *ApJLett.* **294**, L79

ZDZIARSKI, A.A. & LIGHTMAN, A.P. 1987 Pair production and Compton scattering in compact sources and comparison to observations of active galactic nuclei *ApJ* **319**, 643

ZDZIARSKI, A.A. 1985 Power-law X-ray and gamma-ray emission from relativistic thermal plasmas *ApJ* **289**, 514

ZHANG, J. & FANG, L. 1978 Studia Astrophys. Sinica, 1, 8 (English transl. in 1979, Chinese Astroph. 3, 141)

Accretion disc-corona models and X/γ-ray spectra of accreting black holes

By JURI POUTANEN

Stockholm Observatory, SE - 133 36 Saltsjöbaden, Sweden

We discuss properties of thermal and hybrid (thermal/non-thermal) electron-positron plasmas in the pair and energy equilibria. Various accretion disc-corona models, recently proposed to explain properties of galactic as well as extragalactic accreting black holes, are confronted with the observed broad-band X-ray and γ-ray spectra.

1. Introduction

It was realized quite early that broad-band X/γ-ray spectra of Galactic black holes (GBHs) can be explained in terms of successive Compton scatterings of soft photons (Comptonization) in a hot electron cloud. The Comptonizing medium was assumed to be thermal with a given temperature, T_e, and a Thomson optical depth, τ_T. The theoretical spectra were computed by analytical (Shapiro, Lightman & Eardley 1976; Sunyaev & Titarchuk 1980) and Monte-Carlo methods (Pozdnyakov, Sobol' & Sunyaev 1983).

The problem with such an approach is that in any specific geometry arbitrary combinations of (τ_T, T_e) are not possible. Both GBHs and Seyfert galaxies show a hardening of the spectra at ~ 10 keV, which is attributed to Compton reflection (combined effect of photo-electric absorption and Compton down-scattering) of hard radiation from a cold material (White, Lightman & Zdziarski 1988; George & Fabian 1991). Hard radiation, reprocessed in the cold matter, can form a significant fraction of the soft seed photons for Comptonization. The energy balance of the cold and hot phases determine their temperatures and the shape of the emerging spectrum (Haardt & Maraschi 1991, 1993; Stern et al. 1995b; Poutanen & Svensson 1996).

The situation becomes more complicated when a notable fraction of the total luminosity escape at energies above ~ 500 keV. Then hard photons can produce $e^{\pm}$ pairs which will be added to the background plasma. Electrons (and pairs) Comptonize soft photons up to γ-rays and produce even more pairs. Thus, the radiation field, in this case, has an influence on the optical depth of the plasmas, which in its turn produces this radiation. This makes the problem very non-linear.

Another complication appears when the energy distribution of particles starts to deviate from a Maxwellian. In the so called non-thermal models, relativistic electrons are injected to the soft radiation field. The steady-state electron distribution should be computed self-consistently, balancing electron cooling (e.g., by Compton scattering and Coulomb interactions) and acceleration, together with the photon distribution. The pioneering steps in solving this problem were done by Stern (1985, 1988) using Monte-Carlo techniques and by Svensson (1987), Fabian et al. (1986), Lightman & Zdziarski (1987), Coppi (1992) using the method of kinetic equations (see Stern et al. 1995a; Pilla & Shaham 1997; Nayakshin & Melia 1998, for recent developments). Non-thermal models have been used extensively in the end of 1980s and beginning of 1990s for explaining the X-ray spectra of active galactic nuclei (see, e.g., Zdziarski et al. 1990; Svensson 1994), while recently pure thermal model were preferred, since the data show spectral cutoffs at ~ 100 keV in both GBHs and Seyferts (Grebenev et al. 1993, 1997; Johnson et al. 1997).

However, power-law like spectra extending without a cutoff up to at least ~ 600 keV, observed in some GBHs in their soft state (Grove *et al.* 1997a,b), give new strength to the undeservedly forgotten non-thermal models.

Spectral fitting with multi-component models following simultaneously energy balance and electron-positron pair balance, give stronger constraints on the physical condition in the X/γ-ray source, its size and geometry, presence of $e^{\pm}$ pairs, and give a possibility to discriminate between various accretion disc models. In this review, we first describe thermal as well as non-thermal pair models that have been used recently for spectral fitting of GBHs and Seyferts. We discuss spectral properties of $e^{\pm}$ plasmas in energy and pair equilibria for various geometries of the accretion flow. Separately for GBHs and Seyferts, we briefly review X/γ-ray observations. Then, we consider physical processes responsible for spectral formation and confront phenomenological models of the accretion discs with data. We restrict our analysis to "radiative" models where radiative processes and radiative transfer in realistic geometries are considered in details while heating and acceleration mechanisms are not specified.

2. Spectral models

2.1. *Thermal (pair) plasmas*

Since spectra of both Seyferts and GBHs in their hard states cutoff sharply at ~ 100 keV (Grebenev *et al.* 1993, 1997; Zdziarski *et al.* 1996a,b, 1997; Grove *et al.* 1997a,b), thermal models are in some preference.

2.1.1. *General properties*

First, we consider properties of an electron (-positron) plasma cloud in energy and pair equilibria without assuming any specific geometry of the accretion flow. There are four parameters that describe the properties of hot thermal plasmas: (i) $l_h \equiv L_h \sigma_T / (m_e c^3 r_c)$, the hard compactness which is the dimensionless cloud heating rate; (ii) the soft photon compactness, l_s, which represents the cold disc luminosity that enters the hot cloud (corona); (iii) τ_p, the proton (Thomson) optical depth of the cloud (i.e., the optical depth due to the background electrons); and (iv) the characteristic temperature of the soft photons, T_{bb}. Here r_c is the cloud size, σ_T is the Thomson scattering cross-section. In this simplified description, energy and pair balance equations have been solved using analytical and numerical methods (Zdziarski 1985; Ghisellini & Haardt 1994; Pietrini & Krolik 1995; Coppi 1992; Stern *et al.* 1995a).

For sufficiently high compactnesses, the total optical depth can be significantly larger than τ_p. In that case, increase in the cloud heating rate results in the corresponding increase of the cloud total optical depth, τ_T, due to $e^{\pm}$ pairs produced, and in decrease of the cloud temperature, T_e. The ratio l_h/l_s (the amplification factor) has a one-to-one correspondence with the Kompaneets y-parameter (e.g., Rybicki & Lightman 1979), which can be related to the spectral index of the emitted spectrum (see Fig. 1). The spectral index stays approximately constant for a constant Kompaneets y-parameter ($y = 4\Theta\tau_T$, for parameters of interest, here $\Theta \equiv kT_e/m_e c^2$).

Pietrini & Krolik (1995) proposed a very simple analytical formula that relates the observed X-ray spectral energy index to the amplification factor:

$$\alpha \approx 1.6 \left(\frac{l_h}{l_s}\right)^{-1/4}. \tag{2.1}$$

Though the exact coefficient of proportionality depends on the electron temperature and

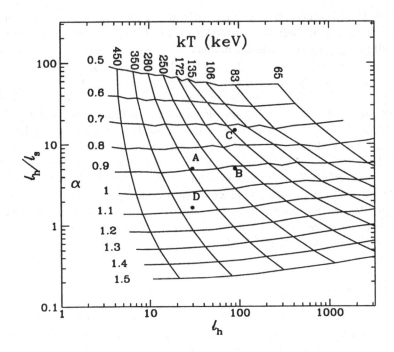

FIGURE 1. Mapping of $(l_h, l_h/l_s)$ to (α, T_e) (from Ghisellini & Haardt 1994). The compactnesses are defined as follows: $l_{h,s} \equiv L_{h,s}\sigma_T/(m_ec^3r_c)$, where r_c is the size of the hot cloud, L_h is the heating rate of the hot cloud, and L_s is the luminosity of the seed soft photons that cool the plasma. This relation holds for $e^{\pm}$ pairs dominated plasmas. For electron-proton dominated plasmas, (α, T_e) depend on $(\tau_T, l_h/l_s)$. One sees that ratio l_h/l_s almost uniquely defines the spectral energy index, α.

the energy of seed photons, this dependence is rather weak. When l_h/l_s increases (the source becomes more "photon starved"), the observed spectrum becomes harder.

2.1.2. *Geometry*

We consider here various geometrical arrangements of the hot plasma cloud and the source of soft photons (see Haardt 1997 for a recent review). The observed spectra of GBHs and Seyferts correspond to $y \approx 1$. This fact does not have a direct explanation in the accretion disc framework. If soft seed photons for Comptonization are produced by reprocessing hard X/γ-ray radiation, then the geometry will define the amplitude of feedback effect and the spectral slope (Liang 1979). What would be the most probable geometry?

Sandwich. The simplest solution is to assume that a hot corona covers most of the cold disc (a sandwich, or a slab-corona model). The radiative transfer in such a geometry was considered by Haardt & Maraschi (1991, 1993) who showed that in the extreme case, when all the energy is dissipated in the corona, the emitted spectra resemble those observed in Seyfert galaxies. Dissipation of energy in the cold disc (with subsequent additional production of soft photons) would produce too steep spectra in disagreement with observations. Even harder spectra observed in GBHs cannot be reconciled with the slab-corona model (Dove *et al.* 1997; Gierliński *et al.* 1997a; Poutanen, Krolik & Ryde 1997), and alternative models with more photon starved conditions and smaller feedback of soft photons are sought.

Magnetic flares. A patchy corona (Galeev, Rosner & Vaiana 1979; Haardt, Maraschi & Ghisellini 1994), where the cold disk is not covered completely by hot material, has certainly a smaller feedback, and the resulting spectra are harder. A patchy corona can be described by a number of active regions above the cold accretion disc. Spectral properties of an active region in the energy and pair balance have been computed recently by Stern *et al.* (1995b) (see also reviews by Svensson 1996a,b for more details). Both patchy and slab-corona models predict an *anisotropy break* (i.e. a break in the power-law spectrum due to the anisotropy of the seed photons) that should appear at the energy corresponding to the second scattering order.

Cloudlets. Another possible solution of the photon starvation problem is to assume that the cold disc within the hot corona is disrupted into cold dense optically thick clouds (Lightman 1974; Celotti, Fabian & Rees 1992; Collin-Souffrin *et al.* 1996; Kuncic, Celotti & Rees 1997) that are able to reprocess hard X/γ-ray radiation and produce soft seed photons for Comptonization. If the height-to-radius ratio of the hot cloud is small, we can approximate this geometry by a plane-parallel slab. We assume further that the cold material is concentrated in the central plane of the hot slab and has a covering factor f_c. Compton reflection comes from these cold clouds (cloudlets) as well as from the outer cooler disc. The seed soft photon radiation is much more isotropic and the emerging high energy spectrum does not have an anisotropy break. The covering factor defines the amplitude of the feedback effect. The total soft seed luminosity (with corresponding compactness, l_s) is the sum of the reprocessed luminosity and the luminosity intrinsically dissipated in the cold disc (with corresponding compactness, l_s^{intr}). For a slab geometry, the heating rate, L_h, of a cubic volume of size h determines the hard compactnesses $l_h \equiv L_h \sigma_T/(m_e c^3 h)$ (where h is the half-height of the slab). Other compactnesses are defined is a similar way.

Figure 2 shows the dependence of the electron temperature and the optical depth on parameters of the cloudlets model, and Figure 3 gives a few selected spectra. Using the method of Poutanen & Svensson (1996), we solve the energy and pair balance equations coupled with the radiative transfer accounting for Compton scattering (exact redistribution function is employed, see, e.g., Nagirner & Poutanen 1994), pair production and annihilation, and Compton reflection. We should point out that in the case of pair dominated plasmas, increase in the amount of soft photons does not necessarily imply a decrease in the plasma temperature. The optical depth decreases rapidly with increase of l_s^{intr}/l_h, and the average energy available per particle can even increase. In the case of electron-proton plasmas, $\tau_T \approx \tau_p$ and T_e decreases with increasing internal dissipation in the cold disc.

"Sombrero". In this model, the cold disc penetrates only a short way into the central coronal region (see, e.g., Bisnovatyi-Kogan & Blinnikov 1977, and Poutanen *et al.* 1997 for recent applications). We can assume that the X/γ-ray source can be approximated by a homogeneous spherical cloud of radius r_c situated around a black hole (probably, a torus geometry for a hot cloud would be more physically realistic, but then it would be more difficult to compute the radiative transfer). The inner radius, r_{in}, of the cold geometrically thin, infinite disc is within the corona ($r_{in}/r_c \leq 1$). This geometry is also similar to the geometry of the popular advection dominated accretion flows (see the article by R. Narayan, R. Mahadevan, and E. Quataert in this volume). Spectra from the sombrero models are almost identical to the spectra expected from the cloudlets model, with the only difference that the amount of Compton reflection would be a bit larger for the same configuration of the outer cooler disc. From the observational point of view, these models are almost indistinguishable.

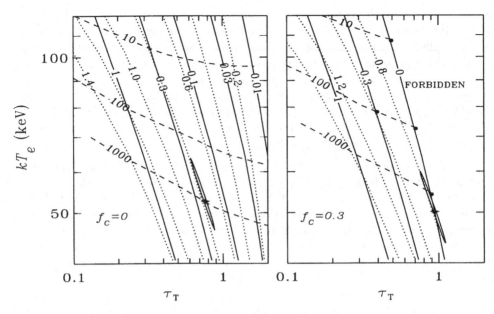

FIGURE 2. Relation between total optical depth, τ_T, of the half-slab and electron temperature, kT_e, for the cloudlets model. Here cold matter is assumed to be concentrated in the central plane of the hot slab. Temperature of cloudlets is fixed at $kT_{bb} = 0.25$ keV. The covering fraction of the cold material is taken to be $f_c = 0$ (left panel) and $f_c = 0.3$ (right panel). For $f_c = 0$, all seed photons are external. Solid curves represent solutions for constant l_s^{intr}/l_h. These relations are the same for pure pair or pure electron-proton plasmas. The dashed curves are the solutions for a constant l_h (assuming pair dominated plasmas, i.e., $\tau_p \ll \tau_T$, and thermal electron distribution). Dotted curves give solutions for a constant intrinsic (without Compton reflection) spectral index α in 2–18 keV range. l_s^{intr} can mean here both intrinsic internally (not reprocessed) and externally produced soft photon luminosity. Region to the right of $l_{s'}^{intr}/l_h = 0$ curve is forbidden (the energy balance cannot be reached). Stars represent best fits to the simultaneous *Ginga* and OSSE data of GX 339-4 in September 1991 (Zdziarski *et al.* 1998) and elongated ellipsoids are the contours plots at 90 per cent confidence level for two interesting parameters ($\Delta\chi^2 = 4.61$). For GX 339-4, external or internally produced soft luminosity (not reprocessed) entering hot slab can be $\sim 23\%$ of the heating rate L_h, if $f_c = 0$. If $f_c = 0.3$, the only solution describing data of GX 339-4 is possible when there are *no* internally generated (except reprocessed) or external soft photons. Cooling is provided by reprocessed radiation only. Spectra for the solutions marked by filled circles are shown in Figure 3.

2.2. *Hybrid thermal/non-thermal pair plasmas*

There are reasons to believe that in a physically realistic situation, the electron distribution can notably deviate from a Maxwellian. A significant fraction of the total energy input can be injected to the system in form of relativistic electrons (pairs). In the so called hybrid thermal/non-thermal model, the injection of relativistic electrons is allowed in addition to the direct heating of thermal electrons.

The most important input parameters of the model are: (i) the thermal compactness, l_{th}, which characterizes the heating rate of electrons (pairs); (ii) the analogous non-thermal compactness, l_{nth}, which characterizes the rate of injection of relativistic electrons, (iii) the soft photon compactness, l_s; (iv) Γ_{inj}, the power-law index of the non-thermal electron injection spectrum, (v) τ_p, the proton (Thomson) optical depth; and (vi) T_{bb}. Compton reflection adds a few more parameters (e.g. the amplitude R, the ionisation parameter ξ) and can be accounted for using angular dependent Green's functions (Magdziarz & Zdziarski 1995; Poutanen, Nagendra & Svensson 1996). By

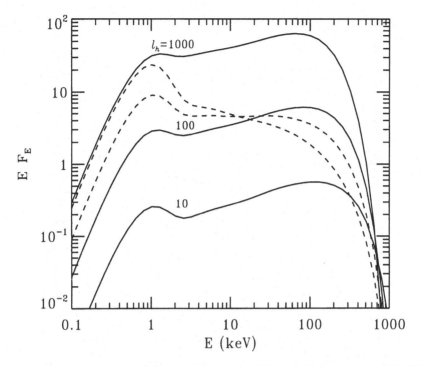

FIGURE 3. Spectra from the hot slab - cold clouds model for solutions represented by filled circles in Figure 2. Solid curves correspond to the solutions with no internal dissipation in the cold clouds ($l_s^{\mathrm{intr}} = 0$) and various hard compactness l_h. Dotted curves correspond to $l_h = 100$, and $l_s^{\mathrm{intr}}/l_h = 0.3$ and 1. With increasing l_s^{intr}/l_h spectra become steeper (softer), but the electron temperature does not decrease (rather slowly increases). In the case of the electron-proton plasmas, T_e would decrease rapidly with constant $\tau_T \approx \tau_p$.

$l_h = l_{th} + l_{nth}$, we denote the total hard compactness. For spectral fitting, we use the code of Coppi (1992) (see also Coppi *et al.* 1998) which is incorporated into the standard X-ray data analysis software XSPEC.

The electron distribution is computed self-consistently balancing electron cooling (by Compton scattering and Coulomb interactions), heating (thermal energy source), and acceleration (non-thermal energy source). The self-consistent electron (-positron) distribution can be characterized by a Maxwellian of the equilibrium temperature, T_c, plus a non-thermal (generally not a power-law) tail. The spectrum of escaping radiation then consists of the incident blackbody, the soft excess due to Comptonization by a *thermal* population of electrons and a power-law like tail due to Comptonization by a *non-thermal* electron (pair) population.

As a first example, we consider how spectra from the hybrid plasmas change with the hard compactness when keeping the ratio of the soft-to-hard compactness constant (this gives an almost constant α) and fixing the non-thermal efficiency (l_{nth}/l_h) at 10 per cent (see solid curves in Figure 4). The electron temperature behaves exactly as in the pure thermal case (it decreases when compactness increases), since relatively small non-thermal efficiency, that we have chosen, does not change the energy balance significantly. The electron distribution is Maxwellian with a weak high energy tail.

Next, we consider how the spectra change as a function of l_h/l_s, while keeping the other parameters constant. The dashed curves in Figure 4 show the evolution of the spectrum with increasing soft seed photon luminosity. For large l_h/l_s, most of the spectrum is

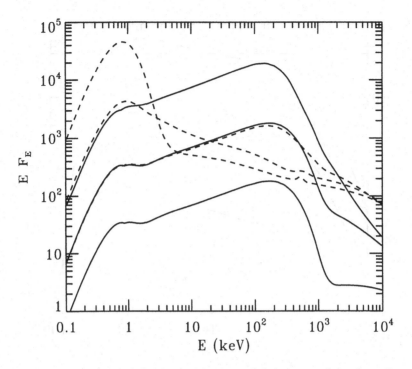

FIGURE 4. Spectra from the hybrid pair plasmas. Solid curves show dependence on hard compactness l_h. Other parameters: $l_h/l_s = 10$, $l_{nth}/l_h = 0.1$, $\tau_p = 1$, $\Gamma_{inj} = 2.5$, $T_{bb} = 0.2$ keV. The resulting electron temperature and optical depth are (kT_e, τ_T)=(126 keV, 1.0002), (123 keV, 1.02), and (82 keV, 1.47) for $l_h = 1, 10, 100$, respectively (l_h increases from the bottom to the top of the figure). For a higher compactness, the spectrum has a sharper cutoff at energies above 1 MeV due to larger optical depth for photon-photon pair production. These spectra are similar to the spectra of GBHs in their hard state (see Fig. 5). Dashed curves show dependence on l_h/l_s. Here we fixed $l_h = 10$, $l_{nth}/l_h = 0.5$. The resulting electron temperature and optical depth (kT_e, τ_T) are (104 keV, 1.07), (34 keV, 1.02), and (5 keV, 1.01) for $l_h/l_s = 10, 1, 0.1$, respectively. Increase in l_s results in a more pronounced blackbody part of the emerging spectrum. The blackbody is modified by Comptonization on thermal electrons.

produced by Comptonization off a *thermal* population of electrons (pairs), while the tail at energies above $m_e c^2$ is produced by *non-thermal* electrons. For low l_h/l_s, the electron temperature drops. Most of the pairs are in the thermal bump, but Kompaneets y-parameter is very small since the electron temperature is small. The resulting spectrum is produced by a *single* Compton scattering off non-thermal electrons. The Maxwellian part of the electron distribution produces a weak power-law tail to the blackbody bump. The annihilation line is quite weak for relatively small compactnesses, and would not be detectable by modern detectors. The cutoff energy at a few MeV is anti-correlated with the compactness.

Quantitatively, the behaviour of the electron distribution with changing of the amount of soft photons is easy to understand. A break between thermal and non-thermal parts of the electron distribution appears where the thermalization timescale due to Coulomb scattering is equal to the Compton cooling timescale (we neglect here thermalization by synchrotron self-absorption, see, e.g., Ghisellini & Svensson 1990; Ghisellini, Haardt, & Svensson 1998). Compton cooling timescale is simply $t_{Compton} = \pi r_c/(\gamma c l_s)$, while Coulomb thermalization operates at $t_{Coulomb} \approx \gamma r_c/(\tau_T c \ln \Lambda)$ timescale (see, e.g., Dermer & Liang 1989; Coppi 1992; Ghisellini, Haardt, & Fabian 1993; here $\ln \Lambda$ is the usual

Coulomb logarithm, typically ~ 15, and γ is the electron Lorentz factor). These relations define the Lorentz factor of the break

$$\gamma_{break} \approx \left(\pi \ln \Lambda \frac{\tau_T}{l_s} \right)^{1/2}. \tag{2.2}$$

Increase in Compton cooling causes the break to shift towards lower energies.

3. Galactic black holes

3.1. *Hard state of Galactic black holes*

3.1.1. *Observations and interpretation*

Galactic black holes (GBHs) are observed in a few different spectral states that can be generally classified as soft and hard. A spectrum in the hard state is characterized by a power-law with the energy spectral index $\alpha \approx 0.4 - 0.9$ with a cutoff at energies ~ 100 keV (Grebenev *et al.* 1993, 1997; Tanaka & Lewin 1995; Phlips *et al.* 1996; Zdziarski *et al.* 1996a, 1997; Grove *et al.* 1997a,b). The presence of an iron line at ~ 6.4 keV and an iron edge at ~ 7 keV, together with a spectral hardening around 10 keV, was interpreted as a signature of Compton reflection of the intrinsic spectrum from relatively cold matter (Done *et al.* 1992; Ebisawa *et al.* 1996b; Gierliński *et al.* 1997a). The amount of Compton reflection $R \equiv \Omega/2\pi \approx 0.3 - 0.5$ ($R = 1$ corresponds to an isotropic X/γ-ray source atop an infinite cold slab). An excess at energies $\lesssim 1$ keV is interpreted as radiation from the accretion disc (Bałucińska & Hasinger 1991; Bałucińska-Church *et al.* 1995) with a characteristic temperature, T_{bb}, of order $0.1 - 0.3$ keV (usually quite poorly determined, due to strong interstellar absorption in that spectral range). Observations by BATSE and COMPTEL revealed also a presence of a high energy excess at $\gtrsim 500$ keV in some GBHs (Cyg X-1: McConnell *et al.* 1994; Ling *et al.* 1997; GRO J0422+32: van Dijk *et al.* 1995). The monochromatic luminosity, EL_E, peaks at about ~ 100 keV. A characteristic spectrum of the hard state GBH is shown in Figure 5.

GBHs show variability in X/γ-rays on all possible time scales, from milliseconds to years. The size of the emitting region cannot be much larger than the minimum variability time scale $\times c \approx 1$ ms $\times c = 300$ km. In the context of accretion onto a black hole it is 10 gravitational radii, $R_g \equiv GM/c^2$, (for a 10 $M_\odot$ black hole), i.e., the inner part of the accretion disc where most of the energy is liberated. Since, in the hard state, GBHs radiate a big fraction of the energy in the hard X-rays/soft γ-rays (see Figs. 5, 8), the region responsible for the production of this radiation lies within $\sim 20 - 50 R_g$. The most efficient cooling mechanism responsible for formation of the spectra is probably thermal Comptonization of soft photons in the $\sim 50 - 100$ keV electron cloud and $\tau_T \sim 1$ (Shapiro *et al.* 1976; Zdziarski *et al.* 1996a, 1997; Gierliński *et al.* 1997a). High sensitivity OSSE observations in the 50-500 keV range allow to determine the electron temperature to within 10 per cent. High energy ($\gtrsim 300$ keV) excesses can be explained only if one introduces an additional spectral component. This component can be produced either in a spatially separated, much hotter region by thermal Comptonization (Liang & Dermer 1988; Liang 1991; Ling *et al.* 1997), or in the same region by a non-thermal tail of the electron distribution (Li, Kusunose & Liang 1996a,b; Poutanen & Coppi 1998). In the former (thermal) case, the very hot ~ 400 keV plasma cloud has to be kept far from the sources of soft photons to avoid cooling, and it is not clear whether one can physically separate it from the rest of the accretion disc. Having a very hot cloud close to the ~ 100 keV inner disc, could also be a problem since radiative conduction would smooth out large temperature gradients. On the other hand, non-thermal tails can be created by

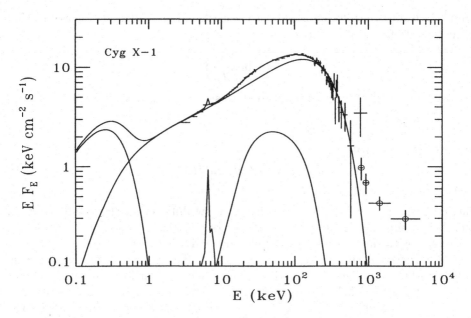

FIGURE 5. The hard state of Cygnus X-1. Simultaneous *Ginga*, OSSE and COMPTEL data from June 1991. Different components correspond to the soft blackbody radiation, thermal Comptonized spectrum, and its Compton reflection. A high energy excess at $\gtrsim 1$ MeV cannot be described by thermal models.

magnetic processes which likely operate in the accretion disc environment (see Dermer, Miller & Li 1996; Li *et al.* 1996a; Li & Miller 1997 and references therein).

3.1.2. *Geometry*

Observations of the Compton reflection feature, together with a fluorescent iron line, suggest the presence of a rather cold weakly ionized material in the vicinity of the X/γ-ray source. From the amplitude of Compton reflection one can derive that the cold matter occupies $\Omega/2\pi \sim 1/3$ solid angle as viewed from the hard photon source (Ebisawa *et al.* 1996b; Gierliński *et al.* 1997a; Życki, Done & Smith 1997a). Further constraints on the geometry, e.g., the covering fraction of the hot cloud as viewed from the soft photon source, can be derived from the observed spectral slopes if the *observed* soft luminosity can be reliably estimated (Poutanen *et al.* 1997). There are no evidence that the cold material extends close to the black hole. (The observed iron lines are quite narrow, and the iron edges are quite sharp implying weakness of gravitational and Doppler effects.)

Sandwich. As we pointed out in Section 2.1.2, an accurate treatment of the radiative transfer and Compton scattering rule out slab-corona (sandwich) models for sources with hard spectra (see Poutanen *et al.* 1997; Dove *et al.* 1997 where the case of Cyg X-1 is considered, and Zdziarski *et al.* 1998 for the interpretation of GX 339-4 data). Even if all the energy is dissipated in the corona, the predicted spectra are too steep and cannot be reconciled with observations. Energy dissipation in the cold disc and the corresponding increase of the amount of seed soft photons worsen the discrepancy. The anisotropy break expected at a few keV in the sandwich model was never observed.

Flares. Magnetic flares (active regions) on the surface of the cold accretion disc also predict anisotropy break in disagreement with observations. Active regions atop the cold disc give the right amount of Compton reflection (since at $\tau_T \sim 1$, a big fraction of it is scattered away), but produce too steep spectra as in the case of the sandwich model.

Detached active regions (Svensson 1996a,b) produce spectra with the right spectral slopes while predicting too much reflection ($R \sim 1$).

Cloudlets. Zdziarski *et al.* (1998) found this model giving the best description of *Ginga* and OSSE data of GX 339-4. In this case, most of the Compton reflection occurs in an outer cold disc. The covering factor of cold clouds within the hot inner disc cannot be larger than $f_c \sim 0.3$ due to the energy balance requirements (see Fig. 2).

Sombrero. This geometry is consistent with the amount of Compton reflection observed in GBHs in their hard state (Dove *et al.* 1997; Gierliński *et al.* 1997a; Poutanen *et al.* 1997). In the case of Cyg X-1, a solution with $r_{in}/r_c = 1$ is energetically possible, but the intrinsic soft luminosity should be rather large in order to produce enough soft photons for Comptonization. The energy balance constrains the inner radius of the cold disc to be larger than $\sim 0.7 r_c$. Probably, $r_{in}/r_c = 0.8 - 0.9$ would satisfy all the observational requirements (see Poutanen *et al.* 1997). Similarly, in case of GX 339-4 (which has a steeper spectrum in the hard state than Cyg X-1), $r_{in}/r_c \geq 0.7$ required.

Concluding, models with the central hot cloud surrounded by a cold disc give the best description of the data.

3.1.3. *Spectral variability and $e^{\pm}$ pairs*

As it was already mentioned, Galactic black holes show variability on different time scales (see recent review by Van der Klis 1995). Here we just consider spectral variations on the time scale of hours. It was shown by Gierliński *et al.* (1997a) that the spectral shape of Cyg X-1 in the *Ginga* spectral range does not vary much when luminosity changes within a factor of two (Fig. 6), while there is evidence that the cutoff energy increases when luminosity drops (best fit curves cross each other at ~ 500 keV). Such a behaviour implies almost a constant ratio l_s/l_h (see Figs. 1 and 2) and a constant Kompaneets y-parameter. We can conclude that the transition radius between the hot and the cold discs does not change much (otherwise, the ratio of the intrinsic energy dissipation in the cold disc to the heating rate of the hot cloud would change, causing spectral slope changes). Alternatively, the transition radius is sufficiently large that most of the seed soft photons are provided by reprocessing hard radiation from the central cloud.

For $e^{\pm}$ pair dominated plasmas, variation in the heating rate would cause variations in the optical depth by pair production and corresponding changes of the cloud temperature. On the other hand, the same behaviour is expected when no pairs are present. Now, variations in the accretion rate cause variation in optical depth and escaping luminosity, and the temperature adjusts to satisfy the energy balance.

It is probably not possible to determine the pair content directly from observations, since spectra from the *thermal* $e^{\pm}$ plasma are indistinguishable from spectra of the electron-proton plasma (annihilation line is much too weak to be observed, see Maciołek-Niedźwiecki, Zdziarski & Coppi 1995; Stern *et al.* 1995b). It is in principle possible to determine the compactness parameter from observations. For example, in the case of Cyg X-1, $l_h \sim 20$ assuming a unitary (i.e. not broken into many pieces) source, while $l_h \sim 400$ is required to make the source pair dominated (Poutanen *et al.* 1997). If the energy dissipation is extremely inhomogeneous (which would be the case if magnetic reconnection is responsible for the energy dissipation), then at a given moment most of the luminosity is produced by a smaller fraction of the cloud and the effective compactness can be much higher. Presence of non-thermal particles in the source would increase pair production and explain the high energy excess at ~ 1 MeV at the same time. However, the quality of the data is not sufficient to make any definite conclusions.

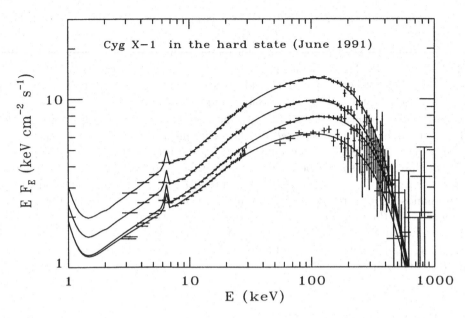

FIGURE 6. Spectral variation of Cyg X-1 on the time scales of hours in the hard state as observed by *Ginga* and OSSE in June 1991 (data are adapted from Gierliński *et al.* 1997a).

3.2. *Soft state of Galactic black holes*

3.2.1. *Observations and interpretation*

Unlike in the hard state, most of the luminosity in the soft state is carried by a blackbody like component with a characteristic temperature, $kT_{\rm bb} \approx 0.5 - 1$ keV. Until recently, there were not so many broad-band data with high spectral resolution and high signal-to-noise ratio (see, e.g., Tanaka & Lewin 1995; Grebenev *et al.* 1993, 1997) that a detailed spectral analysis would be possible. Having *ASCA, RXTE*, and *CGRO* in orbit at the same time changes the situation. Cygnus X-1 was observed by all these observatories simultaneously on May 30, 1996 and by *RXTE* and *CGRO* on June 17-18, 1996, when it was in the soft state. Figure 7 gives an example of the soft state spectrum.

A soft component (energies less than ~ 5 keV) cannot be represented neither by a multicolor disc spectrum, nor a modified blackbody. It is clear that at least two components are required to fit it (e.g., blackbody and a power-law, or two black bodies, Cui *et al.* 1997a,b; Gierliński *et al.* 1997b). One can imagine that the soft blackbody comes from the accretion disc, but the nature of the additional component peaking at ~ 3 keV is not so clear. Gierliński *et al.* (1997b) interpreted it as due to thermal Comptonization of a disc blackbody in a plasma with $kT_e \approx 5$ keV and $\tau_T \approx 3$.

GBH spectra in the hard X-rays/soft γ-rays can be well represented by a power-law which does not have an observable break, at least, up to energies of order $m_e c^2$ (Phlips *et al.* 1996; Grove *et al.* 1997a,b, 1998). COMPTEL has detected Cyg X-1 and GRO J1655-40 at energies up to ~ 10 MeV (A. Iyudin, private communications), and it seems that the power-law at MeV energies is just a continuation of the hard X-ray power-law. Although signatures of Compton reflection are also observed in this state (e.g. Tanaka 1991), its amplitude, R, is much more difficult to determine since it depends on the assumed distribution of ionisation and detailed modelling of the continuum, which is rather curved in the spectral region around the iron edge (see Fig. 7). For example, Gierliński *et al.* (1997b) give $R \approx 0.6 - 0.8$ for Cyg X-1 in the soft state observed on

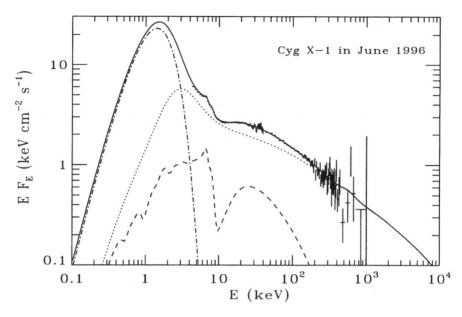

FIGURE 7. The soft state of Cygnus X-1 observed by *RXTE* and OSSE in June 1996 and the best fit hybrid thermal/non-thermal pair model (corrected for interstellar absorption). The solid curve represents the total spectrum, the dashed curve gives the Compton reflection spectrum, and the dotted curve represents the Comptonized continuum. The disc blackbody is shown by the dot-dashed curve. The parameters of the fit are: $l_s = 20$ (frozen), $l_h/l_s = 0.3$, $l_{nth}/l_h = 0.95$, $\tau_p = 0.3$, $kT_{bb} = 0.36$ keV, $\Gamma_{inj} = 3.0$, $R = 0.4$, $\xi = 3.7 \cdot 10^3$, giving a $\chi^2/\text{dof}=168/167$. The temperature of the Maxwellian part of the electron distribution is $kT_e = 30$ keV, and the total Thomson optical depth (including pairs) $\tau_T = 0.32$. The iron edge and the iron line appear to be smeared (requiring to account for rotation of the relativistic accretion disc, see Życki *et al.* 1997a,b; Gierliński *et al.* 1997b) and the reflector to be ionised.

May 30, 1996, while Cui *et al.* (1997) get $R \approx 0.15$ (restricting themselves to a much narrower energy range). Both, the iron line and the iron edge, appear to be smeared due to probably gravitational redshift and Doppler effect, implying that the cold disc extends very close to the central black hole.

The origin of the steep power-law was interpreted in terms of bulk Comptonization in a converging flow (Ebisawa, Titarchuk, & Chakrabarti 1996; Titarchuk, Mastichiadis, & Kylafis 1997). This model predicts a cutoff at $\lesssim m_e c^2$ which does not appear to be the case. The power-law can be produced by Comptonization of soft photons from the accretion disc by a non-thermal corona (the base of the jet?) which is optically thin and covers much of that disc (Mineshige, Kusunose & Matsumoto 1995; Li *et al.* 1996a,b; Li & Miller 1997; Liang & Narayan 1997; Poutanen & Coppi 1998). In that case, the turnover is expected at a few MeV due to absorption by photon-photon pair production.

3.2.2. *Hybrid pair plasma model*

The steep power-law in the γ-ray spectral region can be interpreted as a Comptonization (or, in fact, a single scattering) by *non-thermal* electrons ($e^\pm$ pairs), and an additional component peaking at ~ 3 keV as a *thermal* Comptonization by rather low temperature electrons. It is natural to assume that both components are produced in the same spatial region by electrons having a non-Maxwellian distribution.

The soft state data are shown on Figure 7 together with the model spectrum. Unfortunately, data in the γ-rays are not good enough to determine the compactnesses

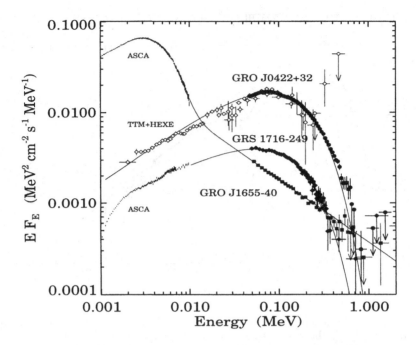

FIGURE 8. Different spectral classes of Galactic black holes (from Grove *et al.* 1998).

unambiguously (one needs very accurate estimates of the amplitude of the annihilation line, as well as shape of the cutoff at $\sim 5 - 10$ MeV). It is worth mentioning once more that the whole spectrum from soft X-rays to γ-rays can be represented by a hybrid model where the electron distribution is determined self-consistently by balancing heating, acceleration and cooling.

3.3. *Spectral state transitions*

Some Galactic black holes have been observed to always be in one of the states (either in the hard, or in the soft), while others have shown transitions between states (see Fig. 8, and Sunyaev *et al.* 1991; Grove *et al.* 1997b; Grebenev *et al.* 1997). The nature of the state transitions is not fully understood yet, and none of the dynamical accretion disc models can fully describe them. Probably, the most developed model presently available is the advection dominated accretion disc model (see Esin *et al.* 1997, 1998, and the article by R. Narayan *et al.* in this volume). This model is able to explain general spectral behaviour in the X-ray range, during the transition. However, restricted to pure thermal plasma, it is not able to explain the γ-ray data.

We restrict our consideration here to the observations and spectral modelling of Cyg X-1. During the transition from the hard to the soft state, the γ-ray luminosity drops, the spectrum becomes steeper, while the cutoff energy increases (Phlips *et al.* 1996; Poutanen 1998). The luminosities of the different spectral components change dramatically in expense of each other, while the bolometric luminosity hardly changes (Zhang *et al.* 1997).

In the hard state, most of the power is deposited, through the thermal channel, to heat the plasma of the inner accretion disc. The non-thermal supply is relatively small, resulting in a weak observed tail at MeV energies. The soft photon input to the system

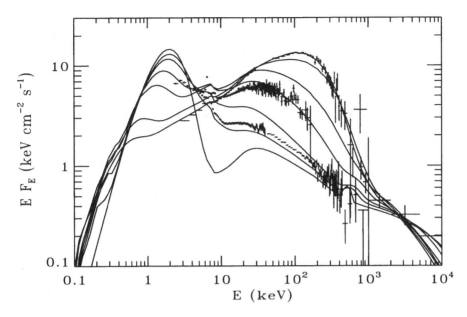

FIGURE 9. Spectral states of Cyg X-1 and the state transition as predicted by the hybrid pair model.

is also rather small. Thus the system is "photon starved" and produces a hard spectrum. Soft photons that are Comptonized to X/γ-rays by thermal electrons are most probably produced by reprocessing of the X/γ-rays in the cold material. This feedback effect fixes the spectral index at a value that is defined by the geometry of the system.

In the soft state, the soft photon luminosity (from the accretion disc) exceeds all other energy injection rates to the system, while thermal energy dissipation (electron heating in the corona/hot disc) is negligible. The non-thermal energy injection rate is $\sim 1/4$ of the total energy output of the system. This almost pure non-thermal model reproduces the broad-band soft state data (see Fig. 7 and Gierliński *et al.* 1998).

Poutanen & Coppi (1998) showed that by using simple scaling laws for the luminosity of the cold disc, the thermal dissipation/heating rate in the corona, and the rate of energy injection from a non-thermal source, all as functions of radius of the corona, the hard-to-soft transition can be explained as the result of a decrease in the transition radius between the inner hot disc (corona) and the outer cold disc by a factor ~ 5. They assumed that the sum of the soft luminosity from the disc, $L_s \propto 1/r$, and the thermal dissipation rate in the corona, $L_{th} \propto 1 - 1/r$, as well as the non-thermal power, L_{nth}, remain approximately constant during transition. This idea is somewhat similar to the proposal by Mineshige *et al.* (1995) (although they fixed the ratio l_{nth}/l_{th}). The model gives a sequence of spectra, shown in Figure 9 together with the data of Cyg X-1. When transition radius decreases, the ratio l_h/l_s (and the spectral index) does not change until the internally generated soft photon luminosity becomes comparable to the reprocessed one. After that the spectrum changes dramatically since l_h/l_s decreases. The model explains the pivoting behaviour of the spectra at ~ 10 keV, and predicts a pivoting behaviour at $\gtrsim 1$ MeV. Smaller soft γ-ray luminosity in the soft state results in a decrease of the pair production opacity, giving a higher cutoff energy. The predicted behaviour in the MeV range can be compared with observations only after the launch of INTEGRAL.

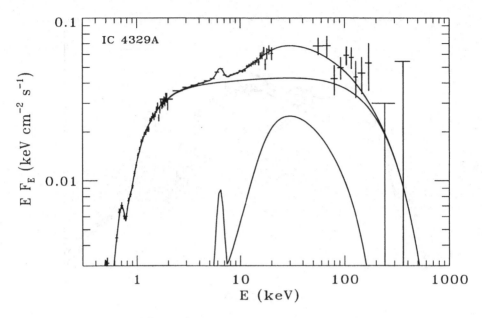

FIGURE 10. The ROSAT/*Ginga*/OSSE broad-band spectrum of Seyfert 1 galaxy, IC4329A (data from Madejski *et al.* 1995). The dashed line is the thermal Comptonization spectrum from a spherical cloud of electron temperature $kT_e = 85$ keV and radial optical depth $\tau = 1.1$. The temperature of the soft seed photons, kT_{bb}, is fixed at 35 eV. Calculation are made for the geometry of the X/γ-ray source where cold accretion disc is situated outside of a hot central spherical cloud (see Poutanen *et al.* 1997; Dove *et al.* 1997). The amount of Compton reflection is factor of 3 larger than expected from an infinite geometrically thin cold slab.

4. Seyfert galaxies

Broad-band spectral properties of Seyfert (Sy) galaxies are discussed in recent reviews by Zdziarski *et al.* (1997), Johnson *et al.* (1997) (see also a review by G. Madejski in this volume).

4.1. *"Normal" Seyfert 1 galaxies*

4.1.1. *Observations*

X-ray/gamma-ray spectra of Sy 1s are very similar to the hard state of GBHs (see Fig. 10), having a power-law spectral index $\alpha \approx 1$ and cutoff energies $E_c \approx 300$ keV (e.g., Gondek *et al.* 1996). The spectral hardening at ~ 10 keV is also detected in Seyferts (Nandra & Pounds 1994; Weaver, Arnaud & Mushotzky 1995). An important difference is that the amount of Compton reflection is generally found to be somewhat higher, $R \sim 0.8$ (Gondek *et al.* 1996; Zdziarski *et al.* 1997). Probably, the only exception is NGC 4151 which has almost an identical spectrum to the one of the GBHs GX 339-4 (Zdziarski *et al.* 1998), and $R \approx 0.4$ (Zdziarski, Johnson & Magdziarz 1996). Radio quiet Sy have not been detected in the γ-ray spectral range (Maisack *et al.* 1995). Spectral similarities with GBHs support the attribution of the X/γ spectra to a scale invariant process, such as Compton scattering. Best fits with thermal Comptonization models to the broad-band data give $kT_e \approx 100$ keV and $\tau_T \approx 1$.

4.1.2. *Geometry*

The *slab-corona* model (Liang 1979; Haardt & Maraschi 1991, 1993) predict slopes of the Comptonized spectrum $\alpha \sim 1$, if all the energy is dissipated in the corona. Harder

X-ray spectra can be achieved only if the coronal temperature is high enough ($\gtrsim 300$ keV) to make an anisotropy break in the observed band (Stern *et al.* 1995b; Svensson 1996a,b). Let us consider as an example one of the brightest Sy 1s, IC 4329A. Spectral fitting to the broad-band data (Madejski *et al.* 1995) gives $kT_e = 80$ keV and $\tau_T = 0.7$ (assuming $kT_{bb} = 35$ eV). However, this model can be ruled out by the requirement of energy balance. For a given τ_T, the reprocessed flux from the cold slab would cool down the corona to 45 keV. Internal dissipation in the cold slab would worsen the discrepancy. Similar conclusion can be drawn for the averaged spectrum of Sy 1s (Zdziarski *et al.* 1997).

Flares. Since in Seyferts, the amplitude of Compton reflection, R, is closer to 1 than in GBHs, one could argue that active region (magnetic flares) model is more viable here. Indeed, having less soft photon returning to the active region, the spectra (harder than in the slab-corona case) satisfy the energy balance condition, and produce the needed amount of Compton reflection. The anisotropy break (never seen in Seyferts) shifts to smaller energies, out of the observed X-ray band, due to smaller (than in GBHs) characteristic energies of soft seed photons.

Cloudlets. The hardness of the spectrum of IC 4329A, provides limits on the covering factor of the cold clouds. The energy balance can be satisfied for $f_c \lesssim 1/3$ (similar to GX 339-4, see Fig. 2; note that curves corresponding to a constant l_h/l_s for $kT_{bb} = 35$ eV are shifted a little to the left). Larger f_c would cool down the hot slab by the reprocessed radiation. For $f_c = 1/3$ (no energy dissipation in clouds is allowed), the best fit to the data gives $kT_e = 70$ keV and the optical thickness of the half-slab $\tau_T = 0.6$. This model requires Compton reflection to be produced externally.

Sombrero. Models with a hot inner disc (modelled as a sphere) and a cold outer disc, give acceptable fits. They are energetically possible if $r_{in}/r_c \gtrsim 0.7$. For $r_{in}/r_c = 0.7$ the best fit gives $kT_e = 75$ keV and $\tau_T = 1.5$, and for $r_{in}/r_c = 1$, $kT_e = 85$ keV and $\tau_T = 1.1$ (see Fig. 10). Both, the sombrero model (where the cold disc is modelled as an infinite, geometrically thin, slab) and the cloudlets model predict too little Compton reflection. Probably, it can be provided by a material far away from the center and plane of the disc (molecular torus, see Krolik *et al.* 1994; Ghisellini *et al.* 1994). This can be checked by the response of the reflection component on changes in the amplitude of the continuum.

4.2. *Seyfert 2s*

Seyfert 2 galaxies have generally higher column densities of the absorber towards the nuclei, than Sy 1s do, so that their intrinsic (nuclear) radiation is not always directly observable (see, e.g. Smith, Done & Pounds 1993; Done, Madejski & Smith 1996; Matt *et al.* 1997). On the other hand, hard X-ray/soft γ-ray spectra of Sy 2s and Sy 1s are very similar (Zdziarski *et al.* 1997; Johnson *et al.* 1997) as predicted by the unification schemes of Seyfert galaxies (Antonucci 1993), and there are no really strong arguments that the intrinsic spectra of Sy 2s are somewhat different from that of Sy 1s.

4.3. *Narrow line Seyfert 1 galaxies*

Another class of Seyfert galaxies, the ultra soft narrow-line Seyfert 1 galaxies (NLSy 1), is characterized by a large soft X-ray excess (very steep spectrum in the soft X-rays) and somewhat steeper (softer) spectra in the standard (1-10 keV) X-ray range (see Turner *et al.* 1993; Pounds, Done & Osborne 1995, Brandt, Mathur & Elvis 1997). Their hard X-ray properties are unknown. Pounds *et al.* (1995) interpreted these objects as Seyferts in their soft (high) state. One would think that in that case, broad fluorescent iron lines should appear in these objects more often than in "normal" Seyferts, since the inner edge of the cold disc should be closer to the central black hole in order to produce large soft X-ray luminosity. One can expect some indications of the correlations between the spectral index and the width of the iron line. We certainly need better data to make further progress in understanding these objects.

5. Final remarks

One of the most intriguing developments during recent years, is the understanding that X/γ-ray spectra of stellar mass black holes (in their hard states) are very similar to the spectra of active galactic nuclei, that are believed to contain $10^6 - 10^8$ solar masses. The spectral fits to the data with the thermal Comptonization models give values of the electron temperature within 50–100 keV and a Thomson optical depth (of the slab) close to unity. These observations give more support to the scale free accretion disc models of the central engine and to Comptonization as the most important scale free radiative process. There are, however, a few questions that have to be addressed.

What is the physical reason for τ_T to be ~ 1? If the plasma is $e^{\pm}$ pair dominated, $\tau_T \sim 1$ is a natural limit just because it is difficult to get compactnesses larger than $\sim 10^3$. Such a compactness is at least an order of magnitude above the estimate, given by the known X/γ-ray luminosity and sizes inferred from the inner edge of the outer cold disc. Significant contribution from non-thermal processes and inhomogeneous energy dissipation probably can remove this discrepancy. On the other hand, $\tau_T \sim 1$ can be achieved in the hot accretion discs radiating at the maximum accretion rate limited by advection (Zdziarski 1998).

What is the geometry of the X/γ-ray producing region in GBHs and Seyferts? Magnetic flares above a cold accretion disc are still a possible solution for Seyferts. We argued that in the case of GBHs, a hot inner disc with a cold disc outside is a more plausible geometry. This can also be a *unifying geometry for both GBHs and Seyferts*. Larger amplitude of the Compton reflection observed in Seyferts can be due to the contribution from the molecular torus. On the other hand, the presence of the gravitationally redshifted fluorescent iron lines in the spectra of some Seyferts (implying inner radius of the cold disc to be at a few GM/c^2, see, e.g., Tanaka *et al.* 1995; Fabian *et al.* 1995; Nandra *et al.* 1997) would be more difficult to explain in such a geometry.

An important observational progress, made during recent years, was the broad-band X/γ-ray data of GBHs in their soft state. Luckily for us, the soft state transition was observed in Cyg X-1 in the summer of 1996. These observations revealed that the soft state spectra cannot be explained by thermal Comptonization. Hybrid thermal/non-thermal model gives an acceptable description of the broad-band data (from soft X-ray to MeV), making predictions for the spectral change in the MeV range. Unfortunately, we will have to wait for INTEGRAL, before we can verify these predictions. This model also successfully reproduces spectral transitions, as a results of redistribution of the energy dissipation between the hot inner cloud and the cold outer disc, with a constant non-thermal energy injection, probably by the base of the jet or non-thermal corona. Thus, the hybrid model can be a *unifying link between hard and soft states of GBHs*. What is the physical reason for the change in the transition radius between hot and cold phases? We do not know the answer yet.

In the soft state, the inner radius of the cold disc moves closer to the central black hole. The profile of the iron line should change notably and smearing of the iron edge is expected due to the Doppler effect and gravitational redshift. Simultaneous data with a high spectral resolution and a broad spectral coverage (to determine continuum unambiguously) are required to quantify the amplitude of these effects. Observationally it is a challenge, since the soft blackbody bump (in GBHs) dominates in the spectral region around the iron line.

In the case of Seyferts, it is possible that those objects that show redshifted iron lines belong to the NLSy 1 class (probably, the "soft state" Seyferts, see, e.g., Lee *et al.* 1998 for the case of MCG-6-30-15). Then, the correlation between the spectral index and the width of the iron line is expected. We can speculate that NLSy 1 are analogous to the soft state GBHs, but observationally this is not well established yet.

This research was supported by grants from the Swedish Natural Science Research Council and from the Anna-Greta and Holger Crafoord's Fund. The author thanks P. Coppi, A. Zdziarski, M. Gierliński, R. Svensson, E. Grove, and F. Haardt for various help during preparation of this review. I also would like to thank the organizers of the Symposium on Non-Linear Phenomena in Accretion Discs around Black Holes for the financial support.

REFERENCES

ANTONUCCI, R. 1993 Unified models for active galactic nuclei and quasars. *Ann. Rev. Astr. Astroph.* **31**, 473–521.

BAŁUCIŃSKA, M. & HASINGER, G. 1991 EXOSAT observations of Cygnus X-1: study of the soft X-ray excess. *Astronomy and Astrophysics* **241**, 439–450.

BAŁUCIŃSKA-CHURCH, M., BELLONI, T., CHURCH, M. J. & HASINGER, G. 1995 Identification of the soft X-ray excess in Cygnus X-1 with disc emission. *Astronomy and Astrophysics* **302**, L5–L8.

BISNOVATYI-KOGAN, G. S. & BLINNIKOV, S. I. 1977 Disk accretion onto a black hole at sub-critical luminosity. *Astronomy and Astrophysics* **59**, 111–125.

BRANDT, W. N., MATHUR, S. & ELVIS, M. 1997 A comparison of the hard ASCA spectral slopes of broad- and narrow-line Seyfert 1 galaxies. *Monthly Not. Roy. Astron. Soc.* **285**, L25–L28.

CELOTTI, A., FABIAN, A. C. & REES, M. J. 1992 Dense thin clouds in the central regions of active galactic nuclei. *Monthly Not. Roy. Astron. Soc.* **255**, 419–422.

COLLIN-SOUFFRIN, S., CZERNY, B., DUMONT, A.-M. & ŻYCKI, P. T. 1996 Quasi-spherical accretion of optically thin clouds as a model for the optical/UV/X-ray emission of AGN. *Astronomy and Astrophysics* **314**, 393–413.

COPPI, P. S. 1992 Time-dependent models of magnetized pair plasmas. *The Astrophysical Journal* **258**, 657–683

COPPI, P. S., ZDZIARSKI, A. A. & MADEJSKI, G. M. 1998 EQPAIR: hybrid thermal/nonthermal model for compact sources. *Monthly Not. Roy. Astron. Soc.* , in preparation.

CUI, W., EBISAWA, K., DOTANI, T. & KUBOTA, A. 1997 Simultaneous ASCA and RXTE observations of Cygnus X-1 during its 1996 state transition. *The Astrophysical Journal* **493**, L75–L78.

CUI, W. *et al.* 1997 Rossi X-ray timing explorer observation of Cygnus X-1 in its high state. *The Astrophysical Journal* **474**, L57–L60.

DERMER, C. D. & LIANG, E. P. 1989 Electron thermalization and heating in relativistic plasmas. *The Astrophysical Journal* **339**, 512–528.

DERMER, C. D., MILLER, J. A. & LI, H. 1996 Stochastic particle acceleration near accreting black holes. *The Astrophysical Journal* **456**, 106–118.

DONE, C., MADEJSKI, G. M. & SMITH, D. A. 1996 NGC 4945: the brightest Seyfert 2 galaxy at 100 keV. *The Astrophysical Journal* **463**, L63–L66.

DONE, C., MULCHAEY, J. S., MUSHOTZKY, R. F. & ARNAUD, K. A. 1992 An ionized accretion disk in Cygnus X-1. *The Astrophysical Journal* **395**, 275–288.

DOVE, J. B., WILMS, J., MAISACK, M. & BEGELMAN M. C. 1997 Self-consistent thermal accretion disk corona models for compact objects. II. Application to Cygnus X-1. *The Astrophysical Journal* **487**, 759–768.

EBISAWA, K., TITARCHUK, L. & CHAKRABARTI, S. K. 1996 On the spectral slopes of the hard X-ray emission from black hole candidates. *Publ. Astron. Soc. Japan* **48**, 59–65.

EBISAWA, K., UEDA, Y., INOUE, H., TANAKA, Y. & WHITE, N. E. 1996 *ASCA* observations of the iron line structure in Cygnus X-1. *The Astrophysical Journal* **467**, 419–434.

ESIN, A. A., McCLINTOCK, J. E. & NARAYAN, R. 1997 Advection-dominated accretion and the spectral states of black hole X-ray binaries: application to Nova Muscae 1991. *The Astrophysical Journal* **489**, 865–889.

ESIN, A. A., NARAYAN, R., CUI, W., GROVE, E. C. & ZHANG, S.-N. 1998 Spectral transitions in Cyg X-1 and other black hole X-ray binaries. *The Astrophysical Journal* submitted (astro-ph/9711167).

FABIAN, A. C., BLANDFORD, R. D., GUILBERT, P. W., PHINNEY, E. S. & CUELLAR, L. 1986 Pair-induced spectral changes and variability in compact X-ray sources. *Monthly Not. Roy. Astron. Soc.* **221**, 931–945.

FABIAN, A. C. *et al.* 1995 On broad iron $K\alpha$ lines in Seyfert 1 galaxies. *Monthly Not. Roy. Astron. Soc.* **277**, L11–L15.

GALEEV, A. A., ROSNER, R. & VAIANA, G. S. 1979 Structured coronae of accretion disks. *The Astrophysical Journal* **229**, 318–326.

GEORGE, I. M. & FABIAN A. C. 1991 X-ray reflection from cold matter in active galactic nuclei and X-ray binaries *Monthly Not. Roy. Astron. Soc.* **249**, 352–367.

GHISELLINI, G. & HAARDT, F. 1994 On thermal Comptonization in $e^{\pm}$ pair plasmas. *The Astrophysical Journal* **429**, L53–L56.

GHISELLINI, G. & SVENSSON, R. 1990 Synchrotron self-absorption as a thermalizing mechanism. In *Physical Processes in Hot Cosmic Plasmas* (ed. W. Brinkmann, A. C. Fabian & F. Giovannelli) pp. 395–400. Kluwer.

GHISELLINI, G., HAARDT, F. & FABIAN, A. C. 1993 On re-acceleration, pairs and the high-energy spectrum of AGN and Galactic black hole candidates. *Monthly Not. Roy. Astron. Soc.* **263**, L9–L12.

GHISELLINI, G., HAARDT, F. & MATT, G. 1994 The contribution of the obscuring torus to the X-ray spectrum of Seyfert galaxies: a test for the unification model. *Monthly Not. Roy. Astron. Soc.* **267**, 743–754.

GHISELLINI, G., HAARDT, F. & SVENSSON 1998 Thermalization by synchrotron absorption in compact sources: electron and photon distribution. *Monthly Not. Roy. Astron. Soc.* **297**, 348–354.

GIERLIŃSKI, M. *et al.* 1997a Simultaneous X-ray and gamma-ray observations of Cyg X-1 in the hard state by Ginga and OSSE. *Monthly Not. Roy. Astron. Soc.* **288**, 958–964.

GIERLIŃSKI, M., ZDZIARSKI, A. A., DOTANI, T., EBISAWA, K., JAHODA, K. & JOHNSON, W. N. 1997b X-ray and gamma-ray spectra of Cyg X-1 in the soft state. In *Proceedings of 4th Compton Symposium* (ed. C. D. Dermer, M. S. Strickman, & J. D. Kurfess). AIP Conference Proceedings, vol. 410, pp. 844–848. AIP.

GIERLIŃSKI, M., ZDZIARSKI, A. A., COPPI, P. S., POUTANEN, J., EBISAWA, K. & JOHNSON, W. N. 1998 Thermal/non-thermal model of Cyg X-1 in the soft state In *Proceedings of the Symposium The Active X-ray Sky* (ed. L. Scarsi, H. Brandt, P. Giommi, & F. Fiore). *Nuclear Physics B Proceedings Suppl.* in press.

GONDEK, D. *et al.* 1996 The average X-ray/gamma-ray spectrum of radio-quite Seyfert 1s *Monthly Not. Roy. Astron. Soc.* **282**, 646–652.

GREBENEV, S. A. *et al.* 1993 Observations of black hole candidates with GRANAT. *The Astrophysical Journal Suppl.* **97**, 281–287.

GREBENEV, S. A., SUNYAEV, R. A., & PAVLINSKY, M. N. 1997 Spectral states of galactic black hole candidates: results of observations with ART-P/Granat. *Adv. Space Res.* **19**, (1)15–(1)23.

GROVE, J. E., GRINDLAY, J. E., HARMON, B. A., HUA, X.-M., KAZANAS, D. & McCONNELL, M. 1997a Galactic black hole binaries: high energy radiation. In *Proceedings of 4th Compton Symposium* (ed. C. D. Dermer, M. S. Strickman, & J. D. Kurfess). AIP Conference Proceedings, vol. 410, pp. 122–140. AIP.

GROVE, J. E., KROEGER, R. A. AND STRICKMAN, M. S. 1997b Two gamma-ray spectral classes of black hole transients. In *The Transparent Universe*. (ed. C. Winkler, T. J.-L. Courvoisier, & Ph. Durouchoux) Proceedings 2nd INTEGRAL Workshop, SP-382, pp. 197–200. ESA.

GROVE, J. E. *et al.* 1998 Gamma-ray spectral states of Galactic black hole candidates. *The Astrophysical Journal* in press.

HAARDT, F. 1997 Models for the X-ray emission from radio quiet AGNs. *Mem. Soc. Astron. Ital.* **68**, 73–80. (astro-ph/9612082).

HAARDT, F. & MARASCHI, L. 1991 A two-phase model for the X-ray emission from Seyfert galaxies. *The Astrophysical Journal* **380**, L51–L54.

HAARDT, F. & MARASCHI, L. 1993 X-ray spectra from two-phase accretion disks. *The Astrophysical Journal* **413**, 507–517.

HAARDT, F., MARASCHI, L. & GHISELLINI, G. 1994 A model for the X-ray and UV emission from Seyfert galaxies and galactic black holes. *The Astrophysical Journal* **432**, L95–L99.

JOHNSON, W. N., ZDZIARSKI, A. A., MADEJSKI, G. M., PACIESAS, W. S., STEINLE, H. & LIN, Y.-C. 1997 Seyferts and radio galaxies. In *Proceedings of 4th Compton Symposium* (ed. C. D. Dermer, M. S. Strickman, & J. D. Kurfess). AIP Conference Proceedings, vol. 410, pp. 283–305. AIP.

KROLIK, J. H., MADAU, P. & ŻYCKI, P. T. 1994 X-ray bumps, iron Kα lines, and X-ray suppression by obscuring tori in Seyfert galaxies. *The Astrophysical Journal* **420**, L57–L61.

KUNCIC, Z., CELOTTI, A., & REES, M. J. 1997 Dense, thin clouds and reprocessed radiation in the central regions of of active galactic nuclei. *Monthly Not. Roy. Astron. Soc.* **284**, 717–730.

LEE, J. C., FABIAN, A. C., REYNOLDS, C. S., IWASAWA, K. & BRANDT, W. N. 1998 An RXTE observation of the Seyfert 1 galaxy MCG-6-30-15: X-ray reflection and the iron abundance. *Monthly Not. Roy. Astron. Soc.* submitted (astro-ph/9805198).

LI, H., KUSUNOSE, M. & LIANG, E. P. 1996a Gamma rays from Galactic black hole candidates with stochastic particle acceleration. *The Astrophysical Journal* **460**, L29–L32.

LI, H., KUSUNOSE, M. & LIANG, E. P. 1996b Non-thermal high energy emission and stochastic particle acceleration in galactic black holes. *Astronomy and Astrophysics Suppl.* **120C**, 167–170.

LI, H. & MILLER, J. A. 1997 Electron acceleration and the production of nonthermal electron distributions in accretion disk coronae. *The Astrophysical Journal* **478**, L67–L70.

LIANG, E. P. T. 1979 On the hard X-ray emission mechanism of active galactic nuclei sources. *The Astrophysical Journal* **231**, L111–L114.

LIANG, E. P. 1991 Structure of thermal pair clouds around gamma-ray–emitting black holes. *The Astrophysical Journal* **367**, 470–475.

LIANG, E. P. & DERMER, C. D. 1988 Interpretation of the gamma-ray bump from Cygnus X-1 *The Astrophysical Journal* **325**, L39–L42.

LIANG, E. P. & NARAYAN, R. 1997 Spectral signatures and physics of black hole accretion disks. In *Proceedings of 4th Compton Symposium* (ed. C. D. Dermer, M. S. Strickman, & J. D. Kurfess). AIP Conference Proceedings, vol. 410, pp. 461–476. AIP.

LIGHTMAN, A. P. 1974 Time-dependent accretion disks around compact objects. II. Numerical models and instability of inner region. *The Astrophysical Journal* **194**, 429–437.

LIGHTMAN, A. P. & ZDZIARSKI, A. A. 1987 Pair production and Compton scattering in compact sources and comparison to observations of active galactic nuclei. *The Astrophysical Journal* **319**, 643–661.

LING, J. C. *et al.* 1997 Gamma-ray spectra and variability of Cygnus X-1 observed by BATSE. *The Astrophysical Journal* **484**, 375–382

MACIOŁEK-NIEDŹWIECKI, A., ZDZIARSKI, A. A. & COPPI, P. S. 1995 Electron-positron pair production and annihilation spectral features from compact sources. *Monthly Not. Roy. Astron. Soc.* **276**, 273–292.

MADEJSKI, G. M. *et al.* 1995 Joint *ROSAT-COMPTON GRO* observations of the X-ray-bright Seyfert galaxy IC 4329A. *The Astrophysical Journal* **438**, 672–679.

MAGDZIARZ, P. & ZDZIARSKI, A. A. 1995 Angle-dependent Compton reflection of X-rays and gamma-rays. *Monthly Not. Roy. Astron. Soc.* **273**, 837–848.

MAISACK, M. *et al.* 1995 Upper limits on the MeV emission of Seyfert galaxies. *Astronomy and Astrophysics* **298**, 400–404.

MATT, G. *et al.* 1997 Hard X-ray detection of NGC 1068 with BeppoSAX. *Astronomy and Astrophysics* **325**, L13–L16.

MCCONNELL, M. L. *et al.* 1994 Observations of Cygnus X-1 by COMPTEL during 1991. *The Astrophysical Journal* **424**, 933–939.

MINESHIGE, S., KUSUNOSE, M. & MATSUMOTO, R. 1995 Low-state disks and low-beta disks. *The Astrophysical Journal* **445**, L43–L46.

NAGIRNER, D. I. & POUTANEN, J. 1994 Single Compton scattering. *Astrophys. Space Phys. Reviews* **9**, 1–83.

NANDRA, K. & POUNDS, K. A. 1994 *Ginga* observations of the X-ray spectra of Seyfert galaxies. *Monthly Not. Roy. Astron. Soc.* **268**, 405–429.

NANDRA, K. *et al.* 1997 *ASCA* observations of Seyfert 1 galaxies. II. Relativistic iron $K\alpha$ emission. *The Astrophysical Journal* **477**, 602–622.

NAYAKSHIN, S. & MELIA, F. 1998 Self-consistent Fokker-Planck treatment of particle distributions in astrophysical plasmas. *The Astrophysical Journal Suppl.* **114**, 269–288.

PHLIPS, B. F. *et al.* 1996 Gamma-ray observations of Cygnus X-1 with oriented scintillation spectrometer experiment. *The Astrophysical Journal* **465**, 907–914.

PIETRINI, P. & KROLIK, J. H. 1995 The inverse Compton thermostat in hot plasmas near accreting black holes. *The Astrophysical Journal* **447**, 526–544.

PILLA, R. P. & SHAHAM, J. 1997 Kinetic of electron-positron pair plasmas using an adaptive Monte-Carlo method. *The Astrophysical Journal* **486**, 903–918.

POUNDS, K. A., DONE, C. & OSBORNE, J. P. 1995 RE 1034+39: a high-state Seyfert galaxy? *Monthly Not. Roy. Astron. Soc.* **277**, L5–L10.

POUTANEN, J. 1998 Modeling X/γ-ray spectra of Galactic black holes and Seyferts. In *Accretion Processes in Astrophysical Systems: Some Like It Hot* (ed. S. S. Holt & T. Kallman). AIP Conference Proceedings, vol. 431, in press. AIP. (astro-ph/9801055).

POUTANEN, J. & COPPI, P. S. 1998 Unification of spectral states of accreting black holes. *Physica Scripta* in press (astro-ph/9711316).

POUTANEN, J. & SVENSSON, R. 1996 The two-phase pair corona model for active galactic nuclei and X-ray binaries: How to obtain exact solutions. *The Astrophysical Journal* **470**, 249–268

POUTANEN, J., KROLIK, J. H. & RYDE, F. 1997 The nature of spectral transitions in accreting black holes: the case of Cyg X-1. *Monthly Not. Roy. Astron. Soc.* **292**, L21–L25.

POUTANEN, J., NAGENDRA, K. N. & SVENSSON, R. 1996 Green's matrix for Compton reflection of polarized radiation from cold matter. *Monthly Not. Roy. Astron. Soc.* **283**, 892–904

POZDNYAKOV, L. A., SOBOL, I. M. & SUNYAEV, R. A. 1983 Comptonization and the shaping of X-ray source spectra – Monte-Carlo calculations. *Sov. Sci. Rev. E Astrophys. Space Phys.* **2**, 189–331.

RYBICKI, G. B. & LIGHTMAN, A. P. 1979 Radiative Processes in Astrophysics. Wiley.

SMITH, D. A., DONE, C. & POUNDS, K. A. 1993 Unified theories of active galactic nuclei: the hard X-ray spectrum of NGC 1068. *Monthly Not. Roy. Astron. Soc.* **263**, 54–60.

STERN, B. E. 1985 On the possibility of efficient production of electron-positron pairs near pulsars and accreting black holes. *Sov.Astr.* **29**, 306–313.

STERN, B. E. 1988 Nonthermal pair production in active galactic nuclei: A detailed radiation transfer model. *Nordita/88-51 A*, preprint.

STERN, B. E., BEGELMAN, M. C., SIKORA, M. & SVENSSON, R. 1995a A large-particle Monte

Carlo code for simulating non-linear high-energy processes near compact objects. *Monthly Not. Roy. Astron. Soc.* **272**, 291–307.

STERN, B. E., POUTANEN, J., SVENSSON, R., SIKORA, M. & BEGELMAN M. C. 1995b On the geometry of the X-ray emitting region in Seyfert galaxies. *The Astrophysical Journal* **449**, L13–L17.

SHAPIRO, S. L., LIGHTMAN, A. P. & EARDLEY D. N. 1976 A two-temperature accretion disk model for Cygnus X-1: structure and spectrum. *The Astrophysical Journal* **204**, 187–199.

SUNYAEV, R. A. & TITARCHUK, L. G. 1980 Comptonization of X-rays in plasma clouds. Typical radiation spectra. *Astronomy and Astrophysics* **86**, 121–138.

SUNYAEV, R. A. *et al.* 1991 Three spectral states of 1E 1740.7-2942: from standard Cygnus X-1 type spectrum to the evidence of the electron-positron annihilation feature. *The Astrophysical Journal* **383**, L49–L52.

SVENSSON, R. 1987 Non-thermal pair production in compact X-ray sources: first order Compton cascades in soft radiation field. *Monthly Not. Roy. Astron. Soc.* **227**, 403–451.

SVENSSON, R. 1994 The nonthermal pair model for the X-ray and gamma-ray spectra from active galactic nuclei. *The Astrophysical Journal Suppl.* **92**, 585–592.

SVENSSON, R. 1996a X-rays and Gamma Rays from Active Galactic Nuclei. In *Relativistic Astrophysics: A Conference in Honour of Professor I. D. Novikov's 60th Birthday* (ed. B. J. T. Jones & D. Markovic), pp. 235–249. Cambridge University Press.

SVENSSON, R. 1996b Models of X-ray and Gamma-Ray Emission from Seyfert Galaxies. *Astronomy and Astrophysics Suppl.* **120C**, 475–480.

TANAKA, Y. 1991 Black hole candidates in binaries. In *Iron line diagnostics in X-ray sources* (ed. A. Treves) Lecture Notes in Physics, vol. 385, pp. 98–110. Springer.

TANAKA, Y. & LEWIN, W. H. G. 1995 Black-hole binaries. In *X-ray binaries* (ed. W. H. G. Lewin, J. van Paradijs, E. P. J. van den Heuvel) Cambridge Astrophysics Series, vol. 26, pp. 126–174. Cambridge University Press.

TANAKA, Y. *et al.* 1995 Gravitationally redshifted emission implying an accretion disk and massive black hole in the active galaxy MCG-6-30-15. *Nature* **375**, 659–661.

TITARCHUK, L., MASTICHIADIS, A. & KYLAFIS, N. D. 1997 X-ray spectral formation in a converging fluid flow: spherical accretion into black holes. *The Astrophysical Journal* **487**, 834–846.

VAN DER KLIS, M. 1995 Rapid aperiodic variability in X-ray binaries. In *X-ray binaries* (ed. W. H. G. Lewin, J. van Paradijs, E. P. J. van den Heuvel) Cambridge Astrophysics Series, vol. 26, pp. 252–307. Cambridge University Press.

VAN DIJK, R. *et al.* 1995 The black-hole candidate GRO J0422+32: MeV emission measured with COMPTEL. *Astronomy and Astrophysics* **296**, L33–L36.

WEAVER, K. A., ARNAUD, K. A. & MUSHOTZKY, R. F. 1995 A confirmation of 2-40 keV spectral complexity in Seyfert galaxies. *The Astrophysical Journal* **447**, 121–138.

WHITE, T. R., LIGHTMAN, A. P. & ZDZIARSKI, A. A. 1988 Compton reflection of gamma-rays by cold electrons. *The Astrophysical Journal* **331**, 939–948.

ZDZIARSKI, A. A. 1985 Power-law X-ray and gamma-ray emission from relativistic pair plasmas. *The Astrophysical Journal* **289**, 514–525.

ZDZIARSKI, A. A., GHISELLINI, G., GEORGE, I. M., SVENSSON, R., FABIAN, A. C. & DONE, C. 1990 Electron-positron pairs, Compton reflection, and the X-ray spectra of active galactic nuclei. *The Astrophysical Journal* **363**, L1–L4.

ZDZIARSKI, A. A., GIERLIŃSKI, M., GONDEK, D. & MAGDZIARZ, P. 1996a The canonical X-ray/gamma-ray spectrum of Seyfert 1s and low state Galactic black hole candidates. *Astronomy and Astrophysics Suppl.* **120C**, 553–558.

ZDZIARSKI, A. A., JOHNSON, W. N. & MAGDZIARZ, P. 1996b Broad-band gamma-ray and X-ray spectra of NGC 4151 and their implications for physical processes and geometry. *Monthly Not. Roy. Astron. Soc.* **283**, 193–206.

ZDZIARSKI, A. A., JOHNSON, W. N., POUTANEN, J., MAGDZIARZ, P. & GIERLIŃSKI, M. 1997 X-rays and gamma-rays from accretion flows onto black holes in Seyferts and X-ray binaries.

In *The Transparent Universe.* (ed. C. Winkler, T. J.-L. Courvoisier, & Ph. Durouchoux) Proceedings 2nd INTEGRAL Workshop, SP-382, pp. 373–380. ESA.

ZDZIARSKI, A. A. 1998 Hot accretion disc with thermal Comptonization and advection in luminous black hole sources. *Monthly Not. Roy. Astron. Soc.* **296**, L51–L55.

ZDZIARSKI, A. A., POUTANEN, J., MIKOŁAEWSKA, J., GIERLIŃSKI, M., EBISAWA, K. & JOHNSON, W. N. 1998 Broad-band X-ray/γ-ray spectra and binary parameters of GX 339-4 and their astrophysical implications. *Monthly Not. Roy. Astron. Soc.* submitted.

ZHANG S. N., CUI W., HARMON B. A., PACIESAS W. S., REMILLARD R. E. & VAN PARADIJS J. 1997 The 1996 soft state transition of Cygnus X-1. *The Astrophysical Journal* **477**, L95–L98.

ZYCKI, P. T., DONE, C. & SMITH, D. A. 1997a Relativistically smeared X-ray reprocessed components in the *Ginga* spectra of GS2023+388. *The Astrophysical Journal* **488**, L113–L116.

ZYCKI, P. T., DONE, C. & SMITH, D. A. 1997b Evolution of the accretion flow in Nova Muscae 1991. *The Astrophysical Journal* **496**, L25–L28.

Emission lines: signatures of relativistic rotation

By A.C. FABIAN

Institute of Astronomy, Madingley Road, Cambridge CB3 0HA, UK

An intrinsically narrow line emitted by an accretion disk around a black hole appears broadened and skewed as a result of the doppler effect and gravitational redshift. The fluorescent iron line in the X-ray band at 6.4 – 6.9 keV is the strongest such line and is seen in the X-ray spectrum of many Seyfert galaxies. It is an important diagnostic with which to study the geometry and other properties of the accretion flow very close to the central black hole. The broad iron line indicates the presence of a standard thin accretion disk in those objects, often seen at low inclination.

1. Introduction

The frequency profile of a line emitted by a finite Keplerian disk has a characteristic double-horned shape and is commonly seen in the spectra of objects as diverse as spiral galaxies and cataclysmic variables. The total line is the sum of the profile from many narrow annuli, each of which has sharp blue and red peaks from the approaching and receding sides, respectively, with a valley in between from the matter moving more across the line of sight. The broadest parts of the final profile originate in the fastest and therefore innermost parts of the disk. A disk around a black hole provides the highest velocities possible and therefore the broadest lines.

Relativistic effects must be taken into account in the computation of a line profile when the velocity is a significant fraction of the speed of light. The emission from the disk is beamed in the direction of motion, which means that the blue horn appears brighter than red one. Transverse doppler shift ('moving clocks run slow') means that the emission from the inner regions of the disk are shifted to the red, thereby skewing the line profile. Finally, gravitational redshift shifts the emission further to the red. The net result is a skewed, broad line.

Mildy skewed lines originating from regions at hundreds of Schwarzschild radii in disks about massive black holes have been tentatively identified in the Balmer lines from some active galaxies (Chen, Halpern & Fillipenko 1989 but see Antonucci, Hurt & Agol 1996).

Broad, skewed lines are clearly seen in the X-ray spectra of most Seyfert 1 galaxies (Mushotzky et al 1995; Tanaka et al 1995; Nandra et al 1997a; Reynolds et al 1997) and Compton-thin Seyfert 2 galaxies (Turner et al 1997). The line profiles indicate that the emission region is at 3–30 Schwarzschild radii and therefore that a relativistic accretion disk is present. It demonstrates that standard thin-disk accretion onto a black hole is taking place. In one Seyfert 1 galaxy, MCG–6-30-15, the line was seen to shift further to lower energies during a minimum state, suggesting that the site of line emission moved closer than 3 Schwarzschild radii (i.e. $3R_s = 6GM/c^2 = 6m$) to the hole (Iwasawa et al 1996b). The most likely explanation is that the black hole is spinning.

No such feature is generally seen in more luminous AGN such as quasars (Nandra et al 1997b) and the evidence in Galactic Black Hole Candidates is ambiguous.

Characteristic Seyfert 1 X—ray Spectrum

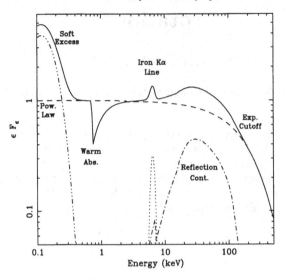

FIGURE 1. Schematic representation of the X-ray spectrum of a Seyfert 1 galaxy.

2. The broad iron line

2.1. *Line production*

The main spectral components in a Seyfert 1 galaxy are shown in Fig. 1. Most of the power is emitted in flares as a hard power-law which is probably due to thermal Comptonization (Zdziarski et al 1994). The flares irradiate the accretion disk, which is relatively cold (i.e. iron is less ionized than FeXVII) producing a reflection component which consists of the backscattered Compton continuum peaking at around 30 keV, and the fluorescent iron line. Weak fluorescent lines are expected from other elements (Fig. 2 and Matt, Fabian & Reynolds 1997) but the fluorescent yield and abundance of iron mean that its line dominates. The combination of doppler and gravitational effects produce a broad, skewed, emission line (Fabian et al 1989). If the black hole is spinning at its maximal rate then the inner edge of the disk moves in close to GM/c^2 and the line is skewed even more to the 'red' (Laor 1991).

Model line profiles are are plotted in Figs 3 and 4. It is seen that the blue wing is most sensitive to the inclination of the disk and the red wing to the inner radius. Note that the line sits on a continuum; model fitting should be carried out with the line and continuum together, rather than on the line alone. This is particularly relevant for the determination of the extent of the red wing.

The fluorescent iron line is produced when one of the 2 K-shell electrons of an iron atom (or ion) is ejected following photoelectric absorption of an X-ray. The threshold for the absorption by neutral iron is 7.1 keV, increasing with energy as the ionization state increases. An L-shell electron can then drop into the K-shell releasing 6.4 keV of energy either as an emission line (34 per cent probability) or an Auger electron (66 per ecnt probability). For cosmic abundances the optical depth of matter at the iron absorption threshold is higher than, but close to, the Thomson depth. The iron line production in an X-ray irradiated surface therefore takes place in the outer Thomson depth. This is only a small fraction of the thickness (say 1 to 0.1 per cent) of a typical accretion disk

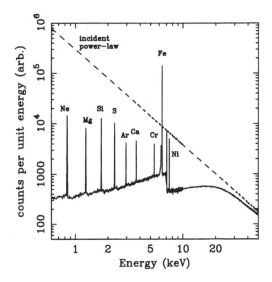

FIGURE 2. The reflection spectrum produced by a power-law of photon index 2 incident onto matter with cosmic abundances.

and it is the ionization state of this thin skin which determines the nature of the iron line.

The reflection continuum is produced by electron scattering and its level relative to the irradiating continuum is approximately the ratio of the Thomson to photoelectric absorption cross-section at that energy. Above observed energies of tens of keV Compton recoil reduces the backscattered photon flux. Although the fluorescent lines are produced after photoelectric absorption, the observed absorption edges are small. This is because the edges are saturated. Iron-K, for example, absorbs most of the incident photons above 7.1 keV but oxygen and iron-L absorb most of the photons below that energy. Only when much of the oxygen is ionized, so it absorbs relatively less, does a large edge become apparent in the reflected spectrum.

The strength of the iron line is usually measured in terms of its equivalent width which essentially means with respect to the direct emission. It varies with the iron abundance, roughly logarithmically (see Matt, Fabian & Reynolds 1997 for formulae). This is because the edges are easily saturated and iron and oxygen both compete for the photons above 7.1 keV and (iron-L and oxygen) absorb the emitted fluorescent photons at 6.4 keV; a low oxygen abundance creates a relatively strong iron line.

Absorption and scattering reduce the strength of the iron line as the surface inclination increases. Ghisellini, Haardt & Matt (1994) give

$$I(\mu) = \frac{I(\mu = 1)}{\ln 2} \mu \log(1 + \frac{1}{\mu}),$$

where $\mu = \cos i$.

Guilbert & Rees (1988) first noted that spectral features due to cold matter are likely from AGN and the reflection continuum was computed by Lightman & White (1988). Monte-Carlo studies of the behaviour of the iron line can be found in George & Fabian (1991) and Matt, Perola & Piro (1991).

X-ray irradiation can photoionize the surface layers of a disk (Ross & Fabian 1993).

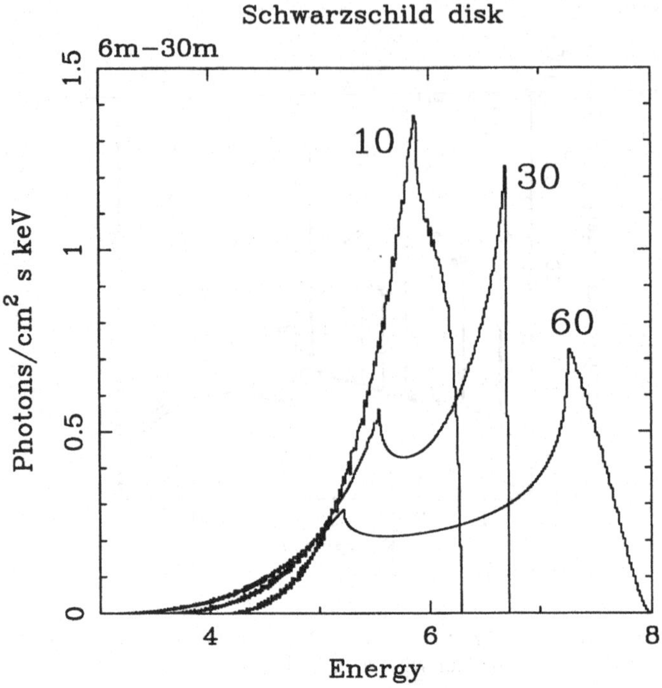

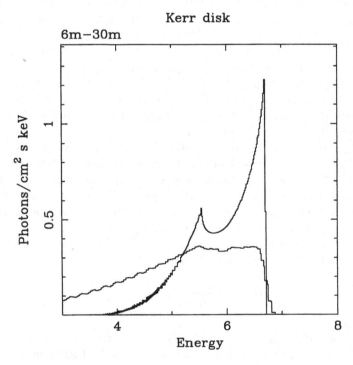

Schwarzschild disk

FIGURE 3. The line profile expected from disk emission around a Schwarzschild black hole at inclinations of 10, 30 and 60 degrees. The inner and outer radii are 6 and $30m$. Note the great sensitivity of the line profile to the disk inclination.

Kerr disk

FIGURE 4. The line profile expected from disk emission around a Kerr black hole, compared with that from a Schwarzschild hole. The inclination is 30 degrees in both cases.

The fluorescent line that the illuminated matter produces depends upon its ionization state and a useful quantity is the ionization parameter $\xi(r) = 4\pi F_x(r)/n(r)$, where $F_x(r)$ is the X-ray flux received per unit area of the disc at a radius r, and $n(r)$ is the comoving electron number density: it measures the ratio of the photoionization rate to the recombination rate (proportional to n). The iron line emission for various ionization parameters has been investigated by Matt et al (1993, 1996) and they conclude the following for various values of ξ.

(i) $\xi < 100$ ergs cm s^{-1} the material produces a 'cold' iron line at 6.4 keV and only a small absorption edge.

(ii) 100 ergs cm s^{-1} $< \xi < 500$ ergs cm s^{-1} does not produce an iron line because photons near the line energy are resonantly trapped and lost due to Auger ejections. There is a moderate absorption edge.

(iii) 500 ergs cm s^{-1} $< \xi < 5000$ ergs cm s^{-1} produces a 'hot' iron line at ~ 6.8 keV with twice the fluorescent yield of the cold line. There is a large absorption edge.

(iv) $\xi > 5000$ ergs cm s^{-1} does not produce an iron line because the iron is completely ionized. There is no absorption edge.

2.2. *Observations of broad iron lines*

The reflection spectrum was first clearly seen in Ginga spectra (Pounds et al 1990; Matsuoka et al 1990). The clearest example of a broad iron line remains that seen from a long ASCA observation of MCG-6-30-15 (Fig. 5; Tanaka et al 1995). The sharp drop seen at about 6.5 keV both demonstrates the good spectral resolution of the CCD detectors and constrains the inclination of the disk to be about 30 deg. If the inclination were greater then this blue edge to the line moves to higher energies (as is seen in the broad line of IRAS 18325-596; Iwasawa et al 1996a). The redward extent of the line constrains the inner radius of the emission to be around $3R_s$ and the overall shape means that most of the emission is peaked within $20R_s$.

Nandra et al (1997a) and Reynolds (1997) have studied the iron line in over 20 Seyfert 1 galaxies and find that most are broader than the instrumental resolution. Nandra et al (1997a) have summed together the lines and a clear extension to low energies is seen. Variability of the iron line in two particular objects is discussed by Nandra et al (1997c) and Yaqoob et al (1996). A narrow component to the line at 6.4 keV may occur in some Seyferts due to reflection from outer structures (e.g. the putative torus).

Nandra et al (1995; 1997b) find no evidence for the iron line or any reflection features in most quasars. In the second paper, it is shown that the equivalent width of the iron line diminishes with source luminosity above about 10^{44} erg s^{-1} and it is suggested that the disk is increasingly ionized, perhaps because the objects are closer to the Eddington limit. This is puzzling because the disk must jump from being 'cold' to completely ionized, otherwise there would be intermediate objects with even larger equivalent widths when the surface iron in the disk is H or He-like (Matt, Fabian & Ross 1993). There should at the same time be a deep broad iron edge which is not seen.

Some smeared edges have been seen in the spectra of Galactic Black Hole Candidates (Ebisawa et al 1996) with little line emission. This may be explained by the surface of the disk being moderately ionized, with the mean value of ξ equal to a few hundred (Ross, Fabian & Brandt 1996).

3. Alternative models for a broad line

Fabian et al (1995) discuss and dismiss some alternative models with which a broad skew iron line can be obtained, including Comptonization by cold electrons. This has

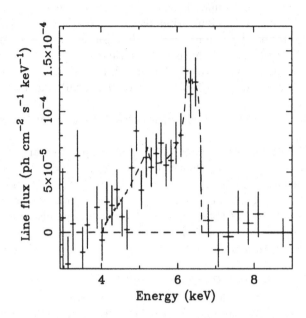

FIGURE 5. The broad iron line seen from the long ASCA observation of MCG–6-30-15 (Tanaka et al 1995). The best-fitting model line is indicated by the dashed line.

been revived recently by Misra & Kembhavi (1997) who argue that a distant, cold ($kT < 0.25\,\mathrm{keV}$), Thomson thick ($\tau > 3$) cloud of gas may be so highly ionized (no iron absorption features can be seen) that it could produce the observed profile by down-scattering. I estimate, using Ferland's CLOUDY code, that the ionization parameter ($\xi = L/nR^2$) at the outer edge of the cloud needs to exceed several times 10^5 for iron to be highly ionized at these low temperatures. The cloud then extends only out to a few tens of R_s for any sub-Eddington mass black hole, and gravitational effects are just as important for the injected line as in the disk model. Comptonization is then an unnecessary complication rather than an alternative to matter close to the black hole.

Skibo (1997) has proposed that energetic protons turn iron in the surface of the disk into chromium and lower Z metals which then enhances their fluorescent emission (see Fig. 2) which, with poor resolution or low signal to noise, might appear as a broad skew line. Apart from issues on the efficiency with which spallation can take place at disk radii where relativistic broadening dominates, it should be noted that the broad line in MCG–6-30-15 (Tanaka et al 1995) is well resolved (the instrumental resolution is about 150 eV there) and it would be obvious if it were due to several separate and well-spaced lines spread over 2 keV. There can of course be doppler-blurring of all the lines, as suggested by Skibo (1997), but it will still be considerable and require that the redward tails be at least 1 keV long.

Finally it is worth noting that the line profile indicates that most of the doppler shifts are due to matter orbiting at about 30 degrees to the line of sight. The lack of any large blue shifted component rules out most jet models. What we cannot determine at present is the geometry in more detail. For example we cannot rule out a 'blobby' disk (Nandra & George 1994). We do however require that any corona be either optically thin or localized, in order that passage of the reflection component back through the corona does not smear out the sharp features.

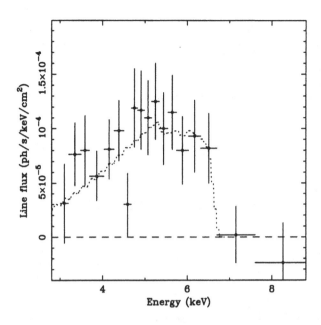

FIGURE 6. The extremely broad iron line seen during the deep minimum of the ASCA long observation of MCG–6-30-15 (Iwasawa et al 1996b).

4. What happens inside $6m$: Kerr black holes

The X-ray emission from MCG–6-30-15 went into a deep minimum state toward the end of the long ASCA observation of 1994, and during that state the line appeared to extend to lower energies (Fig. 6 from Iwasawa et al 1996b, compare with Fig. 5). The only way to increase the redshift of the line is to make the source of emission move within the innermost stable orbit for a non-spinning Schwarzschild black hole (i.e. $3R_s$). The line is indeed well-fitted by the profile of a maximally-spinning Kerr black hole and we tentatively concluded that the line was the first spectroscopic evidence for a Kerr hole (Iwasawa et al 1996b). (It was tentative because of the difficulty in measuring the continuum precisely at that time due to the increase in strength of the warm absorber; Otani et al 1996). Later work by Dabrowski et al (1997) quantified the spin of the hole required as exceeding 95 per cent of its maximal value.

The basic idea is not of course that the spin of the black hole was changing but that the location of the source of hard irradiating X-ray emission changed. As it moves closer to the black hole then, provided that it is not corotating with the disk, more of the continuum falls on the disk by the effect of light bending so what is observed decreases whilst the line equivalent width increases (Martocchia & Matt 1996).

Reynolds & Begelman (1997) point out that although the disk around a Schwarzschild hole ends at $3R_s$ the accretion flow does not immediately become optically thin at smaller radii. The matter falls from the inner edge of the disk and, if illuminated in the right manner, can give rise to an an extremely broad iron line similar to that seen. Young, Ross & Fabian (1997a,b) have computed both the line and continuum structure in detail. The flow is ionized within $3R_s$ leading to the development of a large iron absorption edge (Fig. 7, 8) which is not seen in the data. It appears that the full data can discriminate between matter plunging from $6m$ and a disk extending inward in a Kerr metric.

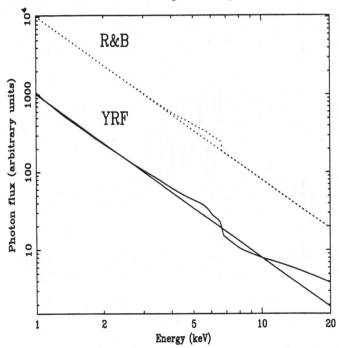

FIGURE 7. Model spectra for the continuum and line alone (upper) and line with reflection component (lower) for flow within $3R_s$. The curves have been displaced apart for clarity.

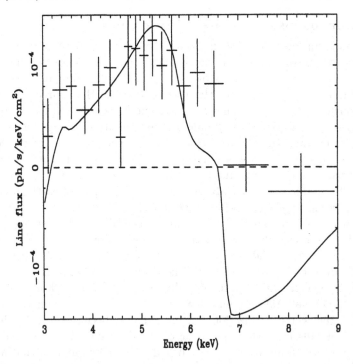

FIGURE 8. Comparison of the predicted reflection spectrum from plunging orbits around a Schwarzschild black hole with the data from MCG–6-30-15, from Young et al (1998). Note that the large predicted edge disagrees with the observed spectrum.

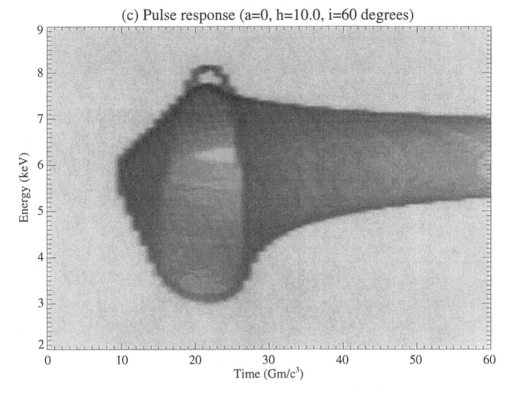

(c) Pulse response (a=0, h=10.0, i=60 degrees)

FIGURE 9. Line response from an impulsive flare occurring at $10m$ along the disk axis above a Schwarzschild black hole, from Reynolds et al (1998). Blackness indicates line intensity. The small loop in the line energy at the top is due to emission from ionized gas plunging from the orbit of marginal stability.

5. The future

Most of the detailed observations of the broad iron line have so far relied on ASCA. This yields only a few hundred line photons per day from the brightest Seyferts. Future missions such as XMM, Constellation-X and XEUS will increase the count rates by orders of magnitude. This will enable the line shape to be defined much more precisely and enable the geometry to be determined in much more detail. For example the inner radius of the emitting matter can be studied with precision. This will tell us whether the black holes are spinning or not, and how fast, and the distribution of the corona and flares above the disk.

Reverberation mapping will be possible, although complex. Unlike for UV studies, the X-ray emission and reflection regions have a similar size so complicating the unravelling of the various effects. Two examples of variations of line profile with time after an on-axis flare are shown in Figs 9 and 10, from Reynolds et al (1998). Time delays enable physical dimensions to be placed on disk models and, with knowledge of the gravitational radii involved from the observed line energies, yield the mass of the black hole.

6. Acknowledgements

I am very grateful to my immediate collaborators, Kazushi Iwasawa, Andy Young and Randy Ross for continuing help, and the Royal Society for support.

(c) Pulse response (a=0.998, h=10.0, i=60 degrees)

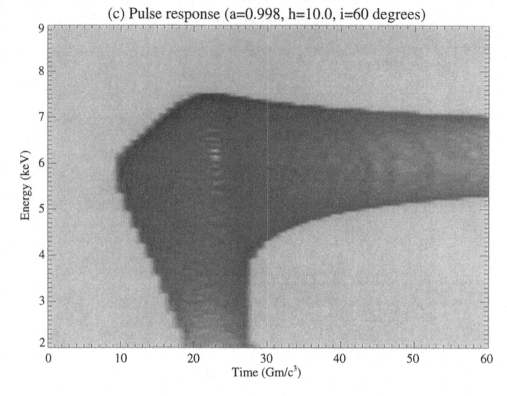

FIGURE 10. Line response from an impulsive flare occurring at 10m along the disk axis above a maximal Kerr black hole, from Reynolds et al (1998). Blackness indicates line intensity.

REFERENCES

ANTONUCCI R., HURT T., AGOL E. 1996, ApJ, 456, L25

CHEN K., HALPERN J.P., FILLIPENKO A.V. 1989, ApJ, 339, 742

DABROWSKI, Y. ET AL. 1997, MNRAS, 288, L11

EBISAWA K., UEDA Y., INOUE H., TANAKA Y., WHITE N.E. 1996, ApJ, 467, 419

FABIAN A.C., NANDRA K., REYNOLDS C.S., BRANDT W.N., OTANI C., TANAKA Y., INOUE H., IWASAWA K. 1995, MNRAS, 277, L11

FABIAN A.C., REES M.J., STELLA L., WHITE N.E. 1989, MNRAS, 238,729

GEORGE I.M., FABIAN A.C. 1991, MNRAS, 249, 352

GHISELLINI G., HAARDT F., MATT G. 1994, MNRAS, 267, 743

GUILBERT P.W., REES M.J. 1988, MNRAS, 233, 475

IWASAWA, K. , FABIAN A.C., MUSHOTZKY R.F., BRANDT W.N., AWAKI H., KUNIEDA H. 1996a, MNRAS, 279, 837

IWASAWA, K. ET AL. 1996b, MNRAS, 282, 1038

LAOR, A. 1991, ApJ, 376, 90

LIGHTMAN A.P., WHITE T.R. 1988, ApJ, 233, 57

MARTOCCHIA A., MATT G. 1996, MNRAS, 282, L53

MATSUOKA M., PIRO L., YAMAUCHI M., MURAKAMI T. 1990, ApJ, 361, 400

MATT G., FABIAN A.C., REYNOLDS C.S. 1997, MNRAS, 289, 175

MATT G., FABIAN A.C., ROSS R.R 1996, MNRAS, 280, 823

MATT G., FABIAN A.C., ROSS R.R. 1993, 262, 179

MATT, G., PEROLA, G.C., PIRO L. 1991, AaA, 247, 25

MISRA R., KEMBHAVI A. 1997, astro-ph 9712327

MUSHOTZKY R.F., FABIAN A.C., IWASAWA K., KUNIEDA H., MATSUOKA M., NANDRA K., TANAKA Y. 1995, MNRAS, 272, L9

NANDRA K., ET AL. 1995, MNRAS, 276, 1

NANDRA K., GEORGE I.M. 1994, MNRAS, 267, 974

NANDRA K., GEORGE I.M., MUSHOTZKY R.F., TURNER T.J., YAQOOB T. 1997a, ApJ, 477, 602

NANDRA K., GEORGE I.M., MUSHOTZKY R.F., TURNER T.J., YAQOOB Y. 1997b, ApJ, 488, 91

NANDRA K., MUSHOTZKY R.F., YAQOOB T., GEORGE I.M., TURNER T.J. 1997c, MNRAS, 284, L7

OTANI, C. 1996, PASJ, 48, 211

POUNDS K.A., NANDRA K., STEWART G.C., GEORGE I.M., FABIAN A.C. 1990, Nature, 344, 132

REYNOLDS, C.S. 1997, MNRAS, 286, 513

REYNOLDS, C.S., WARD, M.J., FABIAN A.C., CELOTTI A. 1997, MNRAS, 291, 403

REYNOLDS, C.S., BEGELMAN, M.C. 1997, ApJ, 488, 109

REYNOLDS, C.S. YOUNG A.J., BEGELMAN M.C., FABIAN A.C. 1998, ApJ submitted

ROSS, R.R., FABIAN, A.C. 1993, MNRAS, 261, 74

ROSS, R.R., FABIAN, A.C., BRANDT W.N. 1996, MNRAS, 278, 1082

SKIBO, J.G. 1997, ApJ, 478, 522

TANAKA, Y. ET AL. 1995, Nature, 375, 659

TURNER T.J., GEORGE I.M., NANDRA K., MUSHOTZKY R.F. 1998, ApJ, 493, 91

YAQOOB, Y., SERLEMITSOS P.J., TURNER T.J., GEORGE I.M., NANDRA K. 1996, ApJ, 470, L27

YOUNG A.J., ROSS R.R., FABIAN A.C. 1997, MNRAS in press

ZDZIARSKI A.A., ET AL 1994, MNRAS, 269, L55

Spectral tests of models for accretion disks around black holes

By JULIAN H. KROLIK

Department of Physics and Astronomy, Johns Hopkins University, Baltimore MD 21218, USA

A review is given of the current state of the art with respect to predicting the output spectra of accretion disks in AGN. Major uncertainties exist due both to unresolved issues in basic disk physics (how does the shear stress depend on local conditions? how is dissipation distributed vertically within the disk?) and to the varying approximations made by different groups in solving the radiation transfer problem (is the ionization balance computed explicitly or in the LTE approximation? are heavy element opacities are included? are all relativistic effects included?). Illustrations are given of how the results depend on the answers are given to these questions.

1. Introduction: the Zeroth Order Picture

To test our ideas about the dynamics of accretion disks around black holes, we compare the radiation we receive from them with the radiative output predicted by our models. Standard models predict a very simple spectrum—a quasi-thermal continuum. However, even the most cursory glance at real accretion disks immediately reveals that, while there often is a spectral component resembling the expected quasi-thermal one, substantial energy is also released in quite different ways—in hard X-rays from "coronal" gas, and also in extremely non-thermal radiation from relativistic jets. In this review I will not discuss how the energy is diverted into coronæ and more strongly non-thermal channels, or how well our models for the radiation from these structures matches what is seen; the *ad hoc* elements are so strong in these models that matching them to observations does not provide strong tests of accretion disk dynamics. Instead, I will concentrate on spectral properties of the quasi-thermal continuum. Here the character of the emergent radiation is directly tied to the structure of the disk, and can in principle provide strong diagnostics of how this structure is dynamically regulated.

It is best to begin with the simplest picture: a quiescent, smooth, geometrically thin disk that is in local thermodynamic equilibrium everywhere from the marginally stable radius out to its outermost boundary. Most of the accretion energy is released at radii near $\sim 10GM/c^2$ (the radius of maximum dissipation is somewhat greater than the marginally stable radius because relativistic effects and the outward transport of energy associated with the angular momentum flux reduce the depostion of heat in the innermost part of the disk). The characteristic temperature at which this energy is radiated is then

$$T_* \sim 2 \times 10^7 \dot{m}^{1/4} m^{-1/4} \text{K}, \tag{1.1}$$

where $\dot{m}$ is the accretion rate relative to the rate which would produce an Eddington luminosity, and m is the mass of the black hole in solar units. Thus, we expect stellar black holes to radiate primarily soft X-rays, while galactic scale black holes (as in active galactic nuclei) should radiate primarily in the ultra-violet.

This description of the spectrum can be easily refined by integrating over the actual distribution of disk dissipation. Assuming LTE, the temperature falls towards larger radius r as $r^{-3/4}$. The sum of all the local black bodies then gives an overall luminosity per unit frequency

$$L_\nu \propto \nu^{1/3} \exp\left(-h\nu/kT_*\right), \tag{1.2}$$

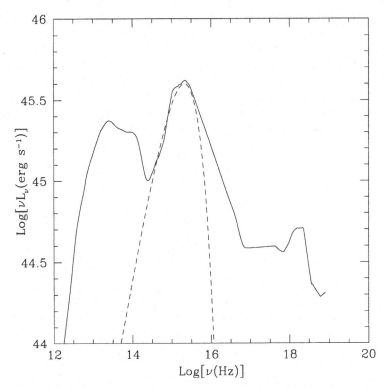

FIGURE 1. This failure of standard accretion disk models is especially prominent in AGN. Here the composite spectrum of radio-quiet quasars assembled by Elvis *et al.* (1994) is shown by the solid line, while a spectrum with the shape of equation 1.2 for $T_* = 7 \times 10^4$ K is shown by the dashed line.

where the characteristic temperature T_* is approximated by the estimate of equation 1.1 (a more careful calculation would be more punctilious about the exact location of the maximum dissipation rate).

The approximate expression given in equation 1.2 can be easily compared with real spectra to gain a sense of perspective about how well our expectations are vindicated in real objects. Ironically, the quality of data at our disposal is much better for AGN than for Galactic black holes. This is because the band in which the thermal emission peaks for Galactic black holes (< 1 keV) is one for which, at least hitherto, spectroscopy has been relatively undeveloped, while in the ultraviolet, where the thermal component of AGN emission peaks, good quality spectroscopy is relatively easy to obtain. As Figure 1 shows, while the UV portion of the spectrum is, to zeroth order, replicated by the simple sum-of-blackbodies spectrum, our simple models of accretion disks don't come close to predicting the extremely broad-band emission seen in these objects.

2. Issues in Detailed Modelling

Given this much success, the next question to ask is whether more detailed predictions are equally successful. However, to make these predictions, we must first take a closer look at the accretion disk model under consideration. In particular, it contains at least two glaring omissions: we have ignored all relativistic corrections, yet by definition most of the light comes from a region whose gravitational potential is relativistically deep; and we have no assurance that the locally radiated spectrum is Planckian.

First consider the latter question. To answer it, we must solve three separate problems: the equilibrium radial structure of the disk (to find the optical depth as a function of radius); the equilibrium vertical structure (to find the temperature and density as a function of altitude at each radius); and the radiation transfer problem (to find the local emergent spectrum as a function of viewing angle). Unfortunately, there are major gaps in our understanding in each of these three areas. As a result, in each of them it is necessary to make some unjustified assumption in order to proceed. Consequently, *every* detailed prediction of accretion disk spectra is model-dependent in important ways.

2.1. *Radial structure*

The fundamental equation governing the radial equilibrium is conservation of angular momentum:

$$-\int dz\, T_{r\phi} \;=\; \frac{\dot{M}\Omega}{2\pi} R_T(r) \tag{2.3}$$

$$= \frac{2\mu_e c^2}{e\sigma_T} \frac{\dot{m}}{x^{3/2}} R_T, \tag{2.4}$$

where $T_{r\phi}$ is the $r - \phi$ component of the stress tensor, $\dot{M}$ is the mass accretion rate, Ω the local orbital frequency, and $R_T(r)$ is a correction factor which accounts for both relativistic effects and the outward transport of angular momentum (Novikov & Thorne 1973). The dimensionless form given in the second line is often the more useful one. In it, μ_e is the mass per electron, e is the radiative efficiency of accretion in rest-mass units, σ_T is the Thomson cross section, and x is the radius in gravitational units (*i.e.*, the unit of length is GM/c^2).

We do not yet know how to make a calculation from first principles for $T_{r\phi}$ in terms of other disk quantities. Instead, it has long been popular to appeal to dimensional analysis and guess that $T_{r\phi} = \alpha p$, where α is a dimensionless constant and p is the local pressure (Shakura & Sunyaev 1973). Recent simulations of the nonlinear development of magneto-rotational instabilities (Stone *et al.* 1996; Brandenburg *et al.* 1996; Brandenburg, this volume) suggest that this may, in fact, not be a bad approximation, and that $\alpha \sim 0.01$ – 0.1, but numerous uncertainties still remain. For example, it is not entirely clear *which* pressure sets the scale of the stress: gas pressure, radiation pressure, or magnetic pressure? Once this decision has been made, an equilibrium may be found if one assumes steady-state accretion and a smooth, geometrically thin structure for the disk. Its surface density is then $\propto \alpha^{-1}$, but it is also very sensitive to which pressure sets the scale. If the stress is proportional to the total pressure, the optical depth of the disk can be much smaller than if it scales with a pressure component that is only a fraction of the total. However, if $T_{r\phi}$ is proportional to the total pressure, when $\dot{m}$ is more than a small fraction of unity, these equilibria are both viscously (Lightman & Eardley 1974) and thermally (Shakura & Sunyaev 1976) unstable. Does this mean that the "α-prescription" breaks down at the level of perturbation theory? Or does it mean that some other equilibrium should be sought?

2.2. *Vertical structure*

Equally troubling questions arise when considering the details of vertical structure. The amount of local dissipation per unit area in an accretion disk is

$$Q = \frac{3}{4\pi} \frac{GM\dot{M}}{r^3} R_R(r) \tag{2.5}$$

$$= \frac{3\mu_e c^5}{GM\sigma_T e} \frac{\dot{m}}{x^3} R_R, \tag{2.6}$$

where R_R is a correction factor that accounts for both relativistic effects, and the outward transport of energy associated with the angular momentum flux (also worked out in

Novikov & Thorne 1973). However, we do not know how this dissipation is distributed with height. Clearly, the run of temperature with altitude will be quite sensitive to the dissipation distribution.

In addition, when radiation pressure contributes significantly to the vertical support of the disk against gravity (as commonly occurs), there can be dramatic differences in structure depending on where the radiation is created. For example, if all the dissipation takes place in the mid-plane, the radiation flux is constant with height, so that the vertical radiation force is simply proportional to the local opacity. On the other hand, if the dissipation is confined to the disk surface, there is no outward radiation flux in the body of the disk, and zero support against gravity. Moreover, if the dissipation takes place mostly within a few optical depths of the surface, its exact distribution will clearly have a major impact on the nature of the emergent radiation.

When radiation pressure dominates the vertical support, even small irregularities in the dissipation distribution can significantly affect the outgoing spectrum. Suppose, for example, that the dissipation in the body of the disk is proportional to the gas density. Then, because the flux increases linearly with altitude in exact balance with the increase of g_z with altitude, the effective vertical gravity is nil, and the density is constant as a function of height. However, at the very top of the disk this balance must be broken, for the density must fall to zero. The details of how sharply the density drops are very sensitive to exactly where the dissipation occurs because the effective gravity within the atmosphere depends on the balance between the small additional gravity gained by a small rise in altitude and the small additional radiation force due to the dissipation within that layer. These details are important because the photosphere generally lies within this region of the density roll-off.

The issue of how the heat deposition varies from place to place becomes still further confused when we consider the evidence that a significant part of the total emisison comes out in hard X-rays, a portion of which may then shine down on the disk. In both AGN and Galactic black hole systems, there are telltale signs of X-ray illumination in the form of the "Compton reflection bump": Soft X-rays shining on cool material suffer strong photoelectric absorption, so the albedo of any accretion disk to photons from 0.5 – 10 keV is generally quite small, unless it is very thoroughly ionized. On the other hand, somewhat harder X-rays ($\sim 10 - 50$ keV) are very readily reflected by electron scattering because the highest energy photoionization edge of any abundant element is that of Fe at 7.1 keV, and photoionization cross sections drop rapidly with increasing energy above the edge. The reflection bump rolls over above 50 keV because higher energy photons lose energy by Compton recoil as they are reflected. These bumps are often so prominent that the corollary absorption must contribute significantly to disk heating.

This external illumination may substantially alter both the radial dependence of the integrated heating rate and the vertical distribution of heating at a fixed radius. Where it is important, the heating is largely confined to a skin whose thickness is at most a few Compton depths (*i.e.* $\sim 10^{24}$ cm^{-2}). Precisely because the energy is absorbed in such a thin layer, its effect there can be very strong. Because the photosphere frequently lies at a comparable depth from the surface, X-ray heating can have a major impact on features in the emergent spectrum.

In the limit that *most* of the heating occurs near the surface, whether due to segregation of the internal dissipation or external illumination, the instabilities endemic to radiation pressure-supported disks are quenched (Svensson & Zdziarski 1994). The reason is that there is then little outgoing radiation flux within the body of the disk, so it collapses to a state of substantially greater density (and also greater total surface density). Increases in the dissipation rate then have no impact on the thickness of the disk, and the feedback which drives the instabilities disappears.

2.3. *Radiation transfer*

While the transfer problem is entirely understood in principle, in practise it is so complicated that in every treatment so far, major approximations have been made. Thus, for technical, rather than conceptual, reasons, the solutions are all, in one way or another, model-dependent. A major part of the art of constructing a good transfer solution therefore lies in adroit choices of approximations.

Certain features must surely be included in any transfer solution. Thomson opacity, for example, is almost always important. Similarly, it is easy to show that free-free opacity always affects at least the lower frequencies. At AGN temperatures, HI and HeII photoionization opacity are certainly significant, but at the higher temperatures of disks around stellar black holes, H and He are equally surely fully stripped.

Beyond this point, however, different workers have made different choices, and some effects may be important in certain ranges of parameter space, but not in others. For example, when computing the H and He photoionization opacity, it is necessary to make some statement about the ionization balance for these elements. The simplest guess is that the ionization fractions are those given by the Saha equation, but it is not obvious that this is always correct. Indeed, one might expect that the intensity of radiation in the H and HeII ionization continua would *couple* to the fractional ionization of these species. If one does wish to compute the actual ionization balance of H and He, there are further choices to be made about which processes are important and which are negligible. Ionizations can take place from excited states as well as the ground state, so the excited state population balance must also be found. How many states must be included? And which processes? In the most recent such calculation (Hubeny & Hubeny 1997), the H atom was permitted 9 different values of the principal quantum number, but all states having the same principal quantum number were assumed to be in detailed balance. In addition (and probably more significantly), only bound-free processes were considered. This may have been a significant oversight because at the densities prevalent in their disk model, the bound-bound transition rates due to electron collisions can be comparable to the bound-free rates. Moreover, because non-LTE effects are very sensitive to density, their character can depend very strongly on the model choices made when calculating the vertical structure of the atmosphere.

Another open question is the possible role of heavy element opacities. In AGN accretion disks, where the temperature might be $\sim 10^5$ K and the density $\sim 10^{14}$ cm^{-3}, the thermal equilibrium H neutral fraction is extremely small, $\sim 10^{-8}$. Partially-ionized stages of the more abundant heavy elements are therefore much *more* common than neutral H; the issue is whether the energy bands in which they have substantial opacity are important to the energy flow. Their ionization continua lie at relatively high energy (at least several tens of eV); their resonance lines have rather lower energy, but the importance of lines depends very strongly on the amplitude of turbulent motions in the disk. This last matter is, of course, central to the uncertainties already discussed regarding shear stress. In disks around stellar black holes, the Saha equation predicts that the dominant ionization stages will have ionization potentials $5 - 10kT$; however, there may be enough representatives of species with ionization potentials factors of a few smaller to significantly contribute to the opacity. This issue, too, also has implications for vertical structure when radiation pressure is an important part of the disk's vertical support.

A further question has to do with the possible effects of Comptonization. Its importance is gauged by the parameter

$$y \equiv (4kT/m_e c^2) \max(\tau_T, \tau_T^2), \tag{2.7}$$

for Compton optical depth τ_T. Here the relevant Compton depth is only that portion of the optical depth above the effective photosphere. When y is at least order unity, repeated Compton scatters may impart significant additional energy to photons before they leave the atmosphere. The magnitude of Comptonization is extremely sensitive to model choices and parameters. For example, if the disk equilibrium is one in which the stress is proportional to the total pressure, the temperature in the atmosphere is close to the effective temperature, free-free absorption is the only absorptive opacity, and the gas pressure scale height in the atmosphere is comparable to the total disk thickness (an assumption likely to overestimate the importance of Comptonization),

$$y \simeq 0.6 \dot{m}^{5/3} m^{-1/6} \left(\frac{x}{10} \right)^{-5/2} R_R^{3/2} R_z^{-2/3} \frac{\omega^2}{(1 - e^{-\omega})^{2/3}}. \tag{2.8}$$

Here $\omega = h\nu/kT$, and R_z is another relativistic correction factor, this one adjusting the vertical component of the gravity (Abramowicz *et al.*1997). The frequency-dependence of y is a consequence of the frequency-dependence of the effective photosphere's location. Unfortunately, at the present state of the art, Comptonization can only be treated within the diffusion equation for radiation transfer. Consequently, studies of Comptonization cannot also deal with questions involving either the angular distribution of the emergent radiation or sharp spectral features such as lines and edges. The second limitation comes about because these features are created by changes in the source function on scales comparable to a scattering length; hence, the diffusion approximation does not give an adequate description.

2.4. *Relativistic effects*

A final element that must be included in any proper model (but which is sometimes forgotten) is a proper accounting for relativistic effects. In addition to the dynamical corrections (encapsulated in R_R, R_T, and R_z), there are also relativistic photon propagation effects (Cunningham (1975)). Disk material close to a black hole moves at speeds close to c; as seen by a distant observer, its radiation is therefore Doppler boosted and beamed. In addition, there are intrinsically general relativistic effects: gravitational redshift, and photon trajectory bending. The last effect must be properly melded with the intrinsic angular radiation pattern of disk material in its own rest frame. Clearly, when (as will certainly happen here), Doppler shifts of order unity occur, there can be a dramatic impact on sharp spectral features such as lines and edges.

3. Results to Date

There have been many efforts over the past fifteen years or more to compute the spectra to be expected from accretion disks around black holes. The methods used include:

- stitching together stellar atmosphere solutions (Kolykhalov & Sunyaev 1984);
- summing local blackbodies, but applying general relativistic photon propagation effects (Sun & Malkan 1989);
- solving the frequency-dependent transfer problem, but assuming that the atmosphere is scattering-dominated at all locations and at all frequencies (Laor & Netzer 1989);
- solving the frequency-dependent transfer problem including non-LTE effects in H and He, but with non-standard stress prescriptions (Störzer *et al.* 1994, Hubeny & Hubeny 1997);

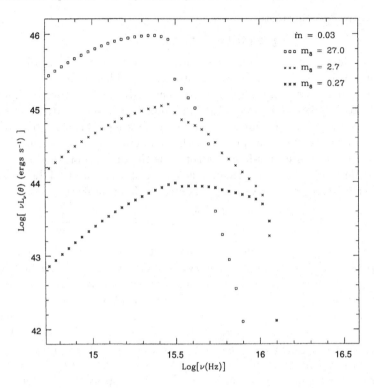

FIGURE 2. The predicted spectra from accretion disks around non-rotating supermassive black holes viewed pole-on. In all three cases shown, $\dot{m} = 0.03$, but the mass varies from $m = 3 \times 10^7$ to $m = 3 \times 10^9$ (from Sincell & Krolik 1998).

- solving the frequency-dependent transfer problem in LTE, but studying the effects of various stress prescriptions (Sincell & Krolik 1998) or external illumination (Sincell & Krolik 1997);
- solving the diffusion/Kompaneets equations in order to study Comptonization (Ross, Fabian & Mineshige 1992; Shimura & Takahara 1993; Dörrer *et al.* 1996).

Because no one calculation is the most nearly complete, I will try to display the range of possibilities by showing a variety of calculations, each highlighting a different aspect of the problem. To set the stage, it's best to begin with how "standard" models depend on parameters. Figure 2 shows the scaling with central mass at fixed accretion rate relative to Eddington according to a calculation which employed full transfer solutions and general relativistic effects, but made the LTE approximation for the H and He ionization balances, and ignored all heavier elements. Although the luminosity increases in proportion to mass at fixed $\dot{m}$, the radiating area increases faster, $\propto m^2$. That is to say, the temperature at the innermost ring scales $\propto (\dot{m}/m)^{1/4}$. Consequently, as this quantity falls, the spectrum becomes softer and the Lyman edge goes steadily deeper and deeper into absorption. If, on the other hand, the central mass is held constant and the accretion rate is varied (Figure 3), the temperature rises, so increasing accretion rate both hardens the overall spectrum and throws the Lyman edge from absorption into emission. In the hottest disks, there can be substantial flux in the HeII continuum, although the edge itself generally stays in absorption for parameters appropriate to AGN.

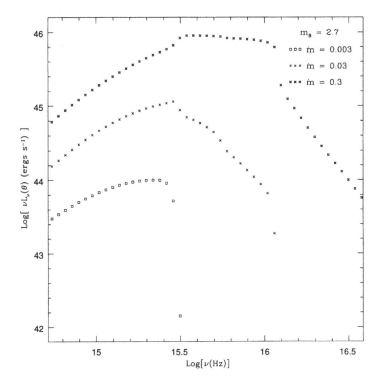

FIGURE 3. The predicted spectra from accretion disks around non-rotating supermassive black holes viewed pole-on. In all three cases shown, $m = 3 \times 10^8$, but the scaled accretion rate varies from $\dot{m} = 0.003$ to $\dot{m} = 0.3$ (from Sincell & Krolik 1998).

3.1. *Dependence on radial structure model*

With this background, we may now inquire into the effects of the systematic uncertainties discussed in the previous section. First, how serious are the effects of the uncertainty in the radial equilibrium? Figure 4 shows what happens when the stress prescription is changed from αp_g to $\alpha \sqrt{p_g p_r}$ to $\alpha(p_r + p_g)$. Greater stress at fixed accretion rate leads to both smaller surface density and smaller volume density. Here the accretion rate is high enough that $p_r \gg p_g$ in the interesting inner rings of the disk. As a result, the temperature in the atmosphere is lowest in the αp_g case and highest when the stress is $\alpha(p_r + p_g)$. At fixed total luminosity, this leads to a generally harder overall spectrum for the disk with the greatest stress, but the qualitative character of the spectrum changes relatively little with different choices for the stress prescription.

3.2. *Dependence on vertical distribution of heating*

The next question to consider is the impact of the vertical distribution of heat deposition. In the most extreme limit, one in which all the heating is concentrated into the disk atmosphere (as, for example, X-ray heating may accomplish), the effects are dramatic. Because radiation pressure no longer supports the disk against the vertical component of gravity g_z, it is geometrically much thinner. Closer to the disk midplane, g_z is also smaller, so the gas pressure scale height (at fixed temperature) is longer. The density at the photosphere is then smaller, even though the density deep inside the disk is much greater. In addition, because the heating rate per unit mass rises upward in the X-ray-heated zone, so does the temperature (although not necessarily monotonically). The

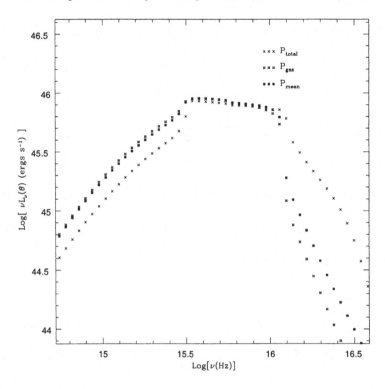

FIGURE 4. The predicted spectrum from an accretion disk around a non-rotating supermassive black hole viewed pole-on. Here $\dot{m} = 0.3$ and $m = 3 \times 10^8$ (from Sincell & Krolik 1998).

result is emergent spectra that are *softer* than the corresponding spectra from internally heated disks at low frequencies, but with more flux at higher frequencies, and ionization edges that are almost always in emission (compare Figure 5 to Figure 2).

3.3. *Implications of the LTE approximation*

Different approximations in the radiation transfer solution are the next issue whose effect we need to evaluate. Is it necessary (or when is it necessary) to compute the departure from LTE of the H and He ionization balances? Störzer *et al.* 1994 argued that when the photospheric pressure was > 100 dyne cm^{-2} (*i.e.*, the density $n \sim 10^{13} T_5^{-1}$ cm^{-3}), non-LTE effects are weak. However, Hubeny & Hubeny (1997) found that the LTE approximation seriously misrepresented the strength of the HI and HeII edges even when the photospheric pressure was several orders of magnitude greater. Part of the problem may be the definition of a non-LTE calculation: Hubeny & Hubeny (1997) included no bound-bound transitions in their evaluation of the H and He excited state balances. Another part of the problem (as already discussed in §2.2) may be that the character of the non-LTE effects is exquisitely sensitive to approximations in the solution of the hydrostatic equilibrium [an issue which affects both Störzer *et al.* 1994 and Hubeny & Hubeny (1997)]. Figure 6 illustrates what the effects of non-LTE *might* be. In this case, both the radiative acceleration and g_z were taken to be constant in the atmosphere, 9 bound levels of HI, 8 of HeI, and 14 of HeII were included, and all bound-bound transitions were neglected. Although the overall shape of the continuum is not very sensitive to non-LTE effects, they can completely change the nature of features at ionization edges.

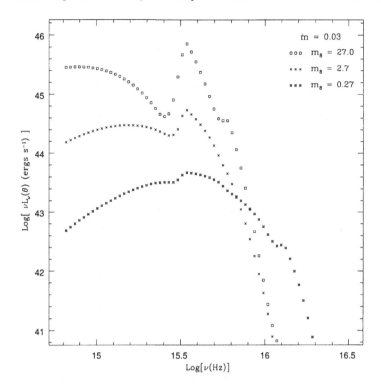

FIGURE 5. The predicted spectra from X-ray-illuminated accretion disks around non-rotating supermassive black holes viewed pole-on. In all three cases shown, $\dot{m} = 0.03$, but the mass varies from $m = 3 \times 10^7$ to $m = 3 \times 10^9$ (from Sincell & Krolik 1997).

3.4. *Influence of heavy element opacity*

Particularly in AGN accretion disks, heavy elements may potentially contribute substantially to the opacity, but these effects have just begun to be explored. To date there has been no calculation of how this added opacity may affect the vertical structure of the disk, and therefore g_z in the atmosphere. Indeed, because including all the necessary atomic data is such a big job, and because the additional frequency resolution required to follow the multitude of heavy element lines adds such a large computational burden, there have been only partial calculations of these effects even assuming a vertical structure computed allowing only for H and He. For this reason, the spectrum shown in Figure 7 (Hubeny, Agol & Blaes 1998) should be regarded merely as a demonstration of the potential impact of heavy element opacity. It was computed on the basis of an atmosphere solution whose vertical structure and temperature profile were determined including only H and He. Moreover, only those heavy element lines falling between 600Å and 1400Å were used. Nonetheless, the effects are large: the integrated outgoing flux is reduced by roughly a factor of two. An atmosphere which includes heavy element opacities in a fully self-consistent fashion will clearly look quite different from one with only H and He.

3.5. *Impact of relativistic effects*

As discussed in §2.4, there are two separate classes of relativistic effects: those entering the structural equations, and those affecting the appearance of the disk to distant observers. Both sorts of corrections are relatively modest when the black hole has little spin

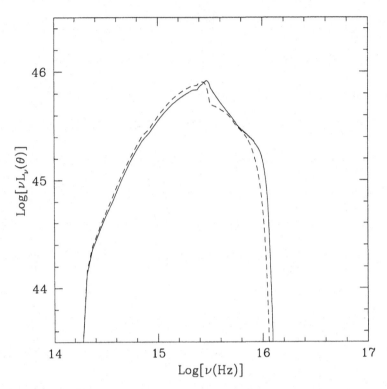

FIGURE 6. LTE (dashed line) and non-LTE (solid line) spectra from accretion disks around supermassive black holes viewed pole-on. In these calculations (Agol 1997), $\dot{m} = 0.072$ and $m = 2 \times 10^9$, and the normalized spin was 0.998.

because the marginally stable orbit, at $x = 6$, is relatively far from the event horizon (at $x = 2$). However, as the spin increases, both the marginally stable orbit and the event horizon move inward, both approaching $x = 1$ as the spin increases toward its maximum value. Figure 8 shows how dramatic these effects can be. All three predicted spectra were calculated using the same physics (stress proportional to $\sqrt{p_r p_g}$, LTE ionization fractions, no heavy elements) and pertain to the same m, $\dot{m}$, and viewing angle. When the central black hole spins rapidly, sharp edge features (as could be seen in the predictions for the spinless black hole) get stretched out over substantial frequency ranges. The HeII edge is stretched even farther than the HI edge because its origin is confined more nearly to the innermost radii. However, because the strongest relativistic effect is the Doppler boosting and beaming of radiation into the direction of orbital motion, these smoothing effects weaken considerably, even for rapidly-spinning black holes, when the viewing angle is nearly along the rotation axis.

4. Comparison to observations

Now that we have seen the range of spectral shapes predicted by various models, we can look at genuine observed spectra to see if any of their features are predicted by the models. Remarkably, the simplest approximation—summing local blackbodies without regard for any of the detailed physics we have discussed—does the best job of reproducing the observations! As shown in Figure 9, its slope comes very close to matching the composite slope (at least for the LBQS sample) at low frequencies, and, of course, it

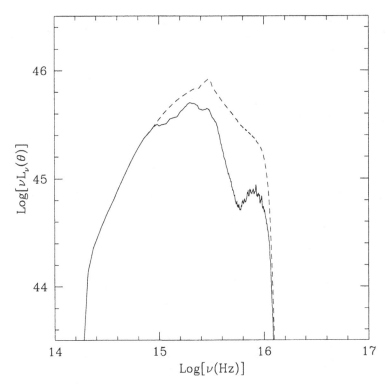

FIGURE 7. Effects of heavy element opacity on the spectrum from an accretion disk around a black hole with the same parameters as in Figure 6. As described in the text, the dashed line is a self-consistent disk spectrum whose opacity tables included only H and He; the solid line was calculated using the disk structure and temperature profile of the pure H and He model, but for predicting the outgoing flux, metal features between 600Å and 1400Å were included. The heavy element ionization fractions were assumed to be in LTE.

is absolutely clean at all ionization edges, in excellent agreement with the numerous observations showing that the Lyman edge is unobservably weak in the great majority of quasars (Koratkar, Kinney & Bohlin (1992)). The zeroth-order approximation still fails, however, to explain the strength of the EUV continuum.

By contrast, the more detailed spectral predictions must be tuned to match the low frequency spectral shape, and nearly always show Lyman edge features, whether in emission or absorption. While it is possible to smooth out the edge with relativistic effects, it seems unlikely that the feature can disappear altogether in composite spectra unless polar views are somehow forbidden.

It is possible that Comptonization can solve two problems—the strength of the EUV continuum and the absence of Lyman edges, but it faces tough prerequisites. As Laor *et al.* (1997) have shown, the slope of the EUV spectrum as seen at the highest frequencies observable in the ultraviolet is a good predictor of the soft X-ray continuum, suggesting that this component extends smoothly all the way across the unobservable EUV. If this is true (the interpolation in Figure 1 assumes this), a significant fraction of the total luminosity of AGN is emitted in the EUV. To make this segment of the spectrum by Comptonization therefore entails putting the same fraction of the total dissipation into the Comptonizing layer. At present there is no known physical mechanism that will do so.

Thus, the interim conclusion regarding how well our models of accretion disks fare

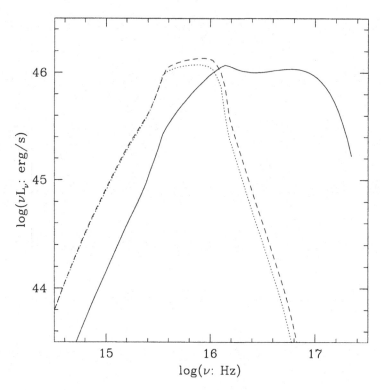

FIGURE 8. Effects of general relativity on the observed spectrum. All three curves are computed for a disk with $\dot{m} = 0.3$ around a black hole of mass $m = 3 \times 10^8$ viewed at an inclination angle of $40°$. However, the dotted curve is the prediction of a model in which the black hole is spinless and the photons travel according to Newtonian rules, while the dashed curve shows how the same disk looks in the real, *i.e.*, relativistic world, and the solid curve shows the spectrum from the same disk if its central black hole had a normalized spin of 0.998.

vis-a-vis observations is one that is frustrating in several respects. On the one hand, there are several major conceptual issues we do not know how to settle that introduce large uncertainties into our predictions; on the other hand, none of the guesses made to date on how to resolve these questions leads to predictions that match observations. It seems quite likely that we are still missing a large piece of the puzzle.

I thank Eric Agol and Mark Sincell for many enlightening conversations, and for supplying figures, sometimes from unpublished work. My research on accretion disks is partially supported by NASA Grant NAG5-3929 and by NSF Grant AST-9616922.

REFERENCES

ABRAMOWICZ, M.A., LANZA, A. & PERCIVAL, M.J. 1997 *Ap. J.* **479**, 179

AGOL, E. 1997 U. of California at Santa Barbara Ph.D. thesis

BRANDENBURG, A., NORDLUND, A., STEIN, R.F., & TORKELSSON, U. 1996 *Ap. J. Letts.* **458**, L45

CUNNINGHAM, C.T. 1975, *Ap.J.* **202**, 788

DÖRRER, T.H., RIFFERT, H., STAUBERT, R., & RUDER, H. 1996 *Astron. & Astrop.* **311**, 69

ELVIS, M., WILKES, B.J., MCDOWELL, J.C., GREEN, R.F., BECHTOLD, J., WILLNER, S.P., OEY, M.S., POLOMSKI, E., & CUTRI, R. 1994 *Ap. J. Suppl.* **95**, 1

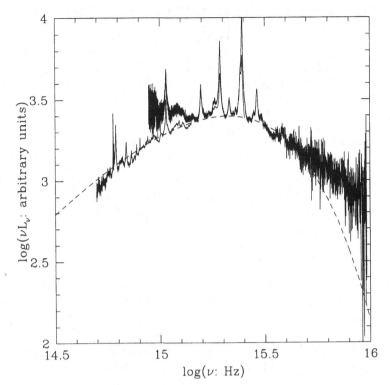

FIGURE 9. Match between a zeroth-order predicted disk spectrum and two composites of quasar spectra. The noisy solid curves are the composites assembled by Francis *et al.* (1991) (from the LBQS sample) and Zheng *et al.*(1997) (from the FOS archive); the dashed curve has the form $\nu^{4/3} \exp\left(-h\nu/kT\right)$ with $T = 7 \times 10^4$ K.

FRANCIS, P.J., HEWETT, P.C., FOLTZ, C.B., CHAFFEE, F.H., & WEYMANN, R.J. 1991 *Ap.J.* **373**, 465

HUBENY, I., AGOL, E. & BLAES, O. 1998 in preparation

HUBENY, I. & HUBENY, V. 1997 *Ap. J. Letts.* **484**, L37

KORATKAR, A.P., KINNEY, A.L. & BOHLIN, R.C. 1992 *Ap.J.* **400**, 435

LAOR, A., FIORE, F., ELVIS, M., WILKES, B.J. & MCDOWELL, J.C. 1997 *Ap.J.* **477**, 93

LIGHTMAN, A.P. & EADLEY, D.M. 1974 *Ap. J. Letts.* **187**, L1

NOVIKOV, I.D. & THORNE, K.S. 1973 In *Black Holes* (eds. C. De Witt & B. De Witt), p. 343. Gordon and Breach.

ROSS, R.R., FABIAN, A.C. & MINESHIGE, S. 1992 *M.N.R.A.S.* **258**, 189

SHAKURA, N.I. & SUNYAEV, R.A. 1973 *Astron. & Astrop.* **24**, 337

SHAKURA, N.I. & SUNYAEV, R.A. 1976 *M.N.R.A.S.* **175**, 613

SHIMURA, T. & TAKAHARA, F. 1993 *Ap. J.* **419**, 78

SINCELL, M.W. & KROLIK, J.H. 1997 *Ap. J.* **476**, 605

SINCELL, M.W. & KROLIK, J.H. 1998 *Ap. J.* in press

STONE, J.M., HAWLEY, J.F., GAMMIE, C.F., & BALBUS, S.A. 1996 *Ap. J.* **463**, 656

STÖRZER, H., HAUSCHILDT, P.H., & ALLARD, F. 1994 *Ap. J. Letts.* **437**, L91

SVENSSON, R. & ZDZIARSKI, A. 1994 *Ap. J.* **436**, 599

ZHENG, W., KRISS, G.A., TELFER, R.C., GRIMES, J.P. & DAVIDSEN, A.F. 1997 *Ap.J.* **475**, 469

Advection-dominated accretion around black holes

By RAMESH NARAYAN
ROHAN MAHADEVAN
AND
ELIOT QUATAERT

Harvard–Smithsonian Center for Astrophysics, 60 Garden Street, Cambridge, MA 02138, USA

This article reviews the physics of advection-dominated accretion flows (ADAFs) and describes applications to several black hole X-ray binaries and galactic nuclei. The possibility of using ADAFs to explore the event horizons of black holes is highlighted.

1. Introduction

Accretion processes around black holes almost inevitably involve rotating gas flows. Consequently, there is great interest in self–consistent solutions of the hydrodynamic equations of viscous differentially–rotating flows. Four solutions are currently known (see Chen et al. 1995 for a discussion). In these solutions viscosity transports angular momentum outward, allowing the accreting gas to spiral in toward the central mass. Viscosity also acts as a source of heat; some or all of this heat is radiated, leading to the observed spectrum.

The most famous of the four solutions is the thin disk model developed by Shakura & Sunyaev (1973), Novikov & Thorne (1973), Lynden–Bell & Pringle (1974) and others (see Pringle 1981 and Frank et al. 1992 for reviews). The accreting gas forms a geometrically thin, optically thick disk, and produces a quasi–blackbody spectrum. The effective temperature of the radiation is in the range $10^5 - 10^7$ K, depending on the black hole mass and the accretion rate ($T_{\text{eff}} \propto M^{-1/4} \dot{M}^{1/4}$). The thin disk solution has been used to model a large number of astrophysical systems.

Shapiro, Lightman and Eardley (1976; hereafter SLE; see also Björnsson & Svensson 1991 and Luo & Liang 1994) discovered a second, much hotter self–consistent solution in which the accreting gas forms a two temperature plasma with the ion temperature greater than the electron temperature ($T_i \sim 10^{11}$K, $T_e \sim 10^8 - 10^9$K). The gas is optically thin and produces a power–law spectrum at X–ray and soft γ–ray energies. The SLE solution is, however, thermally unstable (Piran 1978), and is therefore not considered viable for real flows.

At super–Eddington accretion rates, a third solution is present (Katz 1977; Begelman 1978; Abramowicz et al. 1988; see also Begelman & Meier 1982; Eggum, Coroniti, & Katz 1988), in which the large optical depth of the inflowing gas traps most of the radiation and carries it inward, or "advects" it, into the central black hole. This solution is referred to as an optically thick advection–dominated accretion flow (optically thick ADAF). A full analysis of the dynamics of the solution was presented in an important paper by Abramowicz et al. (1988).

A fourth solution is present in the opposite limit of low, sub–Eddington, accretion rates (Ichimaru 1977; Rees et al. 1982; Narayan & Yi 1994, 1995a, 1995b; Abramowicz et al. 1995). In this solution, the accreting gas has a very low density and is unable to cool efficiently within an accretion time. The viscous energy is therefore stored in the gas as thermal energy instead of being radiated, and is advected onto the central star. The gas is optically thin and adopts a two–temperature configuration, as in the SLE solution. The solution is therefore referred to as an optically thin ADAF or a two–temperature ADAF.

This article reviews ADAFs around black holes, with an emphasis on two–temperature ADAFs.

2. Dynamics of ADAFs

2.1. *Basic Equations*

Consider a steady axisymmetric accretion flow. The dynamics of the flow are described by the following four height-integrated differential equations, which express the conservation of mass, radial momentum, angular momentum, and energy:

$$\frac{d}{dR}(\rho R H v) = 0, \tag{2.1}$$

$$v\frac{dv}{dR} - \Omega^2 R = -\Omega_K^2 R - \frac{1}{\rho}\frac{d}{dR}(\rho c_s^2), \tag{2.2}$$

$$v\frac{d(\Omega R^2)}{dR} = \frac{1}{\rho R H}\frac{d}{dR}\left(\nu \rho R^3 H \frac{d\Omega}{dR}\right), \tag{2.3}$$

$$\rho v T \frac{ds}{dR} = q^+ - q^- = \rho \nu R^2 \left(\frac{d\Omega}{dR}\right)^2 - q^- \equiv f\nu\rho R^2 \left(\frac{d\Omega}{dR}\right)^2, \tag{2.4}$$

where ρ is the density of the gas, R is the radius, $H \sim c_s/\Omega_K$ is the vertical scale height, v is the radial velocity, c_s is the isothermal sound speed, T is the temperature of the gas (mean temperature in the case of a two temperature gas), Ω is the angular velocity, Ω_K is the Keplerian angular velocity, s is the specific entropy of the gas, q^+ is the energy generated by viscosity per unit volume, and q^- is the radiative cooling per unit volume. The quantity on the left in equation (2.4) is the rate of advection of energy per unit volume. The parameter f is thus the ratio of the advected energy to the heat generated and measures the degree to which the flow is advection–dominated. The kinematic viscosity coefficient, ν, is generally parameterized via the α prescription of Shakura & Sunyaev (1973),

$$\nu \equiv \alpha c_s H = \alpha \frac{c_s^2}{\Omega_K}, \tag{2.5}$$

where α is assumed to be independent of R. The steady state mass conservation equation implies a constant accretion rate throughout the flow:

$$\dot{M} = (2\pi R)(2H)\rho|v| = \text{constant}. \tag{2.6}$$

It is useful to rewrite the energy equation (2.4) compactly as

$$q^{\text{adv}} = q^+ - q^-, \tag{2.7}$$

where q^{adv} represents the advective transport of energy (usually a form of cooling). Depending on the relative magnitudes of the terms in this equation, three regimes of accretion may be identified:

• $q^+ \simeq q^- \gg q^{\text{adv}}$. This corresponds to a cooling-dominated flow where all the energy released by viscous stresses is radiated; the amount of energy advected is negligible. The thin disk solution and the SLE solution correspond to this regime.

• $q^{\text{adv}} \simeq q^+ \gg q^-$. This corresponds to an ADAF where almost all the viscous energy is stored in the gas and is deposited into the black hole. The amount of cooling is negligible compared with the heating. For a given $\dot{M}$, an ADAF is much less luminous than a cooling-dominated flow.

• $-q^{\mathrm{adv}} \simeq q^- \gg q^+$. This corresponds to a flow where energy generation is negligible, but the entropy of the inflowing gas is converted to radiation. Examples are Bondi accretion, Kelvin–Helmholtz contraction during the formation of a star, and cooling flows in galaxy clusters.

2.2. *Self-Similar Solution*

Analytical approximations to the structure of optically thick and optically thin ADAFs have been derived by Narayan & Yi (1994). Assuming Newtonian gravity ($\Omega_K^2 = GM/R^3$) and taking f to be independent of R, they showed that equations (2.1)–(2.4) have the following self–similar solution (see also Spruit et al. 1987):

$$v(R) = -\frac{(5 + 2\epsilon')}{3\alpha^2} g(\alpha, \epsilon') \alpha v_{\mathrm{ff}}, \tag{2.8}$$

$$\Omega(R) = \left[\frac{2\epsilon'(5 + 2\epsilon')}{9\alpha^2} g(\alpha, \epsilon')\right]^{1/2} \frac{v_{\mathrm{ff}}}{R}, \tag{2.9}$$

$$c_s^2(R) = \frac{2(5 + 2\epsilon')}{9\alpha^2} g(\alpha, \epsilon') v_{\mathrm{ff}}, \tag{2.10}$$

where

$$v_{\mathrm{ff}} \equiv \left(\frac{GM}{R}\right)^{1/2}, \quad \epsilon' \equiv \frac{\epsilon}{f} = \frac{1}{f}\left(\frac{5/3 - \gamma}{\gamma - 1}\right), \quad g(\alpha, \epsilon') \equiv \left[1 + \frac{18\alpha^2}{(5 + 2\epsilon')^2}\right]^{1/2} - 1. \tag{2.11}$$

γ is the ratio of specific heats of the gas, which is likely to lie in the range 4/3 to 5/3 (the two limits correspond to a radiation pressure–dominated and a gas pressure–dominated accretion flow, respectively). Correspondingly, ϵ lies in the range 0 to 1.

In general, f depends on the details of the heating and cooling and will vary with R. The assumption of a constant f is therefore an oversimplification. However, when the flow is highly advection dominated, $f \sim 1$ throughout the flow, and f can be well approximated as constant. Setting $f = 1$ and taking the limit $\alpha^2 \ll 1$ (which is nearly always true), the solution (2.8)–(2.10) takes the simple form

$$\frac{v}{v_{\mathrm{ff}}} \simeq -\left(\frac{\gamma - 1}{\gamma - 5/9}\right)\alpha, \quad \frac{\Omega}{\Omega_K} \simeq \left[\frac{2(5/3 - \gamma)}{3(\gamma - 5/9)}\right]^{1/2}, \quad \frac{c_s^2}{v_{\mathrm{ff}}^2} \simeq \frac{2}{3}\left(\frac{\gamma - 1}{\gamma - 5/9}\right). \tag{2.12}$$

A number of interesting features of ADAFs are revealed by the self–similar solution. (1) As we discuss below (§4.1.3), it appears that ADAFs have relatively large values of the viscosity parameter, $\alpha \gtrsim 0.1$; typically, $\alpha \sim 0.2 - 0.3$. This means that the radial velocity of the gas in an ADAF is comparable to the free–fall velocity, $v \gtrsim 0.1 v_{\mathrm{ff}}$. The gas thus accretes quite rapidly. (2) The gas rotates with a sub–Keplerian angular velocity and is only partially supported by centrifugal forces. The rest of the support is from a radial pressure gradient, $\nabla P \sim \rho c_s^2/R \sim (0.3 - 0.4)v_{\mathrm{ff}}^2/R$. In the extreme case when $\gamma \to 5/3$, the flow has no rotation at all ($\Omega \to 0$). (3) Since most of the viscously generated energy is stored in the gas as internal energy, rather than being radiated, the gas temperature is quite high; in fact, optically thin ADAFs have almost virial temperatures. This causes the gas to "puff up": $H \sim c_s/\Omega_K \sim v_{\mathrm{ff}}/\Omega_K \sim R$. Therefore, geometrically, ADAFs resemble spherical Bondi (1952) accretion more than thin disk accretion. It is, however, important to note that the *dynamics* of ADAFs are very different from that of Bondi accretion (§2.7). (4) The gas flow in an ADAF has a positive Bernoulli parameter (Narayan & Yi 1994, 1995a), which means that if the gas

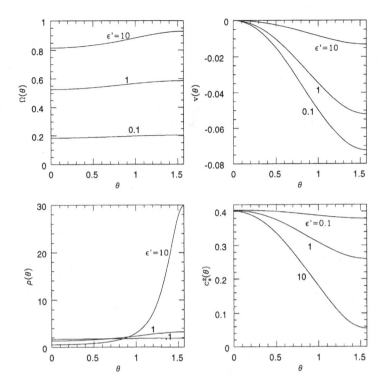

FIGURE 1. Angular profiles for radial self-similar solutions with $\alpha = 0.1$, $\epsilon' = 0.1, 1, 10$. *Top left:* angular velocity Ω/Ω_K as a function of polar angle θ. *Top right:* radial velocity, $v/v_{\rm ff}$. *Bottom left:* density, ρ. *Bottom right:* sound speed squared, $c_s^2/v_{\rm ff}^2$.

were somehow to reverse its direction, it could reach infinity with net positive energy. This suggests a possible connection between ADAFs and jets. (5) The entropy of the gas increases with decreasing radius. ADAFs are therefore convectively unstable (Narayan & Yi 1994, 1995a; Igumenshchev, Chen, & Abramowicz 1996).

2.3. Vertical Structure

The quasi–spherical nature of the gas flow in an ADAF might indicate that the use of height integrated equations is an oversimplification. This led Narayan & Yi (1995a) to investigate the structure of ADAFs in the polar direction, θ. Making use of non–height-integrated equations, they searched for radially self–similar solutions; for example, they assumed that the density scales as $\rho \propto R^{-3/2}\rho(\theta)$, with a dimensionless function $\rho(\theta)$; the radial velocity scales as $v \propto R^{-1/2}v(\theta)$, etc. The results confirmed that ADAFs are quasi–spherical and quite unlike thin disks.

Figure 1 shows the variation of Ω, v, ρ and c_s^2 as a function of θ for some typical solutions ($\alpha = 0.1$, $\epsilon' = 0.1, 1, 10$). For fully advection dominated flows ($f = 1$), different values of ϵ' correspond to different values of γ: the three solutions shown have $\gamma = 1.61, 1.33, 1.06$. If γ is fixed, then increasing ϵ' corresponds to decreasing f. Large values of ϵ' therefore correspond to cooling–dominated thin disk solutions whereas small values of ϵ' correspond to highly advection–dominated flows. From Figure 1, we find that Ω, ρ and c_s^2 in ADAFs ($\epsilon' = 0.1, 1$) are nearly constant on radial shells. The radial velocity v is zero at the poles and reaches a maximum at the equator. Most of the accretion therefore takes place in the equatorial plane. The figure also shows the expected features of a thin disk in the limit of efficient cooling ($\epsilon' = 10$); the density is peaked near the equator, and Ω approaches Ω_K.

Despite the quasi–spherical nature of the flow, Narayan & Yi (1995a) found that the solutions of the exact non–height–integrated equations agree quite well with those of the simplified height–integrated equations (§2.2), provided that "height–integration" is done along θ at constant spherical radius, rather than along z at constant cylindrical radius. The height–integrated equations therefore are a fairly accurate representation of quasi–spherical ADAFs. (Technically, this has been proved only in the self-similar regime.)

2.4. *Pseudo–Newtonian Global Solutions*

The self–similar solution is scale free and does not match the boundary conditions of the flow. To proceed beyond self-similarity it is necessary to solve the full equations (2.1)–(2.4) with proper boundary conditions. For example, far from a central black hole, an ADAF might join on to a thin disk. At this radius, the rotational and radial velocities, as well as the gas density and sound speed, must take on values appropriate to a thin disk. Also, as the accreting gas flows in toward the black hole, it must undergo a sonic transition at some radius R_{sonic}, where the radial velocity equals the local sound speed. In addition, since the black hole cannot support a shear stress, the torque at the horizon must be zero, which is yet another boundary condition to be satisfied by the solutions. (Alternatively, if a causal viscosity prescription is used, there is a boundary condition at the causal horizon; see, e.g., Gammie & Popham (1998)) From the simple scalings of the self–similar solution (2.8)–(2.10), it is evident that the solution does not satisfy the boundary conditions and, therefore, is not a good approximation near the flow boundaries. In fact, the question arises: is the self–similar solution a good approximation to the global flow at any radius?

This question was investigated by Narayan, Kato, & Honma (1997a) and Chen, Abramowicz & Lasota (1997); see also Matsumoto et al. (1985); Abramowicz et al. (1988); Narayan & Yi (1994). Integrating the angular momentum equation (eq. [2.3]) once, one obtains

$$\frac{d\Omega}{dR} = \frac{v\Omega_K(\Omega R^2 - j)}{\alpha R^2 c_s^2}, \tag{2.13}$$

where the integration constant j is the angular momentum per unit mass accreted by the central mass. Recalling that eq. (2.1) integrates to give the constant mass accretion rate, $\dot{M}$, the global steady state problem consists of solving the differential equations (2.2), (2.4), and (2.13), along with proper boundary conditions, to obtain $v(R)$, $\Omega(R)$, $c_s(R)$, $\ln[\rho(R)]$, and the eigenvalue j.

Narayan et al. (1997a) and Chen et al. (1997) obtained global solutions for accretion on to a black hole, using a pseudo–Newtonian potential (Paczyński & Wiita 1980), with $\phi(R) = -GM/(R - R_S)$ and $\Omega_K^2 = GM/(R - R_S)^2 R$, which simulates a Schwarzschild black hole of radius $R_S = 2GM/c^2$. The global solutions agree quite well with the self-similar solutions, except near the boundaries, where there are significant deviations. This means that the self-similar solution provides a good approximation to the real solution over most of the flow.

Figure 2 shows the radial variation of the specific angular momentum and gas pressure in the inner regions of global ADAF solutions for several α. It is seen that low α ADAFs are quite different from high α ADAFs, the division occurring at roughly $\alpha \sim 0.01$. When $\alpha \lesssim 0.01$, the gas in an ADAF is somewhat inefficient at transferring angular momentum outwards, has low radial velocity, and is nearly in hydrostatic equilibrium. The flow is

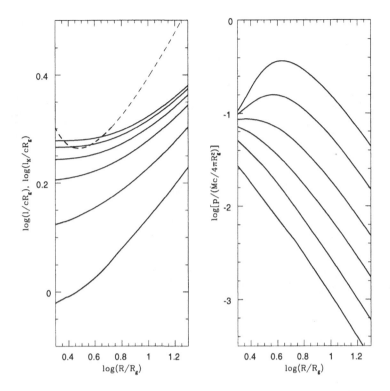

FIGURE 2. *left*: Radial variation of the specific angular momentum, l, for (solid lines, from top to bottom) $\alpha = 0.001, 0.003, 0.01, 0.03, 0.1$, and 0.3. The dashed line shows the Keplerian specific angular momentum, l_K. Note that the low α solutions (0.001 and 0.003) are super-Keplerian ($l > l_K$) over a range of radii, whereas the solutions with larger α have $l < l_K$ for all radii. *right*: Radial variation of the gas pressure for the same α. Note that the low α curves (the two upper ones) have pressure maxima, associated with the super-Keplerian rotation shown in the left panel (taken from Narayan et al. 1997a).

super–Keplerian over a range of radii and has an inner edge near the marginally stable orbit (see fig. 2). The super–Keplerian flow creates a funnel along the rotation axis, which leads to a toroidal morphology. These solutions closely resemble the thick–torus models studied by Fishbone & Moncrief (1976), Abramowicz et al. (1978), Paczyński & Wiita (1980), and others (see Frank et al. 1992), and on which the ion torus model of Rees et al. (1982) is based.

When $\alpha \gtrsim 0.01$, on the other hand, the flow dynamics is dominated by viscosity, which efficiently removes angular momentum from the gas. The gas then has a large radial velocity, $v \sim \alpha v_{ff}$, remains sub–Keplerian at all radii, and has no pressure maximum outside the sonic radius. High α ADAFs therefore do not possess empty funnels and are unlikely to be toroidal in morphology. Instead the flow is probably quasi–spherical all the way down to the sonic radius; in fact, these ADAFs are more akin to slowly rotating settling stars, as illustrated in Fig. 3. As we discuss below (§4.1.3), two–temperature ADAFs have $\alpha \sim 0.2 - 0.3$, which places them firmly in the regime of quasi–spherical, rather than toroidal, flows.

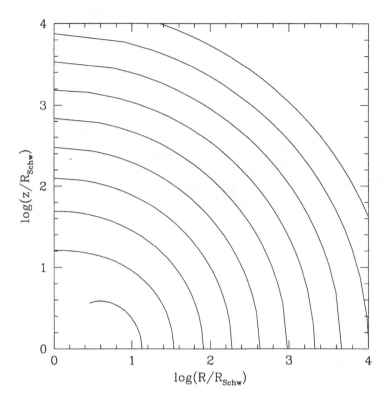

FIGURE 3. Isodensity contours (two contours per decade) in the $R - z$ plane for a global ADAF with $\alpha = 0.3$, $\gamma = 1.4444$ and $f = 1$. The contours are truncated at the sonic radius since the vertical structure is unreliable inside this radius. (Taken from Narayan 1997).

2.5. *Relativistic Global Solutions*

The global solutions described above are based on Newtonian physics. Recently, fully relativistic global solutions in the Kerr geometry have been calculated by Abramowicz et al. (1996), Peitz & Appl (1997) and Gammie & Popham (1998). Gammie & Popham (1998) find that their relativistic solutions are similar to the global solutions of Narayan et al. (1997a) for radii $R \gtrsim 10R_S$, while for $R \lesssim 10R_S$ the solutions differ significantly. They also show that, in the inner regions of the flow, the black hole's spin has a substantial effect on the density, temperature, angular momentum and radial velocity of the accreting gas. This is likely to have an impact on the observed spectrum (Jaroszyński & Kurpiewski 1997). A detailed survey of solutions for various parameters is discussed in Popham & Gammie (1998).

To summarize the last four subsections, a fairly advanced level of understanding has been achieved for the dynamics of ADAFs. There has been progress on two fronts. First, the two-dimensional structure of the flow in $R-\theta$ is understood in the radially self-similar regime. Second, using height-integration (which is shown to be an excellent approximation in the self-similar zone), the radial structure of the flow, including the effects of boundary conditions, has been worked out including full general relativity. Several problems, however, remain. The two-dimensional structure in the boundary regions, where the flow deviates substantially from self-similarity, is not understood. There may be solutions that are qualitatively distinct from the ones discovered so far, e.g., solutions with jets. In addition, three dimensional effects may arise from spiral modes, convective turbulence, etc.. Finally, the introduction of MHD effects in the fluid equations may reveal new phenomena which go beyond the hydrodynamic approximations considered so far.

2.6. *On the Possibility of Radial Shocks*

The global steady state solutions discussed in §2.4 and §2.5 are free of radial shocks. Igumenshchev et al. (1997) have carried out time–dependent simulations of global transonic flows and again find no shocks. However, Chakrabarti and his collaborators (see Chakrabarti 1996 for a review of this work) have claimed that shocks are generic to ADAFs. The clearest statement of his results is found in Chakrabarti & Titarchuk (1995); the authors claim: (i) Low α ($\lesssim 0.01$) flows have shocks and high α ($\gtrsim 0.01$) flows do not. (ii) Low α flows have sub–Keplerian rotation at large radii while high α flows are Keplerian at all radii except very close to the black hole. Neither statement has been confirmed by other workers. In particular, no shocks are seen in any of the global solutions, which span a wide range of parameter values: α ranging from 10^{-3} to 0.3, and γ from 4/3 to 5/3. The reader is referred to Narayan et al. (1997a) and Narayan (1997) for a more detailed discussion of the lack of radial shocks in ADAF solutions.

2.7. *ADAFs versus Pure Spherical Accretion*

Geometrically, ADAFs are similar to spherical Bondi accretion. They are quasi–spherical and have radial velocities which are close to free fall, at least for large α. Also, the gas passes through a sonic radius and falls supersonically into the black hole.

Despite these similarities, it is important to stress that ADAFs are *dynamically* very different from pure spherical accretion. Transport of angular momentum through viscosity plays a crucial role in ADAFs; indeed, there would be no accretion at all without viscosity (recall that $v \propto \alpha$, cf. eq. [2.12]). Bondi accretion, on the other hand, involves only a competition between inward gravity and outward pressure gradient; viscosity plays no role.

In the Bondi solution, the location of the sonic radius is determined by the properties of the accreting gas at large radii. Under many conditions, the sonic radius is quite far from the black hole. In an ADAF, on the other hand, the sonic radius is almost always at a few R_S. Furthermore, the location of the sonic radius depends on the viscosity parameter α (another indication of the importance of viscosity) and the properties of the gas at infinity are irrelevant so long as the gas has enough angular momentum to prevent direct radial infall. In addition, the gas temperature, or energy density, in an ADAF is determined by viscous heating and adiabatic compression, and is generally much higher than in Bondi accretion. Also, the rotational velocities in ADAFs are quite large, reaching nearly virial velocities, whereas Ω is strictly zero in the Bondi solution.

In nature, when gas accretes onto a black hole, it nearly always has too much angular momentum to fall directly into the hole. ADAFs are, therefore, much more relevant than the Bondi solution for the description of accretion flows around black holes.

3. Two–Temperature ADAFs

As mentioned in §1, two types of ADAFs are known, a high–$\dot{M}$ optically thick solution and a low–$\dot{M}$ optically–thin, two–temperature solution. The high–$\dot{M}$ ADAF, though well–understood dynamically, has seen few applications to observations. In particular, the spectral properties of this solution have not been studied in detail (see, however, Szuszkiewicz et al. 1996 for a recent study). In contrast, the low–$\dot{M}$ ADAF has been extensively studied during the last few years and has been widely applied to black hole

X–ray binaries and low–luminosity galactic nuclei. The rest of this article is devoted to this solution.

The two-temperature ADAF solution (though not by this name) was first described in a remarkable paper by Ichimaru (1977). This paper clearly distinguished the ADAF solution from the SLE solution and argued (without proof) that the ADAF solution would be stable. It also suggested that the black hole X-ray binary Cyg X-1, in its so called low state (the hardest spectrum in Fig. 14), corresponds to an ADAF. Unfortunately, this paper was soon forgotten; in fact, its very existence was unknown to later generations of disk theorists until the paper was rediscovered and brought to the attention of the community by Roland Svensson at the present workshop.

Elements of the two-temperature ADAF were described independently by Rees et al. (1982) in their "ion torus" model. This paper, however, though not forgotten, did not have a strong influence on later workers. Indeed, all through the 1980's and the early half of the 1990's, the SLE solution was studied in greater detail than the ion torus model. (Even Svensson did most of his work on the SLE solution!) It appears that most scientists in the community did not recognize that the SLE solution and the ion torus model were distinct, and even those that did were not fully aware of their different stability properties.

The current interest in the two-temperature ADAF solution was initiated by the papers of Narayan & Yi (1994, 1995a, 1995b), Abramowicz et al. (1995), Chen (1995), Chen et al. (1995), and a host of others. These workers discovered the solution for the third time, but this time developed it in considerable detail, studied its properties, and applied it widely to a number of systems.

3.1. Basic Assumptions

Models based on the two–temperature ADAF solution make certain critical assumptions. The validity of these assumptions is not proved and is currently under investigation.

3.1.1. Equipartition Magnetic Fields

It is assumed that magnetic fields contribute a constant fraction $(1 - \beta)$ of the total pressure:

$$p_{\rm m} = \frac{B^2}{24\pi} = (1 - \beta)\rho c_s^2, \tag{3.14}$$

where $p_{\rm m}$ is the magnetic pressure due to an isotropically tangled magnetic field. Note that the usual β of plasma physics is related to the β utilized here by $\beta_{\rm plasma} = \beta/3(1-\beta)$ (the 1/3 arises because the plasma β uses $B^2/8\pi$ for the magnetic pressure, rather than $B^2/24\pi$). The assumption of a constant β is fairly innocuous since, in general, we expect equipartition magnetic fields in most astrophysical plasmas. In particular, Balbus & Hawley (1991) have shown that differentially rotating disks with weak magnetic fields develop a strong linear MHD instability which exponentially increases the field strength to near equipartition values. The exact saturation value of $p_{\rm m}$ is, however, unclear (e.g., Hawley, Gammie, & Balbus 1996). ADAF models assume $\beta = 0.5$ (or $\beta_{\rm plasma} = 1/3$), i.e., strict equipartition between gas and (tangled) magnetic pressure. The presence of equipartition magnetic fields implies that the effective adiabatic index in the energy equation is not that of a monatomic ideal gas. Esin (1997) argues that the appropriate expression is $\gamma = (8 - 3\beta)/(6 - 3\beta)$.

3.1.2. Thermal Coupling Between Ions and Electrons

ADAF models assume that ions and electrons interact only through Coulomb collisions and that there is no non–thermal coupling between the two species. In this case the plasma is two temperature, with the ions much hotter than the electrons. This important assumption may be questioned on the grounds that magnetized collisionless plasmas, such as ADAFs, have many modes of interaction; intuitively, it would seem that the plasma might be able to find a more efficient way than Coulomb collisions of exchanging energy between the ions and the electrons (Phinney 1981). To date, however, only one potential mechanism has been identified. Begelman & Chiueh (1989) show that, under appropriate conditions, large ion drift velocities (due to significant levels of small scale turbulence in the accretion flow) may drive certain plasma waves unstable, which then transfer energy directly from the ions to the electrons. Narayan & Yi (1995b), however, argue that for most situations of interest, the specific mechanism identified by Begelman & Chiueh (1989) is not important for ADAF models.

3.1.3. *Preferential Heating of Ions*

The two-temperature ADAF model assumes that most of the turbulent viscous energy goes into the ions (SLE; Ichimaru 1977; Rees et al. 1982; Narayan & Yi 1995b), and that only a small fraction $\delta \ll 1$ goes into the electrons. The parameter δ is generally set to $\sim 10^{-3} \sim m_e/m_p$, but none of the results depend critically on the actual value of δ, so long as it is less than a few percent. Recently, there have been several theoretical investigations which consider the question of particle heating in ADAFs.

Bisnovatyi–Kogan & Lovelace (1997) argue that large electric fields parallel to the local magnetic field preferentially accelerate electrons (by virtue of their smaller mass), leading to $\delta \sim 1$. In magnetohydrodynamics (MHD), however, an electric field parallel to the magnetic field arises only from finite resistivity effects and is of order $\sim vB/c\mathrm{Re_m}$, where $\mathrm{Re_m} \gg 1$ is the magnetic Reynolds number. Local parallel electric fields are therefore negligible. Bisnovatyi–Kogan & Lovelace (1997) suggest that, instead of the microscopic resistivity, one must use the macroscopic or turbulent resistivity to determine the local electric field, in which case parallel electric fields are large, $\sim vB/c$. The problem with this analysis, as discussed by Blackman (1998) and Quataert (1998), is that the turbulent resistivity is defined only for the large scale (mean) fields. On small scales, the MHD result mentioned above is applicable and so parallel electric fields are unimportant for accelerating particles.

Blackman (1998), Gruzinov (1998), and Quataert (1998) consider particle heating by MHD turbulence in ADAFs. Blackman shows that Fermi acceleration by large scale magnetic fluctuations associated with MHD turbulence preferentially heats the ions in two temperature, gas pressure dominated, plasmas. This is, however, equivalent to considering the collisionless damping of the fast mode component of MHD turbulence (Achterberg 1981), and thus does not apply to the non-compressive, i.e., Alfvenic, component (which is likely to be energetically as or more important than the compressive component).

Following the work of Goldreich & Sridhar (1995) on Alfvenic turbulence, Gruzinov (1998) and Quataert (1998) show that, when the magnetic field is relatively weak, the Alfvenic component of MHD turbulence is dissipated on length scales of order the proton Larmor radius. The damping is primarily by "magnetic" Landau damping (also known as transit time damping, Cherenkov damping, etc.), not the cyclotron resonance (which is unimportant), and most of the Alfvenic energy heats the ions rather than the electrons. For equipartition magnetic fields, Gruzinov argues that Alfvenic turbulence will cascade to length scales much smaller than the proton Larmor radius and heat the electrons. An extension of this work is given in Quataert & Gruzinov (1998).

3.1.4. α *Viscosity*

The viscosity parameter α of Shakura & Sunyaev (1973) is used to described angular momentum transport; α is assumed to be constant, independent of radius. Some authors have proposed that α may vary as a function of (H/R). Since ADAFs have $H \sim R$, no radial dependence is expected, and a constant α appears to be a particularly good assumption (Narayan 1996b). Since the viscous stress is probably caused by magnetic fields, the parameters α and β are likely to be related according to the prescription of Hawley, Gammie & Balbus (1996): $\alpha = 3c(1 - \beta)/(3 - 2\beta)$, where $c \sim 0.5 - 0.6$. This expression for $\alpha(\beta)$ differs from the expression given elsewhere in the literature (e.g., Narayan et al. 1998), $\alpha = c(1 - \beta)$, in properly recognizing that Hawley et al. define magnetic pressure as $B^2/8\pi$, while we use equation (3.14). For equipartition magnetic fields ($\beta = 0.5$), $\alpha \sim 0.4$. The models in the literature usually set α to 0.25 or 0.3.

3.2. *Properties of Two–Temperature, Optically Thin, ADAFs*

For the remainder of this review, we write all quantities in scaled units: the mass is scaled in solar mass units,

$$M = mM_\odot,$$

the radius in Schwarzschild radii,

$$R = rR_S, \qquad R_S = \frac{2GM}{c^2} = 2.95 \times 10^5 m \quad \text{cm},$$

and the accretion rate in Eddington units,

$$\dot{M} = \dot{m}\,\dot{M}_{\text{Edd}}, \qquad \dot{M}_{\text{Edd}} = \frac{L_{\text{Edd}}}{\eta_{\text{eff}} c^2} = 1.39 \times 10^{18} m \quad \text{g s}^{-1},$$

where we have set η_{eff}, the efficiency of converting matter to radiation, equal to 0.1 in the definition of $\dot{M}_{\text{Edd}}$ (cf. Frank et al. 1992).

3.2.1. *Scaling Laws*

Using the self–similar solution (2.8)–(2.10), one can obtain a fairly good idea of the scalings of various quantities in an ADAF as a function of the model parameters. Setting $f \to 1$ (advection–dominated flow) and $\beta = 0.5$ (equipartition magnetic field), one finds (Narayan & Yi 1995b; see also Mahadevan 1997),

$$v \simeq -1.1 \times 10^{10}\,\alpha r^{-1/2} \quad \text{cm s}^{-1},$$

$$\Omega \simeq 2.9 \times 10^4\,m^{-1}r^{-3/2} \quad \text{s},$$

$$c_s^2 \simeq 1.4 \times 10^{20}\,r^{-1} \quad \text{cm}^2\,\text{s}^{-2},$$

$$n_e \simeq 6.3 \times 10^{19}\,\alpha^{-1}\,m^{-1}\dot{m}\,r^{-3/2} \quad \text{cm}^{-3},$$

$$B \simeq 7.8 \times 10^8\,\alpha^{-1/2}\,m^{-1/2}\dot{m}^{1/2}\,r^{-5/4} \quad \text{G},$$

$$p \simeq 1.7 \times 10^{16}\,\alpha^{-1}\,m^{-1}\dot{m}\,r^{-5/2} \quad \text{g cm}^{-1}\,\text{s}^{-2},$$

$$q^+ \simeq 5.0 \times 10^{21}\,m^{-2}\dot{m}\,r^{-4} \quad \text{ergs cm}^{-3}\,\text{s}^{-1},$$

$$\tau_{\text{es}} \simeq 24\,\alpha^{-1}\dot{m}\,r^{-1/2}, \tag{3.15}$$

where n_e is the electron density, p is the pressure (gas plux magnetic), and τ_{es} is the electron scattering optical depth to infinity.

3.2.2. *Critical Mass Accretion Rate*

The optically thin ADAF solution exists only for $\dot{m}$ less than a critical value $\dot{m}_{\rm crit}$ (Ichimaru 1977; Rees et al. 1982; Narayan & Yi 1995b; Abramowicz et al. 1995; Esin et al. 1996, 1997). To derive this result, consider first an optically thin one–temperature gas which cools primarily by free–free emission (Abramowicz et al. 1995). This is a reasonable model at large radii, $r > 10^3$. The viscous heating rate varies as $q^+ \propto m^{-2}\dot{m}r^{-4}$ (eq. 3.15), while the cooling varies as $q^- \propto n_p n_e T_e^{1/2} \propto \alpha^{-2}m^{-2}\dot{m}^2 T_e^{1/2} r^{-3}$, where n_p is the proton number density. For a one–temperature ADAF, $T_e \sim m_p c_s^2/k \sim 10^{12}K/r$, and is independent of $\dot{m}$. By comparing q^+ and q^-, it is easily seen that there is an $\dot{m}_{\rm crit} \propto \alpha^2 r^{-1/2}$, such that for $\dot{m} < \dot{m}_{\rm crit}$, we have $q^+ > q^-$ and a consistent ADAF solution, whereas for $\dot{m} > \dot{m}_{\rm crit}$ no ADAF is possible. Note that $\dot{m}_{\rm crit}$ decreases with increasing r. The decrease is as $r^{-1/2}$ if the cooling is dominated by free–free emission. If other atomic cooling processes are also included, the decrease is more rapid (cf. Fig. 8).

The above argument assumes free–free cooling, which is the dominant cooling mechanism at non–relativistic temperatures. However, once electrons become relativistic, other cooling processes such as synchrotron radiation and inverse Compton scattering take over. Esin et al. (1996) showed that even in this situation it is possible to have a one-temperature ADAF at sufficiently low $\dot{m}$. However, the critical accretion rate $\dot{m}_{\rm crit}$ is very small, $\sim 10^{-6}$ (for $\alpha \approx 0.25$; Esin et al. found $\dot{m}_{\rm crit} \sim 10^{-4}\alpha^2$, but more detailed calculations give a somewhat smaller value). Models with such low mass accretion rates are not of much interest for modeling observed systems.

If we allow a two–temperature plasma, then the ADAF solution can, utilizing the assumptions given in the previous subsections, extend up to larger and more interesting mass accretion rates (Ichimaru 1977; Rees et al. 1982; Narayan & Yi 1995b). By assumption, the viscous energy primarily heats the protons, while the cooling is almost entirely by the electrons. At low densities Coulomb coupling between protons and electrons is very weak and the amount of viscous energy that is transferred to the electrons is very small. Coulomb coupling therefore acts as a bottleneck which restricts the amount of energy that can be lost to radiation.

With increasing $\dot{m}$, Coulomb coupling becomes more efficient, and at a critical density the coupling is so efficient that a large fraction of the viscous energy is transferred to the electrons and is radiated. Above this accretion rate, the flow ceases to be an ADAF; it becomes a standard cooling–dominated thin disk. The critical accretion rate can be estimated by determining the $\dot{m}$ at which the viscous heating, q^+, equals the rate of energy transfer from the ions to the electrons, $q^{\rm ie}$. Alternatively, we can set the ion–electron equilibration time (the time for collisions to force $T_i \approx T_e$), $t_{\rm ie}$, equal to the accretion time, $t_{\rm a}$. The former time scale is given by (Spitzer 1962)

$$t_{\rm ie} = \frac{(2\pi)^{1/2}}{2\,n_e \sigma_T c \ln \Lambda} \left(\frac{m_p}{m_e} \right) (\theta_e + \theta_p)^{3/2},$$

$$\simeq 9.3 \times 10^{-5}\alpha\,\theta_e^{3/2}\,m\,\dot{m}^{-1}\,r^{3/2} \quad {\rm s},$$

where $\ln \Lambda \sim 20$ is the Coulomb logarithm, $\theta_p = kT_p/m_p c^2$, and $\theta_e = kT_e/m_e c^2$. The accretion time is given by

$$t_{\rm a} = \int \frac{dR}{v(R)} \simeq 1.8 \times 10^{-5}\,\alpha^{-1}\,m\,r^{3/2} \quad {\rm s}. \tag{3.16}$$

Setting these two timescales equal gives

$$\dot{m}_{\rm crit} \simeq 5\theta_e^{3/2}\alpha^2 \simeq 0.3\alpha^2, \tag{3.17}$$

where we have used $\theta_e \sim 0.16$, corresponding to $T_e = 10^9$K (Mahadevan 1997). More detailed models (cf. Esin et al. 1997) give $\dot{m}_{\rm crit} \sim \alpha^2$. This value of $\dot{m}_{\rm crit}$ is essentially independent of r out to about $10^2 - 10^3$ Schwarzschild radii. Beyond that, the accreting gas becomes one-temperature and $\dot{m}_{\rm crit}$ decreases with increasing r, as explained above. Figure (8) shows a detailed estimate of the profile of $\dot{m}_{\rm crit}$ vs r. In this review, we will refer to the full profile as $\dot{m}_{\rm crit}(r)$ and refer to $\dot{m}_{\rm crit}(r_{ms})$ as simply $\dot{m}_{\rm crit}$, where r_{ms} is the marginally stable orbit ($r_{ms} = 3$). Thus, $\dot{m}_{\rm crit}$ refers to the maximum $\dot{m}$ up to which an ADAF zone of any size is allowed.

As we discuss later, observations suggest that the two–temperature ADAF solution exists up to $\dot{m}_{\rm crit} \sim 0.05 - 0.1$. This suggests that $\alpha \sim 0.2 - 0.3$ in ADAFs.

3.2.3. *Temperature Profiles of Ions and Electrons*

In a two–temperature ADAF the ions receive most of the viscous energy and are nearly virial (Narayan & Yi 1995b),

$$T_i \simeq 2 \times 10^{12} \beta r^{-1}.$$

The electrons, on the other hand, are heated by several processes with varying efficiencies — Coulomb coupling with the ions, compression, and direct viscous heating (described by the parameter δ) — and cooled by a variety of radiation processes.

To determine the electron temperature profile, consider the electron energy equation (Nakamura et al. 1996, 1997; Mahadevan & Quataert 1997),

$$\rho T_e v \frac{ds}{dR} = \rho v \frac{d\epsilon}{dR} - q^c = q^{ie} + q^v - q^-, \qquad q^c \equiv kTv \frac{dn}{dR}, \qquad (3.18)$$

where s and ϵ are the entropy and internal energy of the electrons per unit mass of the gas, and q^c and q^- are the compressive heating (or cooling) rate and the energy loss due to radiative cooling per unit volume. The total heating rate of the electrons is the sum of the heating via Coulomb collisions with the hotter protons, q^{ie}, and direct viscous heating, $q^v = \delta q^+$. (Recall that δ is usually assumed to be $\sim 10^{-3}$ in ADAF models.)

The electron temperature at radii $r \lesssim 10^2$ in two–temperature ADAFs is generally in the range $10^9 - 10^{10}$K. The exact form of the electron temperature profile depends on which of the three heating terms dominates. Nakamura et al. (1996, 1997) were the first to emphasize the compressve heating term, q^c. Mahadevan & Quataert (1997) showed that compressive heating of the electrons is more important than direct viscous heating so long as $\delta \lesssim 10^{-2}$. Since $\delta \sim 10^{-3}$ in most models, we can ignore direct viscous heating and consider only the following two limits.

$q^{ie} \gg q^c$: This condition is satisfied for $\dot{m} \gtrsim 0.1\alpha^2$. Rewriting equation (3.18) in this regime gives

$$\rho v \frac{d\epsilon}{dR} \simeq q^{ie} + q^c - q^- \simeq q^{ie} - q^-.$$

Since the cooling is efficient enough to radiate away all of the energy given to the electrons, $q^{ie} = q^-$ for $r \lesssim 10^2$, and the internal energy of the electrons does not change with radius, $d\epsilon/dR = 0$. The electron temperature therefore remains essentially constant at $\sim 10^9$K (e.g. Narayan & Yi 1995b). The value of T_e depends weaksly on $\dot{m}$ (Fig. 4).

$q^c \gg q^{ie}$: This condition is satisfied for $\dot{m} \lesssim 10^{-4}\alpha^2$. Rewriting equation (3.18) in this regime gives

$$\rho v \frac{d\epsilon}{dR} \simeq q^c + q^{ie} - q^- \simeq q^c.$$

The radial dependence of the electron temperature is determined by adiabatic compression (Nakamura et al. 1997; Mahadevan & Quataert 1997). The electron temperatures in this regime are slightly higher.

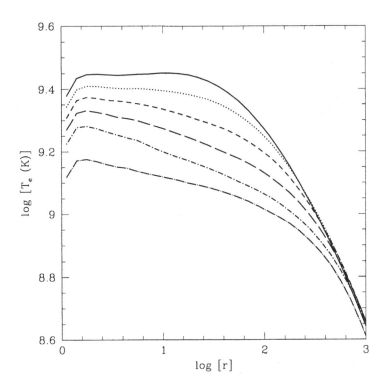

FIGURE 4. Variation of electron temperature T_e with radius r for ADAFs with (from top to bottom) $\log(\dot{m}) = -2, -1.8, -1.6, -1.4, -1.2, -1.1$ (taken from Esin et al. 1997). Note that the electron temperature decreases with increasing $\dot{m}$.

For $10^{-4}\alpha^2 \leq \dot{m} \leq 0.1\alpha^2$, the electron temperature profile lies in between these two extremes, and is determined by solving equation (3.18) without neglecting any of the heating terms.

Figure 4 shows the temperature profile as a function of radius for various $\dot{m}$. The electrons actually become cooler as the accretion rate increases. Since with increasing $\dot{m}$ the energy transferred to the electrons goes up, one might naively expect the electrons to reach higher temperatures in order to radiate the additional energy. In fact, the opposite happens; at high $\dot{m}$, the dominant cooling mechanism is inverse Compton scattering. This is an extremely efficient process when the optical depth approaches unity and its efficiency increases sharply with increasing optical depth, i.e., increasing $\dot{m}$.

Although the electrons achieve relativistic temperatures, pair processes are found to be unimportant in two-temperature ADAFs (Kusunose & Mineshige 1996, Björnsson et al. 1996, Esin et al. 1997). The reason is the low density, which allows very few pair-producing interactions in the medium.

3.2.4. *ADAF Spectra and Radiation Processes*

The spectrum from an ADAF around a black hole ranges from radio frequencies $\sim 10^9$Hz to gamma–ray frequencies $\gtrsim 10^{23}$Hz, and can be divided into two parts based on the emitting particles: (1) The radio to hard X–ray radiation is produced by electrons via synchrotron, bremsstrahlung and inverse Compton processes (Mahadevan 1997). (2) The gamma–ray radiation results from the decay of neutral pions created in proton–

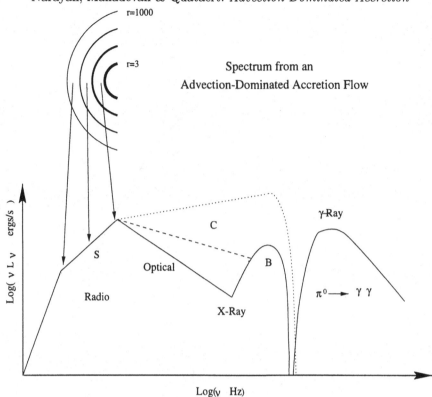

FIGURE 5. Schematic spectrum of an ADAF around a black hole. S, C, and B refer to electron emission by synchrotron radiation, inverse Compton scattering, and bremmstrahlung, respectively. The solid line corresponds to a low $\dot{m}$, the dashed line to an intermediate $\dot{m}$, and the dotted line to a high $\dot{m} \sim \dot{m}_{\mathrm{crit}}$. The γ–ray spectrum is due to the decay of neutral pions created in proton-proton collisions.

proton collisions (Mahadevan, Narayan & Krolik 1997). Figure 5 shows schematically the various elements in the spectrum of an ADAF around a black hole.

The low energy end of the spectrum, labeled S in the figure, is due to synchrotron cooling by semi–relativistic thermal electrons (Mahadevan et al. 1996). The synchrotron emission is highly self–absorbed and is very sensitive to the electron temperature ($\nu L_\nu \propto T_e^7$; Mahadevan 1997). The emission at the highest (peak) frequency comes from near the black hole, while that at lower frequencies comes from further out. The peak frequency varies with the mass of the black hole and the accretion rate, roughly as $\nu_{\mathrm{peak}}^S \propto m^{-1/2} \dot{m}^{1/2}$ (Mahadevan 1997; see figure 6).

The soft synchrotron photons inverse Compton scatter off the hot electrons in the ADAF and produce harder radiation extending up to about the electron temperature ~ 100 keV ($h\nu_{\mathrm{max}}^C \approx kT_e$). The relative importance of this process depends on the mass accretion rate. At high $\dot{m}$, when the Compton y-parameter is large (because of the increased optical depth), the inverse Compton component dominates the spectrum, as shown by the dotted line labeled C in figure 5.

As $\dot{m}$ decreases, Comptonization becomes less efficient and the inverse Compton component of the spectrum becomes softer and less important (dashed and solid lines labeled C in figure 5). At low $\dot{m}$, the X-ray spectrum is dominated by bremsstrahlung emission, which again cuts off at the electron temperature ($h\nu_{\mathrm{max}}^B \approx kT_e$, the curve labeled B in figure 5).

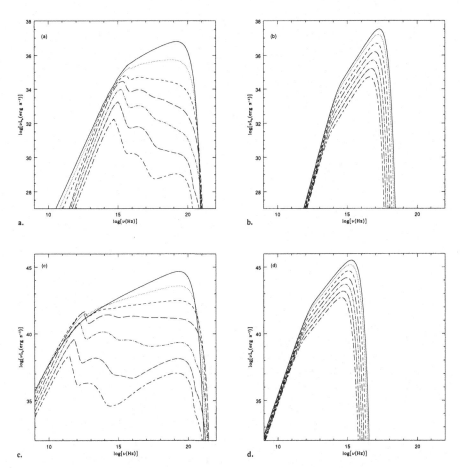

FIGURE 6. (a) Spectra from an ADAF around a 10 solar mass black hole for (from top to bottom) $\log(\dot{m}) = \log(\dot{m}_{\mathrm{crit}}) \approx -1.1, -1.5, -2, -2.5, -3, -3.5, -4$. (b) Spectra from a thin disk at the same accretion rates. Figures (c) and (d) show the corresponding spectra for a 10^9 solar mass black hole. Note that these spectra are for pure disk and pure ADAF models. In practice, real systems are often modeled as an ADAF surrounded by a thin disk (see §3.3). In such composite models, the ADAF part of the spectrum is essentially unchanged, but a dimmer and softer version of the thin disk is also present in the spectrum.

Mahadevan et al. (1997) have studied gamma-ray emission from an ADAF via the decay of neutral pions produced in proton-proton collisions. The results depend sensitively on the energy spectrum of the protons. If the protons have a thermal distribution, the gamma-ray spectrum is sharply peaked at ~ 70 MeV, half the rest mass of the pion, and the luminosity is not very high. If the protons have a power–law distribution, the gamma-ray spectrum is a power–law extending to very high energies (see figure 5), and the luminosity is much higher; the photon index of the spectrum is equal to the power–law index of the proton distribution function.

Figure 6 shows four sequences of spectra, two corresponding to ADAFs (excluding the gamma-ray component) and two to thin disks. Figures (6a) and (6b) are for a 10 solar mass black hole while Figures (6c) and (6d) are for a 10^9 solar mass black hole. All sequences extend over the same range of $\dot{m} = 10^{-4} - 10^{-1.1} \approx \dot{m}_{\mathrm{crit}}$. Note how very different the thin disk and ADAF spectra are, suggesting that it should be easy to tell from spectral observations whether a system has an ADAF or a thin disk.

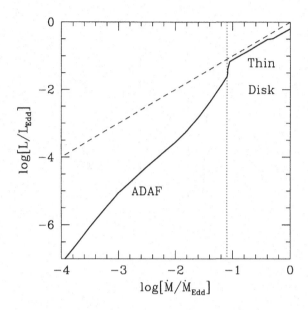

FIGURE 7. The bolometric luminosity vs. mass accretion rate according to the model developed by Esin et al. (1997). The vertical dotted line corresponds to $\dot{m}_{\rm crit}$ (for $\alpha = 0.3$). Above this $\dot{m}$, the accretion is via a thin disk and $L \propto \dot{M}$. Below $\dot{m}_{\rm crit}$, the accretion is via an ADAF at small radii and a thin disk at large radii (cf. §3.3). Here $L \propto \dot{M}^2$ because much of the viscously generated energy is advected into the black hole. The dashed line corresponds to $L = 0.1\dot{M}c^2$.

Figure 6 shows that for high $\dot{m}$, the Compton component of an ADAF spectrum is roughly a power law while for lower $\dot{m}$ distinct Compton peaks are present. This is primarily due to the increase of the electron temperature with decreasing $\dot{m}$ (see fig. 4). A larger T_e yields a larger energy boost per Compton scattering. At low $\dot{m}$, the energy gain due to Compton scattering exceeds the width (in energy) of the input synchrotron photons, resulting in distinct Compton peaks. For increasing $\dot{m}$, however, T_e, and thus the mean energy gain due to Compton scattering, decreases. The Compton peaks therefore "blur" together, yielding an effective power law.

Another striking feature seen in Fig. 6 is that ADAFs are much less luminous than thin disks at low values of $\dot{m}$. This is because most of the energy in an ADAF is advected, rather than radiated, leading to a low radiative efficiency. In fact, the luminosity of an ADAF scales roughly as $\sim \dot{m}^2$ (i.e., the radiative efficiency scales as $\dot{m}$). A thin disk, on the other hand, has a constant efficiency $\sim 10\%$ and the luminosity scales as $\dot{m}$. This is illustrated in Fig. 7 which shows results of a detailed model described by Esin et al. (1997).

3.2.5. *The Particle Distribution Function in ADAFs*

In determining the radiation processes in ADAFs, the electrons are assumed to be thermal, while the protons could be thermal or non–thermal. It is clear that the spectrum will depend significantly on the energy distribution of the particles.

Mahadevan & Quataert (1997) considered two possible thermalization processes in ADAFs: (1) Coulomb collisions and (2) synchrotron self–absorption. Using an analytic Fokker–Planck treatment, they compared the time scales for these processes with the accretion time, and determined the accretion rates at which thermalization is possible. In the case of the protons they found that, for all accretion rates of interest, neither Coulomb collisions nor synchrotron self-absorption leads to any significant thermalization. The proton distribution function is therefore determined principally by the characteristics of the viscous heating mechanism, and could therefore be thermal or non–thermal. Quataert (1998) has argued that Alfvenic turbulence does *not* lead to strong non-thermal features in the proton distribution function; fast mode turbulence (Fermi acceleration), however, may generate a power-law tail.

The electrons exchange energy quite efficiently by Coulomb collisions for $\dot{m} \gtrsim 10^{-2}\alpha^2$, and are therefore thermal at these accretion rates. At lower $\dot{m}$ the emission and absorption of synchrotron photons allows the electrons to communicate with one another, and therefore to thermalize, even though the plasma is effectively collisionless (Ghisellini & Svensson 1990, 1991). At a radius r in the accretion flow, this leads to thermalization for $\dot{m} \gtrsim 10^{-5}\alpha^2 r$ (Mahadevan & Quataert 1997). For still lower $\dot{m}$, the electron distribution function is somewhat indeterminate, and is strongly influenced by adiabatic compression (Mahadevan & Quataert 1997). Detailed spectra of ADAFs at such low $\dot{m}$, which take into account the nonthermal distribution function of the electrons, have not been calculated (see, however, Fujita et al. (1998) for preliminary models of isolated black holes accreting at very low $\dot{m}$ from the interstellar medium).

3.2.6. *Stability of the ADAF Solution*

Thin accretion disks suffer from thermal and viscous instabilities under certain conditions (e.g., Pringle 1981, Frank et al. 1992). What are the stability properties of ADAFs?

Narayan & Yi (1995b) and Abramowicz et al. (1995) have shown that ADAFs are stable to long wavelength perturbations (see Narayan 1997 for a qualitative discussion of the relevant physics), while Kato et al. (1996, 1997) and Wu & Li (1996) showed that a one temperature ADAF may be marginally unstable to small scale perturbations. Using a time–dependent analysis, Manmoto et al. (1996) confirmed that density perturbations on small scales in a one-temperature ADAF do grow as the gas flows in, but not sufficiently quickly to affect the global validity of the solutions. They suggest that such perturbations may account for the variable hard X–ray emission which is observed in AGN and stellar mass black hole candidates. Recently, Wu (1997) has considered the stability of two temperature ADAFs to small scale fluctuations, and has shown that these flows are both viscously and thermally stable under most reasonable conditions.

3.3. *ADAF Plus Thin Disk Geometry: The Transition Radius*

Section 4 describes several applications of the ADAF model to X-ray binaries and AGN. Many of these applications utilize the following geometry, proposed by Narayan, McClintock, & Yi (1996). In this model, the accretion flow consists of two zones separated at a transition radius, r_{tr}. For $r < r_{tr}$, there is a two-temperature ADAF, whose properties are described in the previous two subsections. For $r > r_{tr}$, the accretion occurs partially as a thin accretion disk, and partially as a hot corona; the corona is modeled as an ADAF (Narayan, Barret, & McClintock 1997; Esin, McClintock, & Narayan 1997). The geometry is very similar to that proposed by SLE (see also Wandel & Liang 1991); the main difference is that the hot phase is taken to be a two-temperature ADAF, rather than the SLE solution.

Several proposals have been made to explain why the inflowing gas might switch from a thin disk to an ADAF at the transition radius. Meyer & Meyer–Hofmeister (1994) proposed, for cataclysmic variables, a mechanism in which the disk is heated by electron conduction from a hot corona; this evaporates the disk, leading to a quasi-spherical hot

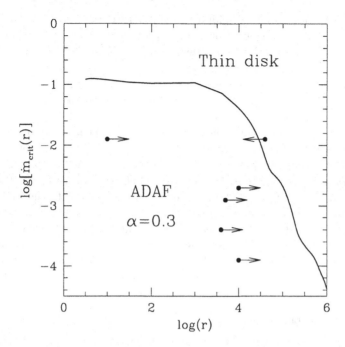

FIGURE 8. The solid line shows the estimated variation of $\dot{m}_{\rm crit}(r)$ with r. The horizontal segment up to $r = 10^3$ is from Esin et al. (1997) and the curve beyond that is based on a detailed model of optically thin cooling in a one-temperature gas (Menou et al. 1998). According to the "strong ADAF proposal" of Narayan & Yi (1995b), the curve also shows the variation of r_{tr} with $\dot{m}$. Regions above and to the right of the curve correspond to the thin disk solution, while regions below and to the left correspond to the ADAF. The dots and arrows show observationally derived limits on r_{tr}. The lower limits on r_{tr} are derived from fitting the spectra of various systems (from above: NGC 4258, V404 Cyg, GRO J1655-40, A0620-00, and Sgr A*), and also requiring the thin disk to be thermally stable in the thermal limit cycle model. The upper limit on r_{tr} in NGC4258 corresponds to the lowest radius at which maser emission has been detected. The calculated curve is consistent with the various constraints shown, but it is clear that fairly large changes are allowed, especially for $\dot{m} > 10^{-2.5}$.

accretion flow. Honma (1996) suggested that turbulent diffusive heat transport from the inner regions of the ADAF produces a stable hot accretion flow out to large radii, which then joins to a cool thin disk. Narayan & Yi (1995b) suggested that small thermal instabilities in the optically-thin upper layers of a thin disk might cause the disk to switch to an ADAF. Other ideas are discussed by Ichimaru (1977) and Igumenshchev, Abramowicz, & Novikov (1997). It is not clear which, if any, of these mechanisms is most important.

It is generally believed that the transition radius r_{tr} is determined principally by $\dot{m}$, but the exact form of $r_{tr}(\dot{m})$ is not known. Narayan & Yi (1995b) suggested that, whenever the accreting gas has a choice between a thin disk and an ADAF, the ADAF configuration is chosen. According to this (rather extreme) principle (the "strong ADAF principle"), $r_{tr}(\dot{m})$ is the maximum r out to which an ADAF is allowed for the given $\dot{m}$; equivalently, it is that r at which $\dot{m}_{\rm crit}(r) = \dot{m}$ (see Fig. 8). This prescription suggests that at low $\dot{m} \ll \dot{m}_{\rm crit} \sim \alpha^2$, r_{tr} will be very large even though a thin disk is perfectly viable at all radii. The evidence to date from quiescent X-ray binaries and galactic nuclei is consistent with this prediction (§4 and Fig. 8).

Even if the "strong ADAF proposal" of Narayan & Yi (1995b) is correct, it does not allow a reliable determination of $r_{tr}(\dot{m})$ because the precise form of $\dot{m}_{\mathrm{crit}}(r)$ is not known. The plot shown in Fig. 8 is based on fairly detailed computations by Esin et al. (1997) and Menou et al. (1998), but it still makes use of simplifying assumptions that may significantly influence the results.

4. Applications

ADAF models have been applied to a number of accreting black hole systems. They give a satisfying description of the spectral characteristics of several quiescent black hole binaries (Narayan et al. 1996, 1997b; Hameury et al. 1997) and low luminosity galactic nuclei (Narayan et al. 1995, 1998; Manmoto et al. 1997; Lasota et al. 1996; Reynolds et al. 1996; DiMatteo & Fabian 1997a) which are known to experience low efficiency accretion. ADAF models have also been applied successfully to more luminous systems which have higher radiative efficiencies (Esin et al. 1997, 1998).

The basic ADAF model has one adjustable parameter, $\dot{m}$; in principle, the transition radius r_{tr} is a second parameter, but most often the results are very insensitive to the choice of r_{tr}. In the future, the Kerr rotation parameter of the black hole will be an additional parameter of the models, but current models assume a Schwarzschild black hole. The mass of the black hole and inclination of the equatorial plane to the line of sight are estimated from observations, while the parameters describing the microphysics of the accretion flow are set to their canonical values (§3.1): $\alpha = 0.25$ or 0.30, $\beta = 0.5$, $\delta = 0.001$. (Any value of δ between 0 and 0.01 gives virtually identical results; see §3.2.3).

The X-ray flux is very sensitive to the density of the plasma and therefore to the accretion rate. For this reason, $\dot{m}$ is usually adjusted to fit the observed X-ray flux. The models described below are generally in good agreement with the remaining data that are not used in the fit: namely, the X-ray spectral slope, the optical/UV data, and, where available, the radio and infrared observations. In contrast, the thin disk model fits the observations very poorly.

The models presented here are the most up to date available, often more advanced than those in the literature; in particular, they include the compressive heating of electrons in the electron energy equation (see §3.2.3) and the full relativistic dynamics for accretion onto a Schwarzschild black hole (the photon transport includes gravitational redshift, but is otherwise Newtonian).

4.1. *Applications to X-Ray Binaries*

4.1.1. *Quiescent Black Hole Transients*

ADAF models have been used to explain the spectra of a number of quiescent black hole soft X-ray transients (SXTs), namely, A0620-00, V404 Cyg, and GRO J1655-40 (Narayan et al. 1996; Narayan et al. 1997b; Hameury et al. 1997; Lasota this volume). SXTs are mass transfer binaries in which the accreting star is often a black hole candidate (though sometimes a neutron star), and the companion star is usually a low-mass main sequence star (van Paradijs & McClintock 1995). Episodically, these systems enter a high luminosity, "outburst," phase, but for most of the time they remain in a very low luminosity, "quiescent," phase. One problem with modeling quiescent SXTs is that a thin accretion disk cannot account for both the observed optical/UV and X-ray flux self-consistently. For example, in the case of A0620-00, a standard thin disk requires an accretion rate of $\dot{M} \sim 10^{-10} M_{\odot} \mathrm{yr}^{-1}$ to explain the optical/UV flux, while the X-ray flux requires an accretion rate of $\dot{M} \sim 10^{-15} M_{\odot} \mathrm{yr}^{-1}$ (McClintock et al. 1995). Another problem is that the optical spectrum resembles a blackbody with a temperature

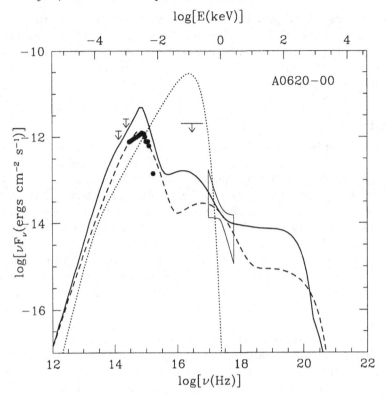

FIGURE 9. Spectrum of an ADAF model of A0620-00 (solid line) at an accretion rate of $\dot{m} = 4 \times 10^{-4}$, compared with the observational data. The dashed line is an ADAF model with $\beta = 0.8$, instead of the standard value of $\beta = 0.5$. The dotted line shows the spectrum of a thin accretion disk with an accretion rate $\dot{m} = 1 \times 10^{-5}$, adjusted to fit the optical flux. (Based on Narayan et al. 1997b)

of $\sim 10^4$K, but a thin disk cannot exist at such a temperature since it would be thermally unstable (Wheeler 1996; Lasota et al. 1996a).

Narayan, McClintock, and Yi (1996) and Narayan et al. (1997b) resolved these problems using an ADAF + thin disk model with $r_{tr} \sim 10^3 - 10^4$. The resulting spectra for A0620-00 and V404 Cyg are shown by solid lines in Figures 9 and 10. The ADAF models of the two sources reproduce the observed X-ray spectral slopes well; they also reproduce the optical/UV fluxes and spectral shapes reasonably well (note especially the good agreement in the position of the optical peak for A0620-00), though the optical flux is generally somewhat too luminous. Hameury et al. (1997) showed that observations of another SXT, GRO J1655-40, are also consistent with the presence of an ADAF in quiescence. The dotted lines in Figures 9 and 10 are steady state thin disk models adjusted to fit the optical flux; these models are clearly ruled out by the data.

The optical emission in the ADAF models of SXTs is due to synchrotron emission from the ADAF. The thin disk itself is very cool and is in a stable "low state." There is thus no difficulty with the thermal instability which a pure thin disk model would face in modeling the optical spectrum (Wheeler 1996; Lasota et al. 1996a). The ADAF model does overproduce the optical flux, but this can be fixed by changing the value of β from 0.5 to 0.8 (see the dashed line in figure 9).

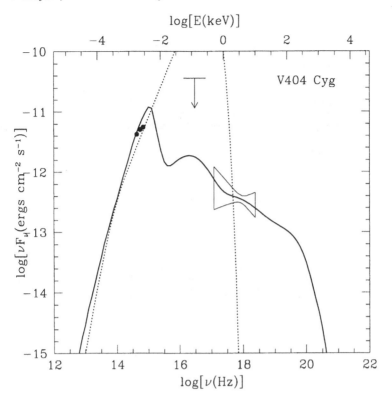

FIGURE 10. Spectrum of an ADAF model of V404 Cyg (solid line) at an accretion rate of $\dot{m} = 2 \times 10^{-3}$, compared with the observational data. The dotted line shows the spectrum of a thin accretion disk with $\dot{m} = 1.8 \times 10^{-3}$ (adjusted to fit the optical flux).

4.1.2. Clues From Outburst Timescales

The above spectral models of quiescent SXTs require an inner ADAF which connects to an outer thin disk at a large transition radius. Lasota et al. (1996a) showed that a large transition radius can also be inferred from the outburst timescales of SXTs. In addition, optical and X-ray observations of the black hole SXT GRO J1655–40 showed an outburst in April of 1996. The optical outburst preceded the X-ray outburst by roughly 6 days (Orosz et al. 1997), which can be understood using an ADAF + thin disk model (but not using a pure thin disk extending down to the last stable orbit). See Lasota (this volume) for a detailed discussion of the implications of SXT outbursts for ADAF models.

4.1.3. Spectral Models of Luminous X-Ray Binaries

Narayan (1996a) proposed that the many different spectral states observed in black hole X-ray binaries can be understood as a sequence of thin disk + ADAF models with varying $\dot{m}$ and r_{tr}. These ideas have been worked out in more detail by Esin et al. (1997; 1998). A schematic diagram of their model is shown in Figure 11.

(a) Quiescent state: This lowest luminosity state has $\dot{m} \lesssim 10^{-2}$ and is discussed above (§4.1.1). Due to the low accretion rate, Comptonization is weak, and the X-ray flux is much lower than the optical flux. The radiative efficiency is very low and the systems are extremely dim (cf. Figs. 6 and 7).

(b) Low state: For $\dot{m}$ above 10^{-2} and up to $\sim 10^{-1}$, the geometry of the accretion flow is similar to that of the quiescent state, but the luminosity and radiative efficiency are

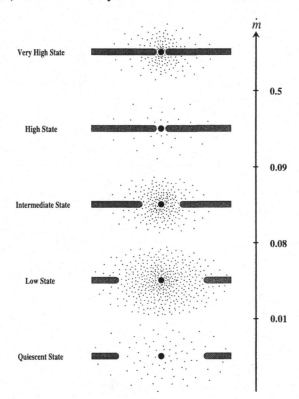

FIGURE 11. The configuration of the accretion flow in different spectral states shown schematically as a function of the total mass accretion rate $\dot{m}$ (from Esin et al. 1997). The ADAF is indicated by dots and the thin disk by the horizontal bars. The lowest horizontal panel shows the quiescent state which corresponds to a low mass accretion rate (and therefore, a low ADAF density) and a large transition radius. The next panel shows the low state, where the mass accretion rate is larger than in the quiescent state, but still below the critical value $\dot{m}_{\mathrm{crit}}$. In the intermediate state (the middle panel), $\dot{m} \sim \dot{m}_{\mathrm{crit}}$ and the transition radius is smaller than in the quiescent/low state. In the high state, the thin disk extends down to the last stable orbit and the ADAF is confined to a low-density corona above the thin disk. Finally, in the very high state, it has been suggested that the corona may have a substantially larger $\dot{m}$ than in the high state, but this is very uncertain.

larger (and increase rapidly with $\dot{m}$). A low state spectrum of GRO J0442+32 is shown in figure 12. Comptonization becomes increasingly important, giving rise to a very hard spectrum which peaks around 100 keV.

(c) Intermediate state: At still higher accretion rates, $\dot{m}$ approaches $\dot{m}_{\mathrm{crit}} \sim 0.1$, the ADAF progressively shrinks in size, the transition radius decreases, and the X-ray spectrum changes continuously from hard to soft. This occurs at roughly constant bolometric luminosity. In this state, the thin disk becomes radiatively comparable to, or even brighter than, the ADAF.

(d) High state: At still higher accretion rates, $\dot{m} > \dot{m}_{\mathrm{crit}}$, the ADAF cannot exist as an independent entity at any radius, and the thin disk comes all the way down to the last stable orbit; there is, however, a weak corona above the thin disk which is modeled as an ADAF (with a coronal $\dot{m} \lesssim \dot{m}_{\mathrm{crit}}$). A characteristic high state spectrum resembles a standard thin disk spectrum, with a power law tail due to the corona.

This model accounts convincingly for the characteristic spectral state variations, from quiescence to the high state, seen in black hole X-ray binaries.

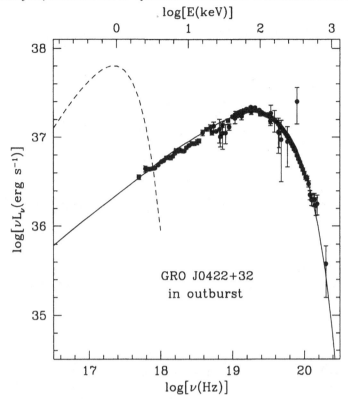

FIGURE 12. An ADAF model of J0442+32 (solid line) in the low state, compared with the observational data (dots and errorbars). The dashed line shows a thin disk model at the same accretion rate, $\dot{m} = 0.1$.

Esin et al. (1997; 1998) applied their model to the 1991 outburst of the SXT Nova Muscae and to the high-low/low-high state transition of Cyg X-1. Nova Muscae went into outburst during the fall of 1991, and was extensively studied in the optical and X-ray bands (Ebisawa et al. 1994; see also Brandt et al. 1992; Lund 1993; Goldwurm et al. 1992; Gilfanov et al. 1993). Its luminosity evolution is summarized in figure 13. The model light curves agree quite well with the data.

Figure 14 shows the broadband simultaneous RXTE (1.3-12 keV) and BATSE (20-600 keV) spectra of Cyg X-1 observed during the 1996 low-high (upper panel) and high-low (middle panel) state transitions. The bottom panel shows a sequence of ADAF + thin disk models, which are in good agreement with the observations. In particular, the model reproduces the range of photon indices seen in the data, the anti-correlation between the soft and hard X-ray flux, the "pivoting" around 10 keV, and the nearly constant bolometric luminosity throughout the transition.

One important result, highlighted by Narayan (1996), and confirmed by the more detailed calculations of Esin et al. (1997), is that the application of ADAFs to luminous black hole X-ray binaries requires a fairly large value of $\alpha \sim 0.25$. (The preliminary work of Narayan 1996 actually suggested $\alpha \sim 1$, but this was revised in Esin et al. 1997.) If α is much smaller, then the ADAF solution cuts off at a very low mass accretion rate (recall that $\dot{m}_{\rm crit} \sim \alpha^2$), and the observed luminosities of low state systems cannot be explained. As discussed in §3.1, α will likely be large if the accreting gas has magnetic fields of equipartition strength.

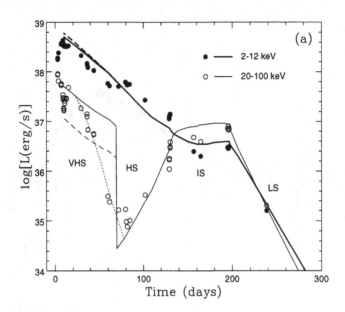

FIGURE 13. Soft and hard X-ray light curves of Nova Muscae 1991. Filled and open circles are data from Ebisawa et al. (1994) corresponding to the 2–12 keV and 20–100 keV bands, respectively. The heavy and thin lines are the model predictions (from Esin et al. 1997). The symbols VHS, HS, IS, LS correspond to the very high state, high state, intermediate state and low state, respectively. In the very high state, the solid, dotted, and dashed lines correspond to different prescriptions for viscous dissipation in the corona (see Esin et al. 1997 for details).

Some black hole X-ray binaries, such as Nova Muscae 1991, exhibit an even higher luminosity state, called the very high state, which is significantly harder than the high state. This does not fit readily into the thin disk + ADAF paradigm. Esin et al. (1997) showed that if there is enhanced viscous dissipation of energy in the corona, as suggested by Haardt & Maraschi (1991, 1993) for modeling coronae in AGN, some properties of the very high state can be understood. Esin et al.'s model of the very high state is, however, quite speculative, and does not fit the observations particularly well.

4.2. *Applications to Galactic Nuclei*

4.2.1. *Sgr A**

Dynamical measurements of stellar velocities within the central 0.1 pc of the Galactic Center indicate a dark mass with $M \sim (2.5 \pm 0.4) \times 10^6 M_\odot$ (Haller et al. 1996; Eckart & Genzel 1997). This is believed to be the mass of the supermassive black hole in Sgr A*. Observations of stellar winds indicate that the expected accretion rate is $6 \times 10^{-6} M_\odot$ yr^{-1} $\leq \dot{M} \leq 2 \times 10^{-4} M_\odot$ yr^{-1} (Genzel et al. 1994; Melia 1992), which corresponds to $10^{-4} \leq \dot{m} \leq 3 \times 10^{-3}$. Using a nominal radiative efficiency of 10% these accretion rates imply an accretion luminosity ($\sim 0.1\dot{M}c^2$) between $\sim 10^{40}$erg s^{-1} and $\sim 10^{42}$erg s^{-1}. Observations in the radio to γ–rays, however, seem to indicate a bolometric luminosity of less than 10^{37}erg s^{-1}. This extremely low luminosity has been used to argue against a supermassive black hole in Sgr A* (Mastichiadis & Ozernoy 1994; Goldwurm et al. 1994).

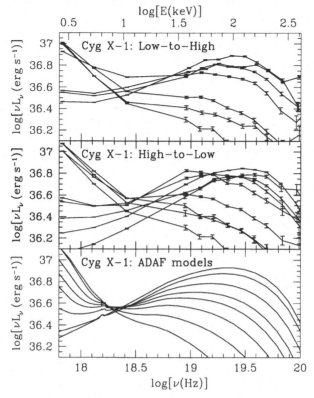

FIGURE 14. The RXTE (1.3-12 keV) and BATSE (20-600 keV) spectra of Cyg X-1 during the 1996 low-high (upper panel) and high-low (middle panel) state transitions. The bottom panel shows a sequence of ADAF + thin disk models which are in good agreement with the observations.

A standard thin disk with the above accretion rate gives rise to a black body spectrum which peaks in the near infrared (Frank et al. 1992). Menten et al. (1997), however, have obtained a strong upper limit on the infrared emission from the Galactic center, and find it to be below $\sim 10^{34}$erg s^{-1}. This effectively rules out any thin disk model of the Galactic Center.

All of the apparently contradictory observations of Sgr A* appear to be naturally accounted for by an optically thin, two temperature, ADAF model. Rees (1982) first suggested that Sgr A* may be advecting a significant amount of energy and Narayan et al. (1995) provided the first spectral model; more detailed recent models (Manmoto et al. 1997; Narayan et al. 1998) confirm that an ADAF model can explain the observations quite well. The solid line in Figure 15a shows the best fit Narayan et al. (1998) model, while the dotted lines correspond to thin disk models, which are easily ruled out by the data. The mass of the central black hole is fixed at its dynamically measured value of $2.5 \times 10^6 M_\odot$ and the accretion rate is varied to fit the X-ray flux. The resulting accretion rate is $\dot{m} \sim 1.3 \times 10^{-4}$, which is consistent with the mass accretion rate estimates from the observations of stellar winds.

The resulting model naturally reproduces the observed spectrum in other wavebands. Synchrotron radiation from the ADAF produces the radio spectrum which cuts off sharply in the sub–mm; the inverse Compton spectrum is consistent with the stringent upper limit on the NIR flux given by Menten et al. (1997); finally, bremsstrahlung radiation is responsible for the X-ray flux which extends up to a few hundred keV. There is, however,

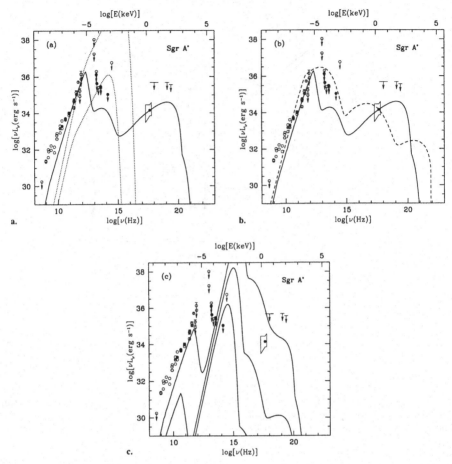

FIGURE 15. (a) Spectrum of a two temperature ADAF model of Sgr A* (solid line). The mass accretion rate inferred from this model is $\dot{m} = 1.3 \times 10^{-4}$, in agreement with an independent observational determination. Dotted lines show the spectra of thin accretion disks, at the same accretion rate (upper) and at $\dot{m} = 10^{-8}$ (lower). (b) The dashed line is a one temperature ADAF model of Sgr A* at $\dot{m} = 3 \times 10^{-7}$ (adjusted to fit the x-ray flux). The solid line is the standars two temperature model. (c) Spectra of ADAF models of Sgr A* where the central star is taken to have a hard surface at $r = 3$ and the advected energy is assumed to be reradiated from the surface as a blackbody. The three spectra correspond to $\dot{m} = 10^{-4}, 10^{-6}$, and 10^{-8} (from top to bottom). All three models violate the infrared limits.

a problem at low radio frequencies, $\lesssim 10^{10}$ Hz, where the model is well below the observed flux. In addition, the γ–ray spectrum from the ADAF is lower than the observed flux by nearly an order of magnitude (not shown in Figure 15; see Figure 1 of Narayan et al. 1998). The latter is not considered to be a serious problem since it is unclear whether the observed γ–rays are in fact from a point or a diffuse source at the Galactic Center.

Sgr A* is one of the few observed systems for which the luminosity is low enough that a one temperature ADAF model can be constructed (recall from §3.2.2 that $\dot{m}_{\rm crit} \sim 10^{-6}$ for a one temperature ADAF so that a source must have a luminosity $\lesssim 10^{-6} L_{\rm edd}$ for a one temperature ADAF model to be possible). This model (dashed line in Figure 15b) is, however, ruled out by the data.

An important feature of the two temperature ADAF model of Sgr A* is that the observed low luminosity is explained as a natural consequence of the advection of energy

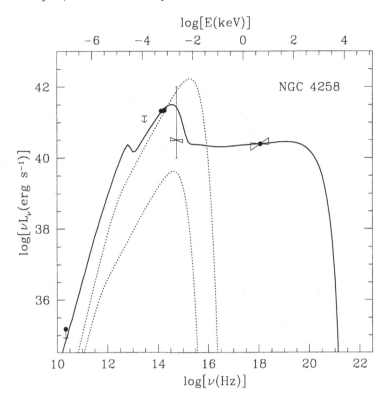

FIGURE 16. Spectrum of an ADAF model of NGC 4258 (solid line), at an accretion rate of $\dot{m} = 9 \times 10^{-3}$ with a transition radius of $r_{tr} = 30$. Dotted lines show the spectra of thin accretion disks at accretion rates of $\dot{m} = 4 \times 10^{-3}$ (upper, adjusted to fit the infrared points) and $\dot{m} = 10^{-5}$ (lower).

in the flow, rather than as a very low accretion rate (the radiative efficiency is very low, $\sim 5 \times 10^{-6}$). The model will not work if the central object has a hard surface and reradiates the advected energy, as demonstrated in Figure 15c. Therefore, the success of the ADAF model implies that Sgr A* is a black hole with an event horizon (§4.3).

4.2.2. *NGC 4258*

The mass of the central black hole in the AGN NGC 4258 has been measured to be $3.6 \times 10^7 M_\odot$ (Miyoshi et al. 1997). Observations of water masers indicate the presence of a thin disk, at least at large radii. Highlighting the fact that the observed optical/UV and X-ray luminosities are significantly sub-Eddington ($\sim 10^{-4}$ and $\sim 10^{-5}$, respectively), Lasota et al. (1996) proposed that the accretion in NGC 4258 at small radii proceeds through an ADAF. In Lasota et al.'s original model, the transition radius was a free parameter, but since then new infrared observations have been made (Chary & Becklin 1997) that constrain the transition radius to be $r_{tr} \sim 30$. The outer thin disk then accounts for the newly observed infrared emission, and the refined model (see Fig. 16, based on Gammie, Narayan & Blandford 1998) is also in agreement with a revised (smaller) upper limit on the radio flux from NGC 4258 (Herrnstein et al. 1998). Maoz & McKee (1997) and Kumar (1997) find, via quite independent arguments, an accretion rate for NGC 4258 in agreement with the ADAF model ($\dot{m} \sim 0.01$). Neufeld & Maloney (1995) estimate a much lower $\dot{m} < 10^{-5}$ in the outer maser disk via a model of the maser emission. The accretion rate close to the black hole cannot be this low; not even an efficiently radiating thin disk can produce the observed infrared and X-ray radiation with such an $\dot{m}$ (the lower dotted line in Fig. 16).

4.2.3. *Other Low Luminosity Galactic Nuclei*

ADAFs have also been used to model a number of other nearby low luminosity galactic nuclei, e.g., M87 (Reynolds et al. 1996) and M60 (Di Matteo & Fabian 1997a) in the Virgo cluster. These and similar elliptical galaxies are believed to have contained quasars with black hole masses $\sim 10^8 - 10^9 M_\odot$ at high redshift (Soltan 1982; Fabian & Canizares 1988; Fabian & Rees 1995). In the case of M87 and several other galaxies there is independent evidence for such dark mass concentrations in their centers (Ford et al. 1997). The unusual dimness of these galactic nuclei is, however, a problem.

Fabian & Canizares (1988) considered six bright nearby ellipticals and, using X-ray gas profiles in the central arcsecond regions, estimated the accretion rates onto the central black holes. For a standard radiative efficiency of 0.1, their estimated accretion rates yielded luminosities which were substantially larger than the observed luminosities. Thus these galactic nuclei appeared to be very underluminous. Fabian & Canizares (1988) highlighted this problem, calling it the dead quasar problem. As suggested by Fabian & Rees (1995), and confirmed by the detailed calculations of Mahadevan (1997), the problem is naturally resolved if the galactic nuclei are presently accreting via an ADAF, rather than a thin disk. ADAFs are naturally underluminous and have low radiative efficiencies. Similarly, Lasota et al. (1996b) suggested that perhaps all LINERs (of which NGC 4258 is an example) have ADAFs.

4.2.4. *AGN Statistics*

Quasars first appear at a redshift $z \sim 5$ and their numbers increase with decreasing z; below a redshift ~ 2, however, the number of quasars decreases rapidly. Quasars are essentially non-existent at the present epoch, $z \sim 0$. In the standard AGN paradigm, all quasars are assumed to be powered by accretion onto supermassive black holes. Yi (1996) has suggested (following Fabian & Rees 1995) that quasars may switch from accretion via a thin disk to accretion via an ADAF at $1 \lesssim z \lesssim 2$; this provides a natural explanation for the decrease in quasar number counts at small z since ADAFs are significantly less luminous and thus much more difficult to detect. What accounts, however, for the change in the accretion mechanism at $z \sim 2$? Yi assumes that at large z, quasars accrete at $\dot{m} \sim 1$, and so the accretion must be via a thin disk. As the quasar evolves, however, two processes lead to decreasing $\dot{m}$: (1) A decrease in the fuel supply, and (2) an increase in the mass of the accreting black hole; since $\dot{m} \propto \dot{M}/M$, both of these cause $\dot{m}$ to decrease and lead to a critical redshift below which $\dot{m} \lesssim \dot{m}_{\rm crit}$; at this point ($z \sim 2$), the accretion flow switches to an ADAF.

4.2.5. *The X-Ray Background*

Di Matteo & Fabian (1997b; see also Yi & Boughn 1998) argue that ADAFs can be used to model the diffuse X-ray background. The spectrum of the diffuse XRB resembles thermal bremsstrahlung in the 3–60 keV range, and has a rollover at ~ 30 keV. The X-ray background is thought to arise from many discrete sources. ADAFs are a natural candidate since (1) they intrinsically produce bremsstrahlung, (2) the electron temperature is in the right range, $\sim 10^9$K, to account for the observed cutoff, and (3) the electron temperature is very insensitive to the parameters of the model. Assuming a modest spectral evolution with redshift, Di Matteo & Fabian (1997b) were able to reproduce the X-ray background fairly well. Since moderately high accretion rates are required to account for the observed flux, however, Comptonization contributes to the

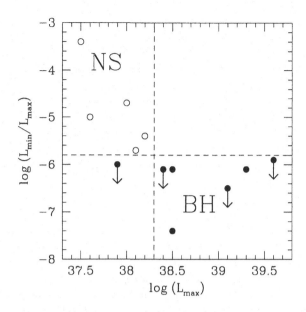

FIGURE 17. A comparison between black hole (BH, filled circles) and neutron star (NS, open circles) SXT luminosity variations (from Narayan et al. 1997c and Garcia et al. 1998). The ratio of the quiescent luminosity to the peak outburst luminosity is systematically smaller for BH systems than for NS systems. This indicates the presence of event horizons at the center of BH candidate systems.

hard X–ray spectrum of the ADAFs. Di Matteo & Fabian avoid this problem by invoking a high degree of clumpiness in the gas so that bremsstrahlung emission dominates over Compton scattering, but this is somewhat ad hoc.

4.3. *Evidence for the Black Hole Event Horizon*

A unique feature of the ADAF applications described above is that the models require the existence of an event horizon in the central object (Narayan et al. 1996; Narayan et al. 1997bc). Since the protons store most of the viscous energy as thermal energy as they fall onto the central object, the luminosity from the ADAF is much less than the viscously generated energy ($\sim 0.1\dot{M}c^2$). If the central object had a hard surface, this thermal energy would be reprocessed and reradiated, resulting in a net luminosity from the object $\sim 0.1\dot{M}c^2$; the spectrum would also differ significantly from that of a pure ADAF (since it would be dominated by the reprocessed radiation; see Figure 15 as an example). If the central object is a black hole, however, the advected thermal energy is carried into the black hole through the event horizon, and is "lost" to the outside observer. The success of ADAF models of the various black hole candidates discussed in the previous sections *without* reprocessed radiation therefore indicates the presence of an event horizon in all of these objects.

Narayan et al. (1997c; see also Yi et al. 1996) highlighted the distinction between black holes and compact stars with surfaces in a different way by comparing the luminosities of black hole and neutron star soft X–ray transients in outburst and quiescence. In outburst, the accretion rates in both neutron star and black hole systems are high, $\dot{m} \sim 1 \gtrsim \dot{m}_{\rm crit}$, and the accretion occurs via a standard thin disk. The observed outburst luminosities, L_{max}, should therefore be proportional to $\dot{M} \propto m$. This is consistent with the observation that black hole SXTs in outburst have a larger L_{max} than neutron star SXTs, as expected because of their larger mass.

In quiescence, the accretion rates are low, $\dot{m} \ll \dot{m}_{\mathrm{crit}}$, and the accretion occurs via an ADAF. Since neutron stars have hard surfaces, however, the advected energy is ultimately reradiated, and the luminosity is still proportional to $\dot{m}$. For an ADAF around a black hole, on the other hand, the advected energy is not reradiated, but is deposited into the black hole. The quiescent luminosities, L_{min}, from black hole candidates should therefore be substantially less, roughly proportional to $\dot{m}^2$ (see Fig. 7). Black hole SXTs should thus experience more substantial luminosity changes from quiescence to outburst than neutron star SXTs. As figure 17 shows, this is exactly what is observed. This provides additional evidence for the existence of event horizons in black hole candidates.

5. Conclusion

The two-temperature ADAF model provides a consistent framework for understanding the dynamics and spectra of black hole accretion flows at low mass accretion rates, $\dot{m} \lesssim 0.1$. By modeling coronae as ADAFs, the model is also being extended successfully to systems with $\dot{m} \gtrsim 0.1$, though not yet to systems close to the Eddington limit ($\dot{m} \sim 1$).

The field is young and there are many open questions. Perhaps the most fundamental issue is whether the assumptions underlying the two-temperature ADAF model are valid (§3.1). Another major question, where much work needs to be done, is the physics underlying the transition radius (§3.3). If this is understood, and the dynamics and thermal structure of the corona can be modeled well, it will be possible to calculate $r_{tr}(\dot{m})$ reliably; this will be an enormous improvement over the present work. Improvements are necessary in the modeling techniques as well. Fully relativistic computations of the coupled dynamics and radiative transfer will be welcome. One needs to include time-dependence in the models in order to understand the complex variations of spectra with time, especially in X-ray binaries such as the SXTs. Ultimately, we expect that time-dependent two-dimensional simulations will elevate the ADAF model to a fully quantitative tool, though this is probably many years away.

Two wider issues should be highlighted. First, the high-$\dot{m}$, optically thick, ADAF (corresponding to radiation trapping) has not been explored at the level of detail of the low $\dot{m}$, two-temperature, ADAF. More quantitative models, especially with respect to their spectral properties, would be worthwhile. It is possible that some of the presently most puzzling sources (e.g. SS433) correspond to this branch of ADAFs. The other issue has to do with jets and outflows. There are tantalizing hints that ADAFs may be particularly efficient at producing outflows (Narayan & Yi 1994, 1995a), but the connection has not been developed in any detail.

We thank Kristen Menou for comments on the manuscript. This work was supported in part by NSF Grant AST 9423209 and NASA Grant 5-2837. EQ was supported in part by an NSF Graduate Research Fellowship.

REFERENCES

ABRAMOWICZ, M., CHEN, X., KATO, S., LASOTA, J. P, & REGEV, O., 1995, *The Astrophysical Journal Letters*, **438**, L37.

ABRAMOWICZ, M., CZERNY, B., LASOTA, J. P., & SZUSZKIEWICZ, E., 1988, *The Astrophysical Journal*, **332**, 646.

ABRAMOWICZ, M., JAROSZYŃSKI, M., & SIKORA, M., 1978, *Astronomy & Astrophysics*, **63**, 221.

ABRAMOWICZ, M., CHEN, X., GRANATH, M., & LASOTA, J.-P., 1996, *The Astrophysical Journal*, **471**, 762.

ACHTERBERG, A. 1981, *Astronomy & Astrophysics*, **97**, 259.

BALBUS, S. A., & HAWLEY, J. F. 1991, *The Astrophysical Journal*, **376**, 214.

BEGELMAN, M. C., 1978, *Monthly Notices of the Royal Astronomical Society*, **243**, 610.

BEGELMAN, M. C., & CHIUEH, T., 1988, *The Astrophysical Journal*, **332**, 872.

BEGELMAN, M. C., & MEIER, D. L., 1982, *The Astrophysical Journal*, **253**, 873.

BISNOVATYI-KOGAN, G.S. & LOVELACE, R. V. E. 1997, *The Astrophysical Journal Letters*, **486**, L43.

BJÖRNSSON, G., ABRAMOWICZ, M. , CHEN, X., LASOTA, J.-P., 1996, *The Astrophysical Journal*, **467**, 99.

BJÖRNSSON, G. & SVENSSON, R., 1991, *The Astrophysical Journal Letters*, **379**, L69.

BLACKMAN, E. 1998, *Physical Review Letter*, in press (astro-ph/9710137).

BONDI, H., 1952, *Monthly Notices of the Royal Astronomical Society*, **112**, 195.

BRANDT, S., CASTRO–TIRADO, A. J., LUND, N., DREMIN, V., LAPSHOV, I., & SUNYAEV, R. A., 1992, *Astronomy & Astrophysics*, **254**, L39.

CELOTTI, A., FABIAN, A. C., & REES, M. J. 1997, *Monthly Notices of the Royal Astronomical Society*, **293**, 239. (astro–ph/9707131).

CHAKRABARTI, S. K. 1990, *Monthly Notices of the Royal Astronomical Society*, **243**, 610.

CHAKRABARTI, S. K., & TITARCHUK, L. G. 1995, *The Astrophysical Journal*, **455**, 623.

CHAKRABARTI, S. K. 1996, Physics Reports, **266**, 229.

CHARY, R. & BECKLIN, E. E., 1997 *The Astrophysical Journal Letters*, **485**, L75.

CHEN, X. 1995, *Monthly Notices of the Royal Astronomical Society*, **275**, 641.

CHEN, X., ABRAMOWICZ, M. A., LASOTA, J.-P., NARAYAN, R., & YI, I. 1995, *The Astrophysical Journal Letters*, **443**, 61.

CHEN, X., ABRAMOWICZ, M. A., & LASOTA, J.-P., 1997, *The Astrophysical Journal*, **476**, L61.

DI MATTEO, T., & FABIAN, A. C., 1997a, *Monthly Notices of the Royal Astronomical Society*, **286**, 50.

DI MATTEO, T., & FABIAN, A. C., 1997b, *Monthly Notices of the Royal Astronomical Society*, **286**, 393.

EBISAWA, K., ET AL., 1994, *Publications of the Astronomical Society of Japan*, **46**, 375.

ECKART, A., & GENZEL, R., 1997, *Monthly Notices of the Royal Astronomical Society*, **284**, 576.

EGGUM, G. E., CORONITI, F. V., & KATZ, J. I., 1988, *The Astrophysical Journal*, **330**, 142.

ESIN, A. A., 1997, *The Astrophysical Journal*, **482**, 400.

ESIN, A. A., MCCLINTOCK, J. E., & NARAYAN, R., 1997, *The Astrophysical Journal*, **489**, 865.

ESIN, A. A., NARAYAN, R., CUI, W., GROVE, J. E., & ZHANG, S-N. 1998, *The Astrophysical Journal*, in press.

ESIN, A. A., NARAYAN, R., OSTRIKER, E., & YI, I. 1996, *The Astrophysical Journal*, **465**, 312.

FABIAN, A. C., & CANIZARES, C. R., 1988, *Nature*, **333**, 829.

FABIAN, A. C., & REES, M. J., 1995, *Monthly Notices of the Royal Astronomical Society*, **277**, L5.

FISHBONE, L. G., & MONCRIEF, V., 1976, *The Astrophysical Journal*, **207**, 962.

FORD, H. C., TSVETANOV, Z. I., FERRARESE, L., AND JAFFE, W. in proc. IAU Symp. 184, *The Central Region of the Galaxy and Galaxies*, ed. Y. Sofue (1997).

FRANK, J., KING, A., & RAINE, D., 1992, in *Accretion Power in Astrophysics*. Cambridge Univ. Press.

FUJITA, Y. ET AL. 1998, *The Astrophysical Journal Letters*, in press.

GAMMIE, C. F., NARAYAN, R. & BLANDFORD, R. 1998, in preperation.

GAMMIE, C. F. & POPHAM, R. G., 1998, *The Astrophysical Journal*, in press (astro-ph/9705117).

GARCIA ET AL. 1998, in *Proc. 13th NAW on CV's*, eds. S. Howell, E. Kuulkers, & C. Woodward.

GENZEL, R., HOLLENBACH, D., & TOWNES, C. H., 1994, *Rep. Prog. Phys.*, **57**, 417.

GHISELLINI, G. & SVENSSON, R. 1990, in *Physical Processes in Hot Cosmic Plasmas*, ed. W. Brinkmann, A.C. Fabian, & F. Giovannelli (NATO AI Ser. C, 305) (Dordrecht: Kluwer), p. 395.

GHISELLINI, G. & SVENSSON, R. 1991, *Monthly Notices of the Royal Astronomical Society*, **252**, 313.

GILFANOV, M., ET AL., 1993, *Astronomy & Astrophysics Supplement*, **97**, 303.

GOLDREICH, P. & SRIDHAR, S. 1995, *The Astrophysical Journal*, **438**, 763.

GOLDWURM, A., ET AL., 1994, *Nature*, **371**, 589.

GRUZINOV, A., 1998, *The Astrophysical Journal*, in press (astro-ph/9710132).

HAARDT, F., & MARASCHI, L., 1991, *The Astrophysical Journal Letters*, **380**, L51.

HAARDT, F., & MARASCHI, L., 1993, *The Astrophysical Journal*, **413**, 507.

HALLER, J., RIEKE, M. J., RIEKE, G. H., TAMBLYN, P., CLOSE, L., & MELIA, F., 1996, *The Astrophysical Journal*, **468**, 955.

HAMEURY, J.-M., LASOTA, J.-P., McCLINTOCK, J. E., & NARAYAN, R., 1997, *The Astrophysical Journal*, **489**, 234 (astro-ph/9703095).

HAWLEY, J. F., GAMMIE, C. F., & BALBUS, S. A., 1996, *The Astrophysical Journal*, **464**, 690.

HERRNSTEIN ET AL., 1998, in preparation.

HUANG, M., & WHEELER, J. C, 1989, *The Astrophysical Journal*, **343**, 229.

HONMA, F., 1996, *Publications of the Astronomical Society of Japan*, **48**, 77.

ICHIMARU, S., 1977, *The Astrophysical Journal*, **214**, 840.

IGUMENSHCHEV, I. V., ABRAMOWICZ, M. A., & NOVIKOV, I. D. 1998, *Monthly Notices of the Royal Astronomical Society*, in press (astro-ph/9709156).

IGUMENSHCHEV, I. V., CHEN, X., & ABRAMOWICZ, M. A. 1996, *Monthly Notices of the Royal Astronomical Society*, **278**, 236.

KATO, S., YAMASAKI, T., ABRAMOWICZ, M. A., & CHEN, X., 1997, *Publications of the Astronomical Society of Japan*, **49**, 221.

KATO, S., ABRAMOWICZ, M. A., & CHEN, X., 1996, *Publications of the Astronomical Society of Japan*, **48**, 67.

KATZ, J., 1977, *The Astrophysical Journal*, **215**, 265.

KUMAR, P. 1998, *The Astrophysical Journal*, in press.

KUSUNOSE, M. & MINESHIGE, S., 1996, *The Astrophysical Journal*, **468**, 330.

LASOTA, J.-P., NARAYAN, R., & YI, I., 1996, *Astronomy & Astrophysics*, **314**, 813.

LASOTA, J.-P., ABRAMOWICZ, M. A., CHEN, X., KROLIK, J., NARAYAN, R., & YI, I., 1996, *The Astrophysical Journal*, **462**, 142.

LUND, N., 1993, *Astronomy & Astrophysics Supplement*, **97**, 289.

LUO, C. & LIANG, E. P., 1994 *Monthly Notices of the Royal Astronomical Society*, **266**, 386L.

LYNDEN–BELL, D., & PRINGLE, J. E., 1974, *Monthly Notices of the Royal Astronomical Society*, **168**, 603.

MAHADEVAN, R., 1997, *The Astrophysical Journal*, **477**, 585.

MAHADEVAN, R., NARAYAN, R., & KROLIK, J., 1997, *The Astrophysical Journal*, **486**, 268.

MAHADEVAN, R., NARAYAN, R., & YI, I., 1996, *The Astrophysical Journal*, **465**, 327.

MAHADEVAN, R., & QUATAERT, E., 1997, *The Astrophysical Journal*, **490**, 605.

MANMOTO, T., TAKEUCHI, M., MINESHIGE, S., MATSUMOTO, R., & NEGORO, H., 1996, *The Astrophysical Journal*, **464**, L135.

MANMOTO, T., MINESHIGE, S., & KUSUNOSE, M. 1997, *The Astrophysical Journal Letters*, **489**, 791.

MAOZ, E. & MCKEE, C. F. 1998, *The Astrophysical Journal*, **494**, 218.

MASTICHIADIS, A. & OZERNOY, L. M., 1994, *The Astrophysical Journal*, **426**, 599.

MATSUMOTO, R., KATO, S., & FUKUE, J. 1985 in *Theoretical Aspects on Structure, Activity, and Evolution of Galaxies III*, eds. S. Aoki, M. Iye, &Y. Yoshii (Tokyo Astr. Obs: Tokyo), p. 102.

MCCLINTOCK ET AL. 1995, *The Astrophysical Journal*, **442**, 358.

MELIA, F. 1992, *The Astrophysical Journal Letters*, **387**, L25.

MENOU, K., NARAYAN, R., & LASOTA, J. P., 1998, in preparation.

MENTEN, K. M., REID, M. J., ECKART, A., & GENZEL, R., 1997, *The Astrophysical Journal Letters*, **475**, L111.

MEYER, F., & MEYER–HOFMEISTER, E., 1994, *Astronomy & Astrophysics*, **288**, 175.

MINESHIGE, S., & WHEELER, J. C., 1989, *The Astrophysical Journal*, **343**, 241.

MIYOSHI, M., MORAN, J., HERRNSTEIN, J., GREENHILL, L., NAKAI, N., DIAMOND, P., & INOUE, M., 1995, *Nature*, **373**, 127.

NAKAMURA, K. E., KUSUNOSE, M., MATSUMOTO, R., & KATO, S. 1997, *Publications of the Astronomical Society of Japan*, **49**, 503.

NAKAMURA, K. E., MATSUMOTO, R., KUSUNOSE, M., & KATO, S. 1996, *Publications of the Astronomical Society of Japan*, **48**, 761.

NARAYAN, R., 1996a, *The Astrophysical Journal*, **462**, 136.

NARAYAN, R., 1996b, in *Physics of Accretion Disks*, (eds. S. Kato, S. Inagaki, S. Mineshige, J. Fukue). p15. Gordan & Breach.

NARAYAN, R., & YI, I., 1994, *The Astrophysical Journal Letters*, **428**, L13.

NARAYAN, R., & YI, I., 1995a, *The Astrophysical Journal*, **444**, 231.

NARAYAN, R., & YI, I., 1995b, *The Astrophysical Journal*, **452**, 710.

NARAYAN, R., YI, I., & MAHADEVAN, R., 1995, *Nature*, **374**, 623.

NARAYAN, R., MCCLINTOCK, J. E., & YI, I. 1996, *The Astrophysical Journal*, **457**, 821.

NARAYAN, R. 1997, in *Accretion Phenomena & Related Outflows, Proc. IAU Colloq. 163 A.S.P. Conf. Series* (eds. D. T. Wickramasinghe, L. Ferrario, G. V. Bicknell).

NARAYAN, R., KATO, S., & HONMA, F. 1997a, *The Astrophysical Journal*, **476**, 49.

NARAYAN, R., BARRET, D., & MCCLINTOCK, J. 1997b, *The Astrophysical Journal*, **482**, 448.

NARAYAN, R., GARCIA, M. R., & MCCLINTOCK, J. 1997c, *The Astrophysical Journal Letters*, **478**, L79.

NARAYAN, R., MAHADEVAN, R., GRINDLAY, J. E., POPHAM, R. G., & GAMMIE, C. 1998, *The Astrophysical Journal*, **492** (astro–ph/9706112).

NEUFELD, D. A. & MALONEY, P. R. 1995, *The Astrophysical Journal Letters*, **447**, L17.

NOVIKOV, I. D., & THORNE, K. S. 1973, in *Blackholes* (ed. C. DeWitt & B. DeWitt), 343. Gordon & Breach.

OROSZ, J. A., REMILLARD, R. A., BAILYN, C. D., & MCCLINTOCK, J. E. 1997, *The Astrophysical Journal*, **478**, 83.

PACZYŃSKI, B., & WIITA, P. J. 1980, *Astronomy & Astrophysics*, **88**, 23.

PEITZ, J., & APPL, S. 1997, *Monthly Notices of the Royal Astronomical Society*, **286**, 681 (astro–ph/9612205).

PHINNEY, E. S. 1981 in *Plasma Astrophysics*, ed. T. Guyenne (ESA SP-161), p. 337.

PIRAN, T. 1978, *The Astrophysical Journal*, **221**, 652.

POPHAM, R. G., & GAMMIE, C. F., 1998, *The Astrophysical Journal*, in press.

PRINGLE, J. E. 1981, *Annual Reviews of Astronomy & Astrophysics*, **19**, 137.

QUATAERT, E., 1998, *The Astrophysical Journal*, in press (astro-ph/9710127).

QUATAERT, E. & GRUZINOV, A. 1998, in preperation.

REES, M. J. 1982, in *The Galactic Center* (ed. G. R. Riegler & R. D. Blandford). AIP, p166. New York.

REES, M. J., BEGELMAN, M. C., BLANDFORD, R. D., & PHINNEY, E. S. 1982, *Nature*, **295**, 17.

REYNOLDS, C. S., DI MATTEO, T., FABIAN, A. C., HWANG, U., & CANIZARES, C. R. 1996, *Monthly Notices of the Royal Astronomical Society*, 1996, **283**, L111.

SHAKURA, N. I., & SUNYAEV, R. A. 1973, *Astronomy & Astrophysics*, **24**, 337.

SHAKURA, N. I., & SUNYAEV, R. A. 1976, *Monthly Notices of the Royal Astronomical Society*, **175**, 613.

SHAPIRO, S. L., LIGHTMAN, A. P., & EARDLEY, D. M. 1976, *The Astrophysical Journal*, **204**, 187 (SLE).

SOLTAN, A. 1982, *Monthly Notices of the Royal Astronomical Society*, **200**, 115.

SPITZER, L. JR. 1962, *Physics of Fully Ionized Gases*, 2nd Ed. John Wiley & Sons, Inc.

SPRUIT, H.C., MATSUDA, T., INOUE, M., & SAWADA, K. 1987, *Monthly Notices of the Royal Astronomical Society*, **229**, 517.

SZUSKIEWICZ, E., MALKAN, M. A., & ABRAMOWICZ, M. A. 1996, *The Astrophysical Journal*, **458**, 474.

VAN PARADIJS, J. & McCLINTOCK, J. E., 1995, in *X-Ray Binaries*, ed. W. H. G. Lewin et al. (Cambridge: Cambridge Univ. Press), p. 58.

WANDEL, A. & LIANG, E. P., 1991, *The Astrophysical Journal Letters*, **376**, 746L.

WHEELER, C. J., 1996, in *Relativistic Astrophysics*, eds. B. Jones & D. Markovic (Cambridge: Cambridge University press)

WU, X. 1997, *Monthly Notices of the Royal Astronomical Society*, **292**, 113 (astro–ph/9707329).

WU, X., & LI, Q. 1996, *The Astrophysical Journal*, **469**, 776.

YI, I., & BOUGHN, S. P. 1998, *The Astrophysical Journal*, in press (astro–ph/9710147).

YI, I. 1996, *The Astrophysical Journal*, **473**, 645.

YI, I., NARAYAN, R., BARRET, D., & McCLINTOCK, J. E. 1996, *Astronomy & Astrophysics Supplement*, **120**, 187.

Accretion disc instabilities and advection dominated accretion flows

By JEAN–PIERRE LASOTA

UPR 176 du CNRS; DARC Observatoire de Paris, 92195 Meudon Cedex, France

In the ADAF + cold accretion disc model of quiescent soft X-ray transients it is the cold part of the flow which is responsible for the outbursts. A dwarf–nova type instability triggers an outburst in the cold disc. A heat front propagates through this part of the flow and brings it to a hot state. The heat front cannot propagate in the ADAF but the the outer disc will diffuse into the ADAF, increasing the accretion rate onto the central black hole. This model of soft X-ray transients is beautifully confirmed by the rise to outburst of the transient system GRO J1655–40.

1. Introduction

Similarities between dwarf novae (DNs) and Soft X-ray Transients (SXTs) were noticed a long time ago (van Paradijs & Verbunt 1984), so it is natural to try to explain properties of both classes of systems by the same mechanism. The 'standard' DN disc instability model (DIM) was applied to SXTs by Cannizzo, Wheeler & Ghosh (1982), Lin & Taam (1984), Huang & Wheeler (1989), Mineshige & Wheeler (1989). Hameury, King & Lasota (1996) suggested that X–ray illumination of the secondary star could trigger a mass transfer instability (MTI), thus reviving the alternative model which was proposed for DN outbursts (Bath 1973) but discarded after a physical mechanism triggering the disc instability was found. This modified version of the MTI did not survive long, however: Gontikakis & Hameury (1993) have shown that it cannot, in general, reproduce the typical timescales associated with the SXT phenomena. Contrary to what is sometimes said (e.g. King & Ritter 1997) this *does not* mean that X–ray irradiation of the secondary can be neglected during the outburst itself (Hameury, King & Lasota 1988).

The only model left was therefore the DIM. The existing version (Mineshige & Wheeler 1989) was far from being perfect since it could not correctly reproduce long recurrence times and the light curves could not be trusted because of the small number of grid points used in the numerical code. One could however expect that improving the numerical methods would improve the situation. Some efforts in this direction were undertaken by Cannizzo, Chen, & Livio (1995), Cannizzo (1998a;b). This work concentrated on the description of the decay from outburst. Some progress was achieved but the value of the results obtained is not clear because they required drastic modification of the standard model, such as 'switching off' the convection in the disc. In any case a SXT event consists of quiescence, rise to outburst and the decay from it, so a description of the decay only cannot be considered a satisfactory model. But even the validity of the decay phase model has been questioned because it does not take into account the effects of X–ray irradiation clearly observed in SXTs. It has been argued convincingly that the form of the light–curves in the systems is determined by X–ray illumination (King & Ritter 1997). The viscosity prescription that was adopted in the DIM models of SXTs (the viscosity parameter is assumed to be $\alpha = \alpha_0 (H/R)^{1.5}$, where H is the the disc semi–thickness), produces (for the assumed value $\alpha_0 = 50$) very long risetimes to outburst, much longer than those observed.

The real, insurmountable problem for the standard DIM of SXTs is due to the properties of the quiescent state. The 'standard' model predicts accretion rates $\sim 10^6$ g s^{-1} (e.g. Mineshige & Wheeler 1989) whereas observations show that the accretion rate is several orders of magnitude higher (McClintock, Horne, Remillard 1995; Lasota 1996a). It is asserted now that 'many authors' noted that 'the standard model fails to account for observations of systems in quiescence' (Cannizzo 1998b), but the first detections of X–rays from quiescent SXTs were hailed as a triumphant confirmation of the 'standard' model (the only exception was probably Shin Mineshige, who had predicted (Mineshige, Ebisawa, Takizawa *et al.*1992), that such detections would put the model in deep trouble) and it was only the present author who pointed out that these detections are in violent contradiction with the 'standard' DIM (Lasota 1996a).

A simple way of solving the problem of observed properties of quiescent SXTs was proposed by Narayan, McClintock & Yi (1996; NMY). According this model (see Narayan this volume) the quiescent accretion flow consists of two components: a cold outer accretion disc and an inner, very hot, advection–dominated accretion flow (ADAF). The parameters of the original model by NMY had to be modified after it was shown (Lasota, Narayan & Yi 1996) that in a self–consistent model the *outer*, cold, disc cannot contribute significantly to optical/UV flux in quiescent SXTs.

The new version of the model (Narayan, Barret & McClintock 1997) provides a very satisfactory and consistent description of the quiescent SXTs A0620–00, V404Cyg and GRO J1655–40. Spectral properties during the outburst phase have been described in this framework by Esin *et al.*(1997). A general scenario for the outburst cycle was proposed by Mineshige (1996) and the rise to outburst has been modeled by Hameury *et al.*(1997; HLMN) who showed that in GRO J1655–40 it is compatible only with the two–zone model of NBM. The present article is mainly devoted to a presentation and discussion of this work.

The ADAF+cold disc model of SXTs is, at present, the only model which provides a complete and consistent description the SXT cycle. The model itself is not without weaknesses (the main one being our lack of understanding of the transition between the two types of flows) and several of its components should be improved, but, contrary to some assertions (Cannizzo 1998; Chen *et al.*1998) it is fully consistent with observations (Narayan, Garcia & McClintock 1997; Lasota & Hameury 1998).

2. GRO J1655–40

GRO J1655–40 is an exceptional BHT because it provides two independent pieces of evidence for a two–component accretion flow in which a geometrically thin accretion disc surrounds an ADAF. One piece of evidence is provided by the fit of the optical and X–ray spectrum of the quiescent system analogous to that used by NBM for V404 Cyg and A0620-00 (GRO J1655–40 is the third BHT X–ray detected in quiescence); the other piece of evidence is given by the properties of the rise to outburst observed by Orosz *et al.*1997(OBRM) in April 1996. As was shown by HLMN, the observed 6–day delay in the rise of the X–rays *and* the shapes of both optical and X–ray light–curves are in perfect agreement with the ADAF + cold disc model and incompatible with models in which (in quiescence) a cold disc extends down to the last stable orbit around the central black hole.

2.1. *Observations*

GRO J1655–40 is not a typical BHT because it has remained active, on and off, for nearly three years. Other BHTs have shown X–ray and optical activity several months after an outburst; however, none of them have sustained their activity for more than about a year (e.g. Tanaka & Shibazaki 1996). The frequent outbursts of GRO J1655–40 in recent years may be due to an enhancement of the mass transfer rate.

The estimate of the mass transfer rate is difficult because, in principle, the secondary of GRO J1655–40 should be in the Hertzsprung gap (KKRF). Systems with companions in the Hertzsprung gap should transfer mass at a very high rate ($\gtrsim 10^{19}$ g s^{-1}), since the secondary expands on a thermal time; this does not seem to be the case now in GRO J1655–40, and no observational determination of the mass transfer rate is available at present.

On longer timescales GRO J1655–40 behaves more like other BHTs, since no outburst of the source has been reported in the previous 25 years. GRO J1655–40 can therefore be considered as a 'bona fide' X-ray transient system.

GRO J1655–40 is also distinguished by its radio outbursts, which are associated with superluminal expansion events (Tingay et al. 1995; Hjellming & Rupen 1995).

About 6 days prior to the most recent X-ray outburst of GRO J1655–40 (in April 1996), a remarkable optical precursor was observed (ORBM). As shown in Figure 1, starting from an initially quiescent state, the optical intensities (BVRI) were observed to rise gradually for several days and to brighten by about 30% before the onset of the X-ray outburst (in Fig. 1 only the V–magnitude is shown). HLMN showed that this delay is naturally explained by an ADAF + cold disc system. One should note here that, the delay in question is between the beginnings of activity (the rise of luminosity) and *not*, as is frequently mistakenly believed, between the maxima of luminosities in various wavelengths.

2.2. *The UV-Delay in Dwarf-Nova Outbursts*

The X-ray delay observed in the outburst of GRO J1655–40 is analogous to the well-known UV delay observed for dwarf novae (e.g. Warner 1995, and references therein). For dwarf novae the rise in the UV flux starts about 5 to 15 hours after the beginning of the optical outburst. In the DIM one interprets the UV-delay seen, as being due to an "outside-in" (or Type A) outburst. According to the DIM (e.g. Cannizzo 1993a, and references therein), a thermal instability in the outer disc creates an inward propagating heat front. This front transforms the disc from a cold (quiescent) state to a hot state. Because the UV flux is mainly emitted close to the white dwarf, one expects a delay in its rise, a delay equal to the time it takes the front to travel from the outer disc to the white dwarf. It is not clear that the DIM in its standard form can account for the observed delay. According to Pringle, Verbunt & Wade (1986) and Cannizzo & Kenyon (1987) the calculated travel time of the front is much shorter than the observed UV-delay. Thus, in its standard form the model fails to explain the UV-delay. There is a dissenting view (see e.g. Cannizzo 1998a) that there is no real problem with the UV–delay in the DIM because its failure to reproduce the observed timescales is due to the use of stellar atmosphere to model the disc's radiation. It is well known that disc radiation is incorrectly represented by stellar atmosphere (see e.g. Shaviv and Wehrse 1991) but the use of the correct models in time–dependent calculations is, for the moment, out of the question because of the the amount of computer time required. Cannizzo (1998a) mentioned that the use of blackbody spectra to represent disc radiation gives an UV–delay close to the observed one. The blackbody approximation is, however, not an accurate representation of the accretion disc radiation (Shaviv & Wehrse 1991). The recently observed 1-day EUV–delay in SS Cyg (Mauche 1996) provides an even better test of the model than the UV–delay because it represent directly the heat front travel–time from the ignition point to the surface of the white dwarf. Unfortunately there is no model to compare this

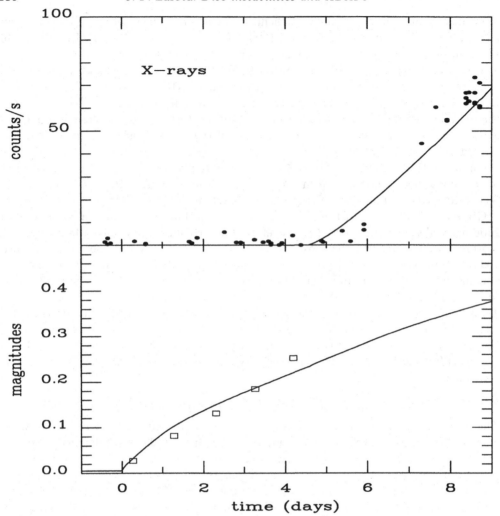

FIGURE 1. Observed optical and 2-12 keV X-ray light curves during the initial phase of the April outburst of GRO J1655–40 (From ORBM) and the X–ray and V lightcurves according to HLMN. For the sake of clarity, only one average data point per night is plotted. The V magnitudes are represented by squares. Time has been set to zero at an arbitrary point close to the onset of the outburst.

observation with: in the 'standard' model of SS Cyg (Cannizzo 1993a) *all* outbursts are of the *inside–out* type. The question of UV–delay in DNs clearly requires more study.

If the DIM is to be rescued; then two solutions are available. Both of them invoke a central "hole" in the accretion disc. At the edge of such a hole, an inward moving heat front would have to stop. The hole would then fill up on a viscous time scale, which is much longer than the heat front propagation time. Livio & Pringle (1992) suggested a mechanism for creating such a hole: they argued that at quiescent mass accretion rates, the magnetic field of a weakly-magnetized white dwarf can disrupt the inner accretion disc. They showed that such a model can reproduce the UV-delay observed in dwarf nova outbursts. This model cannot apply to systems in which the accreting object is a black hole, since a black hole cannot support a permanent magnetic field.

Meyer & Meyer-Hofmeister (1994) proposed a different scheme for quiescent accretion onto a white dwarf that also results in a central hole. They invoke inefficient cooling in the disc's upper layers, which leads to the formation of a hot corona and ultimately to the evaporation of the inner disc. As a result, the inner accretion flow consists of a pure coronal plasma. A similar solution for quiescent SXTs has been independently proposed by NMY.

One should stress that in both cases, the hot inner flow provides a natural explanation for the hard X-ray emission observed in quiescent dwarf novae and SXTs.

Whether a hole is created by magnetic fields or by evaporation, the effect on the outburst of a dwarf nova is similar: when the heat front arrives at the inner edge of the truncated disc it cannot propagate any further; the (surface) density contrast slowly fills up the hole on a viscous time scale, thereby producing the required delay of the UV outburst.

The presence of an inner hole 'hole' in the accretion disc is necessary if one wants to avoid using extremely low values of the viscosity parameter α in the description of the quiecsent state of WZ Sge (Lasota, Hameury & Huré 1996; Hameury, Lasota & Huré 1997).

In the case of SXTs things are simpler: observations of the April 1996 outburst of GRO J1655–40 clearly imply the presence of a two-component accretion flow in this system.

2.3. *The spectral model of GRO J1655-40*

HLMN used the ADAF+cold disc model to fit the spectral data on GRO J1655–40 in quiescence and to estimate the parameters of the quiescent accretion flow. The picture is similar to the two-zone model proposed by NBM for fitting the spectra of V404 Cyg and A0620-00 in quiescence; in that model, the flow inside the transition radius consists of a hot two-temperature ADAF; outside this radius, the accretion flow forms a cold disc that can be subject to the dwarf nova type instability. The thin accretion disc extends only down to a transition radius r_{tr} which, in general, is greater than a few thousand Schwarzschild radii (in what follows r is measured in units of $2GM/c^2$).

In constructing this model, one selects the system parameters for GRO J1655–40 such as the black hole mass, the binary inclination, the distance, and the velocity at the inner edge of the outer thin accretion disc. The mass of the black hole in GRO J1655–40 and the inclination of the system are very well determined: $M_1 = 7\ M_\odot$ and $i = 70^o$ (Orosz & Baylin 1997). The outer radius of the thin accretion disc is taken to be $\log(r_{out}) = 5.0$ but this parameter has very little effect on the results. Based on studies of the radio jets, the adopted distance was D $= 3.2$ kpc (Hjellming & Rupen 1995).

Dynamical and geometrical information about the thin accretion disc can be obtained from studies of the broad, double-peaked Balmer lines (Smak 1981; Horne & Marsh 1986), and from this the transition radius r_{tr} can be constrained. Of interest here is v_{in}, the projected velocity at the inner edge of the thin accretion disc. Estimates of this velocity have been inferred for several SXTs from orbit-averaged profiles of the H_α emission line (see NMY, and references therein).

The width of the H_α provides an estimate $v_{tr} \geq 1045$ km s^{-1}, the circular velocity at the transition radius r_{tr} between the thin disc and the ADAF (HLMN). Then r_{tr} is equal to $r_{tr} = (c \sin i/v_{in})^2/2 \leq 3.6 \times 10^4$. In the spectral models of GRO J1655–40 four values of r_{tr} were used.

The solid lines in Fig. 2 represent the four models, where each model consists of a pure ADAF for radii $\log(r) < \log(r_{tr})$, and a thin accretion disc in the radius range $\log(r_{tr}) \leq \log(r) \leq \log(r_{out})$. The models correspond to $\log(r_{tr}) = 4.5,\ 4.0,\ 3.5$ and 3.0, respectively. The models assume equipartition between gas and magnetic pressure ($\beta = 0.5$ in the notation of NBM) and the viscosity parameter is taken to be $\alpha = 0.3$ in

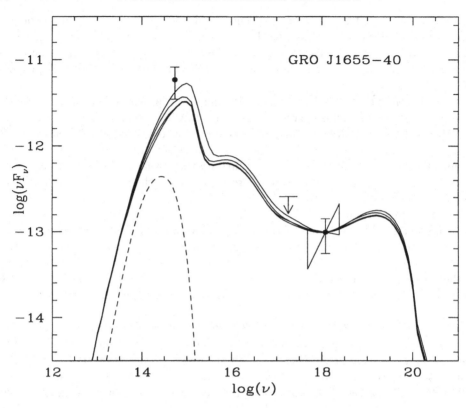

FIGURE 2. Quiescent spectrum of the non-stellar component of GRO J1655–40. The dot and error bar on the left represent the estimated V band flux of the quiescent accretion flow in GRO J1655–40, the arrow shows the upper limit on the soft X-ray flux as measured with ROSAT, and the dot on the right with error bar and "bow-tie" indicates the ASCA constraint on the X-ray flux and spectral index. The solid lines represent model spectra corresponding to an accretion flow consisting of an inner ADAF plus an outer thin disc. In each case, the mass accretion rate has been adjusted to fit the ASCA X-ray flux. From below (in the optical band), the four models have transition radii, $\log(r_{\mathrm{tr}}) = 4.5, 4.0, 3.5, 3.0$, respectively. The spectrum of the outer cold disc is represented by a dashed line.

the ADAF region. It is assumed that a fraction 0.001 of the viscous energy directly heats the electrons in the ADAF (and the corona) while the remaining 0.999 of the energy goes initially into the ions (i.e. $\delta = 0.001$, see NBM for details). The shape and normalization of the computed spectra are quite insensitive to the values chosen for r_{tr}, β, α and δ (see NBM). The dashed line represents the spectrum of the cold disc prior to the outburst.

In each model, only one parameter has been adjusted, namely the mass accretion rate. This has been optimized such that the model flux in the ASCA band agrees with the observed flux. It is worth noting that despite the large range of r_{tr} covered by the four models, the mass accretion rates vary very little from one model to another; the accretion rates range from $\dot{M} = 3.4 \times 10^{16}$ g s^{-1} to $\dot{M} = 3.7 \times 10^{16}$ g s^{-1}. Thus, the spectral models constrain the $\dot{M}$ of GRO J1655–40 in quiescence quite well. Technically, the models determine only the parameter combination $\dot{M}/\alpha$ and so $\dot{M}$ depends on a knowledge of α. However, the value of α in ADAFs is fairly well constrained by the various studies done to date (Narayan 1997), and is unlikely to vary by more than a factor of ~ 3 either way from the value we have assumed, $\alpha = 0.3$. This suggests that the above estimate of $\dot{M}$ in the ADAF is good to about a factor of ~ 3. One should remember that the accretion rate so determined is constant in the ADAF but does not have to be constant in the outer cold accretion disc, so that $\dot{M}(\text{ADAF}) = \dot{M}(r_{\text{tr}})$ in the outer disc but if the outer disc is subject to a DN–type instability the accretion will increase with radius.

The outer disc will be unstable to the dwarf nova instability if it has an effective temperature greater than about 6000 K, and therefore the quiescent disc cannot exceed this temperature at any radius. The four models presented in Fig. 2, with $\log(r_{\text{tr}}) = 4.5, 4.0, 3.5, 3.0$, have maximum effective temperatures in their discs of $T_{\text{crit}} = 1700$, 3700, 8400, 20000 Kelvin respectively. The requirement $T_{\text{crit}} < 6000$ K thus provides the constraint $\log(r_{\text{tr}}) > 3.7$, or $r_{\text{tr}} > 5000$. Just prior to outburst, one expects the cold disc to be very close to the limiting value of T_{crit}. HLMN therefore estimate that GRO J1655–40 had its transition radius at $r_{\text{tr}} \sim 6000$, or $R_{\text{tr}} \sim 10^{10}$ cm, at the time of the April 1996 outburst.

It is obvious that the quiescent data are incompatible with any model which is based only on a thin accretion disc. The ROSAT and ASCA data clearly indicate that (1) the X-ray flux of GRO J1655–40 in quiescence lies below the optical flux, and (2) the X-ray spectrum is quite hard, with a photon index < 2.7 (2 σ). A thin accretion disc model, with either a constant or variable $\dot{M}$ as a function of radius, cannot possibly reproduce such a spectrum. Thus, GRO J1655–40 is similar to A0620-00 and V404 Cyg (see NMY and NBM), where again the quiescent spectra are found to be inconsistent with a pure thin disc model but are fitted well with an ADAF+cold disc model.

3. The rise to outburst in GRO J1655–40

HLMN assumed that the mass transfer rate from the companion star, i.e. the accretion rate at the outer rim of the accretion disc, has the value given by OB, viz. $\dot{M}_{\text{transfer}} = 2 \times 10^{17}$ g s^{-1}, which is higher than the mass transfer in the ADAF. As mentioned above the value of the mass transfer in GRO J1655–40 is not known; OB use an incorrect argument to obtain their value but it seems to be reasonable and the rise to outburst does not depend on its exact value. For this value of $\dot{M}_{\text{transfer}}$, the outer cold disc is unstable to the dwarf nova instability.

The mechanism leading to the cold disc/ADAF transition is not known. The most likely explanation is that the transition between the outer cold accretion disc and the ADAF is due to evaporation into a corona which gradually erodes matter in the disc as the cold inflowing material approaches the transition radius; such a model has been proposed for dwarf novae by Meyer & Meyer-Hofmeister (1994), and by NMY and Esin et al.(1997) for SXTs.

Since a self–consistent description of the evaporation process is not available we must estimate the evaporation rate by an approximate method. Narayan & Yi (1995) have derived that the maximum allowable mass transfer rate in the ADAF at small radii is $\dot{m}_{\text{ADAF,max}} = 0.3\alpha^2$ (in Eddington units) and that $\dot{m}_{\text{ADAF,max}}$ decreases at large radii ($r > 10^3$) (see also Abramowicz et al. 1995). Assuming that the mass transfer rate within the inner ADAF is equal to the maximum, and using $M_1 = 7 M_\odot$, HLMN adopted the the following approximate prescription for evaporation:

$$\dot{M}_{\text{ev}} = \frac{2.8 \times 10^{17}}{(1 + K R_{\text{tr},10}^2)} \text{ g s}^{-1}, \qquad R_{10} \geq R_{\text{tr},10}, \qquad (3.1)$$

where K is a constant which is adjusted so as to give the required value of the transition radius, and $R_{\text{tr},10}$ is the transition radius in units of 10^{10} cm.

The local surface density evaporation rate is then related simply to the derivative of

$\dot{M}_{ev}$ with respect to R, i.e.

$$\dot{\Sigma}_{ev} = \frac{1}{2\pi R}\frac{d\dot{M}_{ev}}{dR} = \frac{9 \times 10^{-4}K}{(1 + KR_{10}^2)^2} \text{ g s}^{-1}\text{cm}^{-2}. \qquad (3.2)$$

This prescription for the evaporation is numerically close to the formula given by Meyer & Meyer-Hofmeister (1994) and a similar formula suggested by Narayan has been recently used by Cannizzo (1998)

The accretion rate in a quiescent disc must satisfy

$$\dot{M}(r) < \dot{M}_{crit}(r) = 9.6 \times 10^3 m_1^{1.73} r^{2.6} \text{ g s}^{-1}, \qquad (3.3)$$

$\dot{M}_{crit}$ from Ludwig et al. (1994) was used. The disc first becomes unstable at its inner edge when $\dot{M}$ in the disc reaches the critical value near the transition radius. This triggers an inside–out outburst. Of course this is an inside–out outburst from the point of view of the cold outer disc. From the point of view of the ADAF *all* outbursts are obviously 'outside- in'. Since most of the mass evaporation occurs close to the transition radius, the condition for the outburst is equivalent to the requirement $\dot{M}_{ev} = \dot{M}_{crit}$. For $KR_{tr,10}^2 > 1$, we then find

$$R_{tr} = 3.3 \times 10^{10}K^{-0.43}m_1^{0.19} \text{ cm}. \qquad (3.4)$$

To fix the value of K one can use the results of the spectral model of GRO J1655–40 in quiescence which requires a transition radius $\sim 10^{10}$ cm. This means that one requires a value of K of the order of 1 – 10. HLMN use $K = 5$, which gives $R_{tr} = 10^{10}$ cm for the quiescent model just before outburst. In this model, the mass transfer rate feeding the ADAF prior to the onset of the instability is found to be 4.6×10^{16} g s^{-1}, which is in excellent agreement with the $\dot{M}$ in the ADAF estimated ($\dot{M} \sim 3.5 \times 10^{16}$ g s^{-1} for $\alpha_{ADAF} = 0.3$) by fitting the spectrum of GRO J1655–40 in quiescence, especially that the transfer rate might have increased during the last month before the April 1996 outburst.

In the presence of evaporation the usual disc equation for mass conservation has to be modified as follows:

$$\frac{\partial \Sigma}{\partial t} + \dot{\Sigma}_{ev} = -\frac{1}{R}\frac{\partial}{\partial R}(R\Sigma v_R), \qquad (3.5)$$

where Σ is the surface column density in the disc, and v_R is the radial velocity. Because the evaporation law is independent of Σ, evaporation results in a disc which is sharply cut off at the transition radius R_{tr}. The inner boundary condition is given by the relation

$$\dot{M}_{disc}(R_{tr}) = 0. \qquad (3.6)$$

In the calculations by HLMN, the disc inclination was taken to be 70°, the outer radius of the disc was taken to be 4×10^{11} cm, and the inner boundary of the grid was set at $R_{in} = 4 \times 10^8$ cm. Thus the inner boundary corresponds to $R_{in} = 194R_S$ rather than $3R_S$, but this is merely for numerical convenience and does not affect any of the results presented here. Once the outburst gets underway and the thin disc extends inward to R_{in}, the time it needs to travel the additional distance to the black hole is quite short compared to the time it took to move from $R = R_{tr} = 10^{10}$ cm down to $R = R_{in} = 4 \times 10^8$ cm. Therefore, very little error is made by truncating the numerical simulation at R_{in}. The outburst was calculated with the implicit, adaptive grid code of Hameury *et al.*(1998) with the viscosity parameter α varying between 0.035 in the cool state and 0.15 in the hot state.

Figure 1 compares the initial phases of an outburst as seen in the numerical calculations with observations. The bottom panel shows the V magnitude variations $-\Delta m = -m +$

m_0, where m_0 accounts for the presence of diluting light originating essentially from the secondary (which dominates over the ADAF). For simplicity, it was assumed that m_0 is constant, and corresponds to a 48 $L_\odot$ secondary with an effective temperature of 6500 K (Orosz & Baylin 1997). During quiescence, the disc is extremely faint—fainter than both the secondary and the ADAF—and contributes less than ~ 1 % of the total light; however, in outburst, its optical luminosity, although still smaller than that of the secondary, dominates the ADAF, justifying the assumption that m_0 is constant.

The B, V, and I magnitudes decrease simultaneously in the calculations; this is independent of the magnitude of the diluting light m_0. The slopes however are directly related to m_0: for large diluting fluxes, the logarithm appearing in the definition of the magnitude can be linearized, and one has ($L(t) \ll L_0$)

$$-\Delta m = 1.09 \frac{L(t)}{L_0}, \tag{3.7}$$

where $L(t)$ is the disc luminosity and L_0 the luminosity of the companion plus the ADAF. The faster rise in the B band is thus simply due to the fact that B–V = 0 for a disc in the hot state, whereas most of the diluting light comes from the secondary with B–V ~ 0.5. This gives a factor of ~ 1.5 between the slopes of the B and V magnitudes, as observed.

The calculations predict that the various optical bands should go into outburst simultaneously, whereas observations suggest a IRVB delay. This problem is discussed in section 3.1.

The top panel in Fig. 1 shows the calculated X-ray light curve. This has been computed assuming that the emission from the ADAF has an efficiency of 0.1%, while the matter flowing through $R_{\rm in}$ in the thin disc has an efficiency of 10%. A conversion factor of 5.3×10^{35} ergs/count was assumed in order to relate ASM counts/s (2-12 kev) to X-ray luminosity (ergs/s). The calculations show that it takes about 5 days for the transition radius to move from its initial value of 10^{10} cm to values small enough that X-rays can be emitted, in excellent agreement with the optical to X-ray time delay observed in the April 1996 outburst of GRO J1655–40 (ORBM). This explanation of the observed delay is the most outstanding success of the ADAF+cold disc model of SXTs..

In the calculations both the quiescent mass transfer rate into the ADAF and the transition radius are solely determined by the evaporation law. On the other hand, the X-ray delay corresponds to the time it takes to rebuild a standard inner disc, and is thus proportional to the mass of the disc, and inversely proportional to the accretion rate at the transition radius. This accretion rate depends essentially on $\alpha_{\rm h}/\alpha_{\rm c}$, the ratio of the Shakura-Sunyaev viscosity parameter in the hot and cold states of the disc. Therefore, an increase in K results in a smaller transition radius and a smaller ADAF luminosity; the corresponding shortening of the X-ray delay can in turn be compensated for by taking a smaller $\alpha_{\rm h}$, i.e. by increasing the viscous time.

The calculations described so far show that a two-component accretion flow model consisting of an outer thin disc and an inner ADAF explains most of the key observations of GRO J1655–40. HLMN showed that a model in which the disc extends down to the last stable orbit around the black hole cannot reproduce the characteristic properties of the rise to outburst of GRO J1655–40 i.e. it cannot reproduce both the X–ray delay *and* the form of the lightcurves.

Two models were considered. The first model describes an inside-out outburst (solid lines in Fig. 3). The surface density in the outer parts of the accretion disc was adjusted so as to reproduce the correct slope of the X-ray and optical light curves, the viscosity being as in the previous model. For this reason, the agreement between the predicted and observed light curves is very good, better in fact than in the case of Fig. 1, for

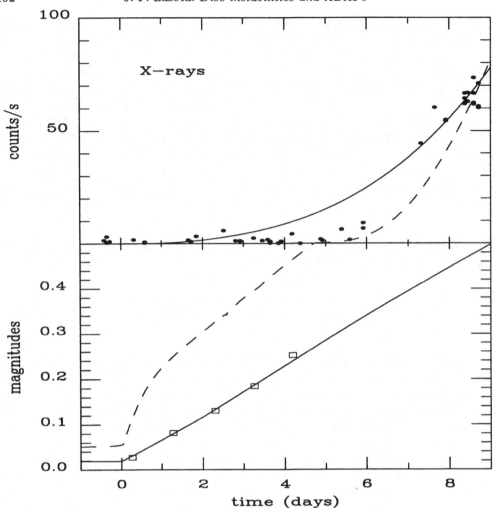

FIGURE 3. Same as Fig. 1 but for outbursts in a thin cold disc extending down to 4×10^8 cm. The solid curve corresponds to an inside-out outburst; the dashed curves to an outside-in situation. In the latter case, matter was artificially added at 8×10^{10} cm to trigger the outburst.

which such a fitting procedure was not performed, since a relaxed case (i.e. after several outbursts) was considered in order to minimize the number of free parameters. The X-ray intensity, which increases simultaneously with the optical flux, is initially quite faint. Thus, the X-ray delay depends on the sensitivity of the X-ray detector; for example, the ASM would be unable to detect the X-rays during the first 1–2 days, and therefore the model predicts an X-ray delay of this order. The delay could be increased further by decreasing the viscosity parameter, which would slow the rise of the X-ray intensity. However, this would cause the rise in the optical flux to be unacceptably slow.

The second model (dashed lines in Fig. 3) describes an outside-in outburst. In order to obtain the longest possible characteristic times, the outburst should start as far out in the accretion disc as possible. In long period systems like GRO J1655–40, the outer disc is always cold and stable; therefore to trigger the outburst we added some matter at $R = 8 \times 10^{10}$ cm. The outburst then began at $R = 7 \times 10^{10}$ cm. The viscosity chosen was such as to reproduce the observed X-ray delay ($\alpha_h = 0.10$): it takes 2.6 days for the heat front to reach the inner edge of the disc, and an additional 3.1 days for the mass accretion rate to reach 10^{16} g s^{-1}, the level at which the X-ray flux becomes detectable. Thus, this model reproduces the observed X-ray delay. However, the optical light curves are not in agreement with observations since most of the disc reaches a hot state before the heat front reaches the inner edge of the disc. Consequently, the optical flux increases too rapidly, on a thermal time scale, and then more slowly, on a viscous time scale. The observed optical light curve does not exhibit such prominent two-phase behaviour, nor does it show such a rapid optical increase, features which are the signatures of an outside-in outburst.

It is therefore difficult to reconcile the observations with a pure disc model. The difficulty of building a viable model describing the rise to outburst of GRO J1655–40 plus the near impossibility of explaining the X-ray flux in quiescence rule strongly against the one component disc model.

3.1. *The 'BVRI delay' – a problem for the DIM*

In addition to the 6–day X–ray delay ORBM report delays between the rise of the I, R, V, B magnitudes, the delay between I and B being of the order of 1 day. One should be very careful in interpreting delays observed between the rise of fluxes in various wavelengths. When one sees that an outburst is rising first in I and then in R and further "propagates" to V and B one is tempted to conclude that one is observing the beginning of an *outside–in* event. Such a conclusion was in fact arrived at by ORBM and they thought that their observation confirmed the validity of the DIM because they assumed that they had observed the propagation of a heat front which was igniting the disc first in the outer, coldest parts (increase in I) then in the hotter regions emitting the R, V and B colours. Such an interpretation appears compelling because the heating front propagation time is

$$t_{\mathrm{fr}} \approx \frac{R}{\alpha c_s} \sim 2.8\alpha^{-1}R_{10}T^{-1/2}\mathrm{hr} \qquad (3.8)$$

so the observed 1–day delay seems to be naturally explained in this fashion.

However, as explained in HLMN, the consequence of these observations is just the opposite: if there really is a ~ 1 day delay between the rise of B and V or V and I, say, then the DIM cannot be the correct explanation of this phenomenon.

The reason is the following: the DIM attributes disc outbursts to a thermal–viscous instability. In quiescence the disc is cold ($T_{eff} \lesssim 6000$ K) and non–steady. Matter transfered from the secondary accumulates somewhere in disc so that both temperature and density increase. At some ring (for *outside–in* outbursts this ring is close to the outer disc radius) the disc becomes unstable. For geometrically thin discs the thermal timescale

$$t_{\mathrm{th}} \sim \frac{H}{\alpha c_s} \qquad (3.9)$$

is (H/R) times shorter than the heat-front propagation time and much shorter than local viscous time

$$t_{\mathrm{vis}} \sim \frac{H}{\alpha c_s} \times \left(\frac{H}{R}\right)^{-2} \qquad (3.10)$$

at which the surface density Σ varies. It follows that the ring will be brought on a thermal time to $T_{eff} \gtrsim 15000$ K. At these temperatures $(B - V) \sim 0$, $(V - I) \sim 0$ so according to the DIM the B, V, I magnitudes must increase simultaneously. This is also true for rings heated up by the propagating heat front created by the initial steep temperature gradient. As the front propagates, it brings, at the local thermal time, the cold disc matter to high temperatures so that a delay of the order of 1 day between the start to rise B, V, I magnitudes cannot be the reflection of the front propagation.

The most likely solution to this problem is the possibility that the mass transfer rate from the secondary may have significantly increased since the previous observation of the source in quiescence, one month prior to the outburst. In fact, it is quite plausible that such an increase could have triggered the thermal instability that caused the April 1996 outburst. In any case, such an increase in the mass transfer rate would not show up in the B-band because the quiescent temperature of the system does not exceed 6500 K.

Observations of the SU UMa type DN WX Hyi by Kuulkers *et al.*(1991), who caught this system at the start of the outburst, show a very similar behaviour to that observed in GRO J1655–40: the V–band flux starts to rise first followed by the B–flux and shorter wavelength fluxes. The rise is faster in B than in V. The timescales in WX Hyi are of the order of 1 h whereas in GRO J1655–40 the are of the order of 1 day, but the size of GRO J1655–40 is ~ 22 times larger than that of WX Hyi. It is interesting to note that there is evidence for an increase of mass transfer rate before SU UMa's superoutbursts (Smak 1997), but Kuulkers *et al.*(1991) observed normal outbursts.

4. Conclusions

The ADAF+cold disc model provides a consistent model for the quiescent state of SXTs and it describes very well the rise to outburst due to a DN–type instability in the cold disc.

REFERENCES

ABRAMOWICZ, M.A., CHEN, X.M., KATO, S., LASOTA, J.P. & REGEV, O. 1995 *ApJ* **438**, L37

BATH, G.T. 1973 *Nature Phys. Sci* **246**, 84

CANNIZZO, J.K. 1993a In *Accretion Disks in Compact Stellar Systems* (ed. J.C. Wheeler), p. 6, World Scientific

CANNIZZO, J.K. 1993b *ApJ* **419**, 318

CANNIZZO, J.K. 1998a *ApJ* submitted

CANNIZZO, J.K. 1993b In *The 13th North American Workshop on CVs and Related Objects*(eds. S.B. Howell, E. Kuulkers & C. Woodward) (ASP Conference Series) in press

CANNIZZO, J.K, CHEN, W. & LIVIO, M. 1995 *ApJ* **454**, 880

CANNIZZO, J.K. & KENYON, S.J. 1987 *ApJ* **320**, 319

CANNIZZO, J.K., WHEELER, J. C. & GHOSH, P. 1982 In *Pulsations in Classical and Cataclysmic Variables Stars* (ed. J.P. Cox & C.J. Hansen), p. 13, University of Colorado Press

CHEN, W. ET AL. 1998 In *Accretion Processes in Astrophysics: Some like it hot* (ed. S. Holt) in press

ESIN, A.A., MCCLINTOCK, J.E. & NARAYAN, R. 1997 *ApJ* ,

GONTIKAKIS, C. & HAMEURY, J.-M. 1993 *A&A* **271**, 118

HAMEURY, J.-M., KING, A.R. & LASOTA, J.-P. 1986 *A&A* **162**, 71

HAMEURY, J.-M., KING, A.R. & LASOTA, J.-P. 1988 *A&A* **192**, 187

HAMEURY, J.-M., LASOTA, J.-P., HURÉ, J.-M. 1997 *MNRAS* **287**, 937

HAMEURY, J.-M., LASOTA, J.-P., MCCLINTOCK, J.E. & NARAYAN, R. 1988 *ApJ* **489**, 234 (HLMN)

HAMEURY, J.-M., MENOU, K., DUBUS, G., LASOTA, J.-P. & HURÉ, J.-M. 1998 *MNRAS* submitted

HJELLMING, R.M. & RUPEN, M.P. 1995 *Nature* **375**, 464

HORNE, K. & MARSH, T.R. 1986 *MNRAS* **218**, 761

HUANG, M. & WHEELER, J. C. 1989 *ApJ* **343**, 229

KING, A.R. & RITTER, H. 1997 *ApJ* in press

KOLB, U., KING, A.R., RITTER, H. & FRANK, J. 1997 *ApJ* **485**, 33

KUULKERS, E., HOLLANDER, A., OOSTERBROEK, T. & VAN PARADIJS, J. 1991 *A&A* **242**, 401

LASOTA, J.-P. 1996a In *Compact Stars in Binaries; IAU Symposium 165* eds. J. van Paradijs E.P.J. van den Heuvel & E. Kuulkers p. 43, Kluwer

LASOTA, J.-P., HAMEURY, J.-M 1998 In *Accretion Processes in Astrophysics: Some like it hot* (ed. S.Holt) in press

LASOTA, J.-P., HAMEURY, J.-M. & HURÉ, J.-M. 1995 *A&A* **302**, L29

LASOTA, J.-P., NARAYAN, R. & YI, I. 1996 *A&A* **314**, 813

LIN, D. N. C. & TAAM, R.E. 1984 In *High Energy Transients in Astrophyics* (ed. S. E. Woosley). AIP Conf. Proc. No. 115, p. 83

LIVIO, M. & PRINGLE, J.E. 1992 *MNRAS* **259**, 23P

LUDWIG, K., MEYER-HOFMEISTER, E. & RITTER, H. 1994 AnA **290**, 473

MAUCHE, C. W. 1996 In *Astrophysics in Extreme Ultraviolet; IAU Coll. 152* (ed. S. Bowyer & R.F. Malina), p. 317, Kluwer

MCCLINTOCK, J.E., HORNE, K. & REMILLARD, R.A. 1995 *ApJ* **442**, 358

MEYER, F. & MEYER-HOFMEISTER, E. 1994 *A&A* **288**, 175

MINESHIGE, S. 1996 *PASJ* **48**, 93

MINESHIGE, S. & WHEELER, J. C. 1989 *ApJ* **343**, 241

MINESHIGE, S., EBISAWA, K., TAKISAWA, M. ET AL. 1992 *PASJ* **44** 117

NARAYAN, R. 1997 In *Accretion Phenomena and Related Outflows; IAU Colloq. 63* (eds. D. Wickramasinghe, L. Ferrario & G. Bicknell), p. , ****

NARAYAN, R. & YI, I. 1995 *ApJ* **452**, 710

NARAYAN, R., BARRET, D. & MCCLINTOCK, J.E. 1997a *ApJ* **482**, 448 (NMB)

NARAYAN, R., GARCIA, M. & MCCLINTOCK, J.E. 1997b *ApJ* **478**, L79 (

NARAYAN, R., MCCLINTOCK, J.E. & YI, I. 1996 *ApJ* **452**, 821 (NMY)

OROSZ, J.A. & BAILYN, C.D. 1997 *ApJ* **477**, 876

OROSZ, J.A., REMILLARD, R. A., BAILYN, C.D. & MCCLINTOCK, J.E. 1997 *ApJ* **478**, L83 (ORBM)

PRINGLE, J.E., VERBUNT, F. & WADE, R.A. 1986 *MNRAS* **221**, 169

SHAVIV, G. & WEHRSE, R. 1991 *A&A* **251**, 117

SMAK, J.I. 1997 *Acta Astron.* **31**, 395

SMAK, J.I. 1997 In *Accretion Phenomena and Related Outflows; IAU Colloq. 63* (eds. D. Wickramasinghe, L. Ferrario & G. Bicknell), p. , ****

TANAKA, Y. & SHIBAZAKI, N. 1996 *ARA&A* **34**, 607

VAN PARADIJS, J. & VERBUNT, F. 1984 In *High Energy Transients in Astrophyics* (ed. S. E. Woosley). AIP Conf. Proc. No. 115, p. 49

WARNER, B. 1995 *Cataclysmic Variable Stars*, CUP

Magnetic fields and multi-phase gas in AGN

By ANNALISA CELOTTI [1,2]
AND MARTIN J. REES [2]

[1] S.I.S.S.A., via Beirut 2-4, 34014 Trieste, Italy

[2] Institute of Astronomy, Madingley Road, Cambridge CB3 0HA, U.K.

We stress the role that magnetic fields in the central regions of AGN can have in creating and maintaining a multi-phase medium. This is expected to be a common process in various environments, such as broad line regions, accretion disk coronae and jets. Even a small amount of thermal gas trapped by the field can re-process the 'primary' radiation and produce observable spectral signatures. We briefly recall the overall scenario and the typical physical parameters involved and then discuss whether similar processes can take place in an accretion flow.

1. Introduction

Observational evidence, from the optical to the γ-ray band, suggests the presence of gas with a range of physical properties in the inner regions of Active Galaxies. Hot/relativistic material co-exists with 'cold' quasi-thermal gas, on a large range of scales in these complex environments: tenuous hot plasma forms a corona above or around relatively cold and dense plasma, in the form of an accretion disk or re-processing clouds; clumps of photoionized gas, responsible for the observed narrow and broad emission lines, are embedded in a tenuous hotter medium.

The formation, confinement and survival of a multi-phase medium has been considered with much detail in the case of thermal re-processing by the material which gives raise to the broad lines. Difficulties associated with the possibility of maintaining the photoionized matter in two phases which are in gas pressure equilibrium, first led to the suggestion that the most likely and effective way of reaching and sustaining temperature and density gradients is given by the presence of magnetic fields in equipartition with other forms of energy (Rees 1987).

Magnetic fields are believed to be a common ingredient in the AGN environment and so can provide effective confinement and insulation of cold gas embedded in a much hotter ambient. In particular, more detailed work, focused on smaller scales, has shown that this is plausibly the case in the magneto-sphere around the central engine (typically on scales $\leq 50R_s$) as well as in relativistic jets, thanks to the combined action of the magnetic and radiation fields (Celotti, Fabian & Rees 1992; Kuncic, Blackman & Rees 1996; Kuncic, Celotti & Rees 1997; Celotti et al. 1998).

Here we briefly recall the global scenario, mainly focusing on the typical physical parameters involved in the above environments. We then discuss whether similar processes can take place in the accretion flow (torus).

2. The picture and its ingredients: field, gas, radiation

2.1. What field strength is expected? Reasonable guesses

Magnetic fields in the central regions ($R \simeq 10R_s$) of AGN in equipartition with energetically relevant components are expected to be of the order of $\sim 10^4 B_4$ G, with $B_4 \simeq 1$. In fact, whether the energetics is dominated by the accreting gas, by the kinetic energy of a jet/outflow, or by the radiation field,

$$B \simeq 4 \times 10^4 (L_{46}\eta_{-1})^{1/2} R_{14}^{-1} \beta_{-1}^{-1/2} \qquad \text{G} \tag{2.1}$$

$$B \simeq 2 \times 10^4 L_{\text{jet46}}^{1/2} (\theta R_{14})^{-1} \Gamma_1^{-1} \qquad \text{G} \tag{2.2}$$

$$B \simeq 10^4 (L_{46})^{1/2} R_{14}^{-1} \qquad \text{G} \tag{2.3}$$

in the three cases, respectively. Here η is the accretion efficiency, βc the gas in–falling velocity, θ the opening angle and Γ the Lorentz factor of the bulk outflow in a jet.

2.2. Dense and cold gas...

This field can then trap some plasma and, as this is likely to have cooled to the Compton temperature, a density contrast as high as $\sim T_{\text{virial}}/T_{\text{Compt}}$ with respect to the external/hot phase medium can be attained. Two body processes start to become very efficient in cooling the dense trapped gas, which can thus fall further towards the equivalent blackbody temperature of the radiation field

$$T_{\text{bb}} \simeq 2 \times 10^5 L_{46}^{1/4} R_{14}^{-1/2} \qquad \text{K.} \tag{2.4}$$

The gas pressure is limited by the balance with the total magnetic field stresses and therefore a maximum density can be achieved, of the order of

$$n \simeq 10^{17} B_4^2 T_5^{-1} \qquad \text{cm}^{-3}. \tag{2.5}$$

2.3. ...in thin sheets/filaments

The spatial distribution of the gas is clearly determined by the field structure and the net external force acting on it. This would limit the scale–height of the gas to a typical value such that the tension of the distorted field (in the perpendicular direction) balances the net effect of gravity, radiation pressure and acceleration. Typically, in presence of loops with a toroidal field component, the action of gravity leads to dimensions $h_{\text{g}} \sim kT/m_{\text{p}}g$. However, other forces can dominate in these environments, such as the gas inertia to dynamical acceleration and the radiation force. In particular the opacity (mostly free–free) of the extremely dense and cold gas can give raise to a radiative acceleration a. This presumably acts in the direction opposite to gravity with an intensity which depends on the incident spectrum, but can reach $a \simeq g \int (\sigma_\nu L_\nu / \sigma_T L_E) d\nu \sim 10^4 g$, where L_E represents the Eddington luminosity. This determines an upper limit to the effective scale–height of the trapped gas (in this direction); the most notable, and perhaps surprising, result is that this scale is only of the order of meters

$$h_{\text{eff}} \sim 6 \times 10^6 \frac{g}{a} M_8^{-1} T_5 R_{14}^2 \sim 6 \times 10^2 M_8^{-1} T_5 R_{14}^2 \qquad \text{cm,} \tag{2.6}$$

and even smaller if the gas is subjected to stronger forces.

It should be stressed that, from the point of view of the large–scale dynamics, these structures of cold and dense gas could be treated as a uniform loading to the global flow, as long as the individual clouds/filaments are small enough that, dragged by the field, they are constrained to follow the same dynamics as the hot material. This requires them to be smaller than the scale–height h_{eff} imposed by the effective local acceleration, which may be dominated by the effect of magnetic stresses or internal random motions (see also § 3).

2.4. The survival of the cold phase

Although measurable by an ordinary tape measure, h_{eff} greatly exceeds, for instance, the characteristic gyro–radii, which when $B_4 \simeq 1$ are below a centimeter for all sub–relativistic electrons, and for cool–phase ions as well. More accurate estimates have

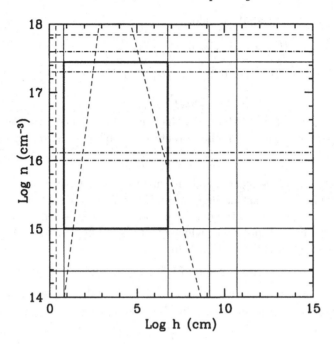

FIGURE 1. The parameter space density n vs dimension h for structures of cold gas, magnetically confined and embedded in a relativistic jet environment. The various lines represent constraints on these two parameters set by energetical, observational considerations and by the requirement that the cold phase can survive over the relevant timescales against conduction and diffusion. More precisely the various lines represent: *(vertical lines, from left to right)* non–thermal diffusion depth transverse to the field lines in a cooling timescale, minimum scalesize of structures unaffected by magnetic diffusion, gravitational pressure scale–height, maximum sound crossing scalesize for pressure equilibrium on a dynamical timescale, cloud dimension corresponding to a filling factor 10^{-8}; *(oblique lines, from left to right)* non–thermal diffusion depth transverse to the field lines in a dynamical timescale, non–thermal diffusion depth in a cooling timescale; *(horizontal lines, from top to bottom, respectively)* jet kinetic power of the order of 10^{46} erg s^{-1}, bremsstrahlung emission comparable to an observed spectral luminosity of 10^{46} erg s^{-1}, thermal pressure balancing the confining magnetic pressure, Comptonization luminosity comparable to an observed spectral luminosity of 10^{46} erg s^{-1}, spectral turnover at optical frequencies due to free–free absorption, neutral hydrogen column density limit of $N_{\rm H} = 10^{21}$ cm^{-2}, minimum density for gas to remain cool in the presence of thermal diffusion, minimum density for 2–body processes to cool gas more efficiently than particle–photon processes. The rectangular area defined by the thick lines is the allowed region of the parameter space. From Celotti et al. (1998).

shown that the strong magnetic field effectively acts as an insulator for the cold gas with respect to thermal conduction by the hot external particles. Magnetic diffusion is also not efficient enough to suppress, on radiative and dynamical time–scales, the existence of such pockets of cold matter, in AGN environments such as magneto–sphere and relativistic jet.

In Fig. 1, we show, as illustration, the limits on the density and dimension of structures of cold material trapped in a relativistic jet in order to both satisfy energetic and observational constraints and survive disruptive effects over a dynamical time–scale (see figure caption).

A further threat to the clouds/filaments survival is however present in turbulent flows. In fact, while in the environments just mentioned the field can be either anchored in

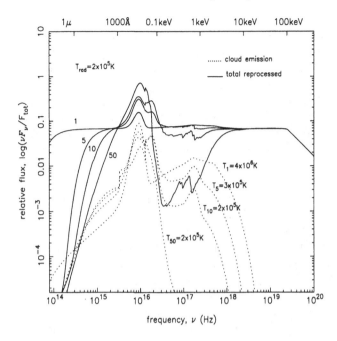

FIGURE 2. Reprocessed spectra emitted by cold gas structures (clouds) located in the central AGN magneto–sphere, for different cold gas and external radiation conditions. The 'primary' incident spectrum is characterized by a low energy cut–off and an equivalent radiation temperature $T_{\rm rad}$. The different lines represent a different number of clouds along the line of sight. Their corresponding temperatures are also indicated. The clouds heating and cooling are generally dominated by free–free plus photoelectrical absorption and collisionally excited recombination line and free–free emission. From Kuncic et al. (1997).

the disk (corona) or have an ordered component (jet), turbulence might cause these structures to be shredded (see § 3).

2.5. *Radiative signatures*

Because of the tiny transversal dimensions, the cold matter occupies a small fraction of the total volume. Also, the mass in cool gas can be less that the hot phase one, provided that the scale–height $< (T_{\rm bb}/T_{\rm virial})R$. Nevertheless, because of the extreme high density, this matter can be very effective in absorbing (mainly through free–free) and re–processing radiation impinging on it.

Furthermore, the 3D structure of the field can allow the matter to accumulate on sheet–like surfaces or filaments in the direction perpendicular to the net force, distributing it with high covering factors. Even though the volume filling factor is small, the surface covering factor can easily be of order unity.

Under these circumstances, the absorbed radiation can be energetically important and the spectral effect relevant. The re–emitted spectrum is expected to be quasi–thermal at the typical gas temperature and any spectral feature would be broadened and shifted by the high velocity motion of the gas.

In Fig. 2, we show the reprocessed spectrum resulting from the detailed radiative transfer calculation (obtained by using the code CLOUDY) in the case of cold matter located in the inner $\sim 50R_{\rm s}$ of an AGN magneto–sphere (see figure caption).

3. Torii

One can envisage a similar picture in ADAFs. Here we would like to stress a couple of points, namely the relevance to the observational properties of ADAFs and the issue posed by the turbulent flow on the clouds survival. More specific work is in preparation.

Because of the comparatively low radiation field in ADAFs, the gas might cool to temperatures much lower than 10^5 K, permitting a density contrast between the two phases as high as 10^9. The low temperature and high density once again cause the trapped gas to be an efficient absorber, even more so as a significant fraction of the ADAF radiation is emitted as cyclo–synchrotron at low frequencies.

Therefore, if these dense filaments existed within an ADAF, with (at least at some radii) a substantial covering factor, they could modify the standard ADAF spectrum. In particular, the radio and mm bump could be absorbed by free–free emission (re–appearing in the optical or UV bands). The actual effect on the radio spectrum would depend on whether the clouds were mainly in the outer part (in which case they could absorb everything) or just in the inner parts (where they would absorb only the higher–frequency sub–mm radiation). We stress here that this is an interesting possible complication in confronting ADAF models with observations.

The radiation pressure could be important, despite the low luminosities, because of the large free–free absorption cross–section. A possibly severe threat to filament survival – as already mentioned – is posed in an environment where the magnetic field provides a high viscosity. In fact, the random motions of field loops and large–amplitude Alfven waves would cause violent local accelerations, which correspondingly reduce the thickness of filaments that could be carried with the flow. The acceleration would be largest on the smallest scales. Let us consider, simply as an illustration, a Kolmogorov spectrum where the velocities associated with a scale h are proportional to $h^{1/3}$: the associated accelerations would then depend on scale as $h^{-1/3}$. This means that the filaments of a given scale can less easily follow the smaller–scale eddies. On the other hand, stretching of field lines would decrease the scale of the filaments.

We draw attention to this issue, emphasizing the uncertainty in the fate of dense filaments that are advected inwards with the flow towards regions of progressively stronger field nearer the hole. One can envisage an element of gas, trapped within a magnetic tube, starting off in the broad line region and moving into the denser and more extreme conditions envisaged in the ADAF region. But can the cool–phase gas survive, or will it be shredded and homogenized? It is unclear whether the disordered motions associated with the magnetic viscosity would shred any filaments into structures so thin that they cannot survive the effects of conductivity and diffusion.

This is an important open question for matter which is coupled to the field lines. However, material decoupled from the field (e.g. ejecta from stars, behaving like diamagnetic clouds) could survive more easily since it could be squeezed by the surrounding field but would not necessarily follow the violent random motion associated with the 'α–viscosity'. Therefore, irrespective of the answer to the question posed above, small clouds or filaments of magnetically–confined cool gas could still have important consequences for the ADAF spectrum.

4. Conclusions

We have stressed the key role that magnetic fields in quasi–equipartition in the central regions of AGN can have in creating and maintaining a multi–phase medium, by confining and insulating cold, dense gas with small filling factor. This is expected to be a common process in these environments (broad line region, accretion disk coronae, jets, accretion torii,...). Even a small amount of thermal gas, thus trapped by the field, can significantly re–process the 'primary' radiation and produce observable spectral signatures.

A. C. thanks the organizers for the last–minute invitation and for providing a scientifically, culturally and geographically interesting environment for this meeting. Gary Ferland is thanked, once again, for the use of CLOUDY. The Royal Society (M. J. R.), MURST and the Institute of Astronomy PPARC Theory Grant (A. C.) are acknowledged for financial support.

REFERENCES

CELOTTI, A., FABIAN, A.C.& REES, M.J. 1992 Dense thin clouds in the central regions of active galactic nuclei. *MNRAS* **255**, 419–422.

CELOTTI, A., KUNCIC, Z., REES, M.J. & WARDLE, J.F.C. 1998 Thermal material in relativistic jets. *MNRAS* **293**, 288–298.

KUNCIC, Z., BLACKMAN, E. & REES, M.J. 1996 Physical constraints on the sizes of dense clouds in the central magnetospheres of active galactic nuclei. *MNRAS* **283**, 1322–1330.

KUNCIC, Z., CELOTTI, A. & REES, M.J. 1997 Dense, thin clouds and reprocessed radiation in the central regions of active galactic nuclei. *MNRAS* **284**, 717–730.

REES, M.J. 1987 Magnetic confinement of broad-line clouds in active galactic nuclei. *MNRAS* **228**, 47–50p.

Supermassive binary black holes in galaxies

By PAWEL ARTYMOWICZ

Stockholm Observatory, SE-13336 Saltsjöbaden, Sweden

There is a strong observational evidence for the presence, even ubiquity, of single supermassive black holes in the nuclei of galaxies. Galaxies, on the other hand, colloide and merge frequently and therefore may contain supermassive binary black holes. We review the current observational evidence of the existence of such objects: double nuclei of galaxies, wiggly jets from active nuclei, double emission lines, and temporal variability of emission. In theory, the observability and frequency of such pairs depends sensitively on the rate of the orbital evolution of the binary under the influence of the surrounding stellar system, as well as the galactic interstellar medium (gas). We consider some important aspects of this interaction, including the 'loss cone' problem leading to the evolutionary deadlock, the possible role of orbital eccentricity in speeding up the eventual merger of the black holes, and the effects of a circumbinary accretion disk. While the observational evidence is generally consistent with theoretical expectations, there is ample room for improvement in both the theory and the observations.

1. Introduction

Single supermassive black holes are objects significantly more massive than stars (i.e., $> 10^3 M_\odot$; In the following, we abbreviate a supermassive black hole as BH). As mentioned in section 2 below, there is now accumulating and persuasive observational evidence for the presence of BHs in galaxies, both normal and active ones, perhaps in almost every galaxy including our own (see also Madejski this volume).

In contrast, binary supermassive black holes (binary BHs) remain elusive, difficult to both detect and identify as such by observation, and to study theoretically. It is in largely because their evolution is almost entirely a matter of external influences (at least until the vary latest stages of evolution, mentioned in section 3.1.), rather than their internal physics. They are thought to form in merging galaxies, both of which happen to have a pre-existing BH. Galaxy mergers are supposed to be frequent enough for the binary BHs to be a fairly common, if not necessarily very long-lived, phenomenon. If found, the massive, close binaries may help test the theory of gravitation for field hundreds of times stronger than in the binary pulsar case. There is already some observational evidence for binary BHs (section 2).

All of the above provides a strong motivation for considering the process of the formation and evolution of such double objects from the complementary perspectives of the theories of stellar dynamics (in section 3) and gas dynamics (in section 4). Among others, we will emphasize the problem of eccentricity evolution of a binary MBH in a galaxy, and review an almost complete circle that the views on this issue, made in the last 2 decades.

2. Galactic mergers and massive binary black holes

2.1. Mergers of galaxies

Mergers, especially at early epochs (redshifts $z = 1 - 2$), played a crucial role in the formation and morphological evolution of galaxies. For instance, one well studied way to form elliptical galaxies is by a merger or otherwise a strong tidal perturbation of spiral galaxies (Barnes & Hernquist 1992, Zepf & Ashman 1993, Moore et al. 1998). On-going mergers are also common (e.g., Schweizer 1986, Hutchings *et al.*1988, Liu & Kennicutt 1995, Schweizer *et al.*1996).

Many details, however, remain to be clarified, for instance the statistics of globular clusters and the stellar density distribution in the centers of ellipticals. The N-body simulations of mergers of spherical isotropic galaxies without central BHs did not reproduce correctly the correlation of core radius and luminosity (or effective radius; Barnes & Hernquist 1992). Ebisuzaki *et al.*(1991) performed galaxy merger calculations assuming the presence of BHs, and found that the interaction of the BH pair with the merger remnant puffs it up kinematically. As required by observations, the core radius increases in each merger, as opposed to the calculations without BHs where the radius stays constant.

The precise frequency of mergers at early and present epochs is not known, but a galaxy like our own may have accreted on the order of one dwarf galaxy per Gyr. Generally, it is found that mergers are rapid, i.e. that they occur on a short dynamical-friction time scale of the bulges of components with respect to each other (e.g., Governato et al. 1994) However, N-body calculations (Smith & Miller 1994) show that small nuclei of galaxies may either merge rapidly or slowly, depending on the background gravitational potential.

2.2. *Black holes in galaxies*

There is a rapidly accumulating evidence for massive, presumably single, BHs in the centers of galaxies. (For reviews and references see Rees 1990, Kormendy & Richstone 1995). The dynamical studies assign masses $\sim 10^6$ to $10^{9.5} M_\odot$ to the central compact objects (most probably BH) in the Galaxy, M31, M32, M87, NGC 3115, NGC 3377, NGC 4258, and NGC 4594. BHs are found in $\sim 20\%$ of nearby E-Sbc galaxies, consistent with predictions based on quasar energetics. BH masses are typically equal to 0.004 times the mass of the bulge of the galaxy.

The two facts, (i) that many galaxies contain BH, and (ii) they often merge, requires a relatively frequent formation of BH pairs. Whether or not a pair disintegrates, merges, or just remains forever (for > Hubble time) as a pair in the merger stellar remnant is a central theoretical and observational question yet to be satisfactorily answered. Below we summarize the main lines of evidence for the existence of supermassive binary BHs.

2.3. *Radio-loud and radio-quiet AGN*

Wilson & Colbert (1995) discussed the two physically distinct classes of AGNs: radio-loud and radio-quiet. They argued that the black hole masses and mass accretion rates in the two classes are not greatly different, because their continua and lines in a wide range of wavelengths are similar. The difference between the classes is likely associated with the spin of the black hole. Since the high spin rates are difficult to acquire in ways other than a merger of two or more BH, Wilson and Colbert proposed that the merger of binary BHs in galaxies is ultimately responsible for the differences between the radio-loud and radio-quiet objects. While it has not been estimated how often successful mergers must occur (the ones in which the binary BH is short-lived compared with Hubble time), it is clear that this explanation can only be valid if the probability that galaxy merger is followed by a BH merger is not much smaller than unity.

2.4. *Black holes in double galactic nuclei?*

While many nuclei of galaxies are double, most are poorly resolved. We concentrate here on just the two best studied objects, M31 and NGC 4486B. Lauer et al. (1993) found that the nucleus of M31 is double. HST images revealed that the brighter of the two nuclear components (P1) is separated by only 0.49" (1.8 pc) from the photocenter of the surrounding bulge, which itself coincides with the less luminous peak (P2). Explain-

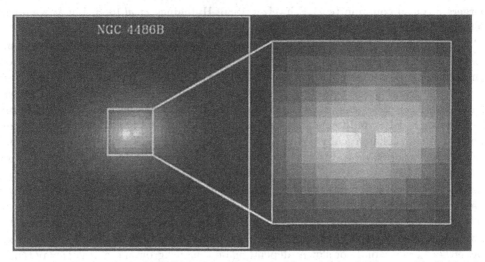

FIGURE 1. Double nucleus of NGC4486B: a single or a binary object? (Source: K. Gebhardt, T. Lauer, and NASA)

ing the double structure by dust absorption or superposition of a foreground object is difficult. P1 could be a dense remnant of a stellar system cannibalized by M31. The authors favored a hypothesis in which P1-P2 system is a supermassive binary with separate stellar systems associated with the two black holes (Lauer et al. 1993).

One problem with this interpretation is that M31 does not appear to have recently undergone a merger, while the decay time for P1's orbit due to dynamical friction is very short ($< 10^6$ yr from N-body simulations assuming nearly circular orbit; Emsellem, Combes 1997). Nevertheless, P1 could still be a separate stellar system that is infalling rather than orbiting the mass center on a thight orbit of less than 20" radius. Moreover, N-body merger calculations of Smith & Miller (1994) indicated that under favorable conditions, such as a 'flat' surrounding gravitational potential that counteracts the tidal disruption, the two cores (without BH) may in some cases survive for as many as thousands of orbits before they eventually merge. This offers a possible reason for the suspected stability of the P1+P2 system.

On the other hand, Tremaine (1995) proposed that the nucleus of M31 contains a thick eccentric disk of stars traveling on aligned, Keplerian orbits around the $10^{7.5} - 10^8 M_\odot$ black hole thought to exist at the center. In this model, P1 is the part of the disk farthest from the black hole, which is bright because the disk stars linger near apocenters.

NGC 4486B, shown in Figure 1, also has a double nucleus similar to that seen in M31 (Lauer et al. 1996). The NGC 4486B nucleus comprises two peaks separated by 12 pc. The double structure is not likely to arise from dust absorption. It also resembles M31 in its kinematics (Kormendy *et al.*1997). Interpretation of the double nucleus of NGC 4486B in terms of the eccentic disk model of Tremaine (1985) is one possibility. Another one is that NGC 486B contains a physical double system containing BHs (cf. section 4.1).

2.5. *Wiggly jets from AGN*

Spiral, apparently precessing, jets accompany many radio galaxies and AGNs (e.g., 4C26.03, NGC 315, NGC 326, Sagittarius A; cf. Ekers et al. 1978, Brown 1982). On large spatial scales of galaxies, the jets may precess under the gravity of companion galaxies, as in 3C 130 (Jaegers & de Grijp, 1985), or in 3C 75, which sports a pair of twin jets originating in the double nucleus of the central galaxy in cluster Abell 400 (Owen et al. 1985).

The wiggles in small-scale jets may be due to Lense-Thirring precession (geodetic precession) of a jet emanating from one of the BHs in a binary system (Begelman et al. 1980, Hunstead et al. 1984, Lu 1990, Katz 1997). In this case, the precession period equals

$$t_{\text{geo}} = 600 \, \text{yr} \, (r/10^{16} \text{cm})^{5/2} (M/10^8 M_\odot)^{-3/2} (M/m) \,, \tag{2.1}$$

where m and M are the masses of the low and high mass components and r is the orbital distance.

Alternatively, wiggles can be directly introduced into the jet by the fact that the point of origin of the jet is different at different times or, more importantly, its velocity is modulated with the binary period by addition of the binary's orbital velocity to the intrinsic jet velocity, which causes jet direction to sweep a cone with the opening angle proportional to orbital/jet velocity ratio. Kaastra & Roos (1992) discussed the applications of this model to M87 and 3C 273. Hardee et al. (1994) presented hydrodynamical models of jet corrugations induced by a periodic disturbance near the origin, and found the conditions leading to the amplification of wiggles.

In the interesting case of the quasar 1928+738 with a superluminal jet, Roos et al. (1993) established that the jet precession cannot be geodetic, because the apparent precession period is short, only 2.9 yr (in observer frame). If interpreted in terms of eq. (2.1) above, $t_{\text{geo}} \approx 3$ yr would imply an extremely short destruction time of the binary due to gravitational radiation emission, of order ~ 10 yr $(10^{16} \text{cm}/r)(m/M)$. The orbital motion, on the contrary, yields a reasonable estimate of the orbital radius,

$$r \approx 1.4 \times 10^{16} \text{cm} (M/10^8 M_\odot)^{1/3} (P/3 \, \text{yr})^{2/3}. \tag{2.2}$$

Gravitational radiation acts on time scale of $\sim$Myr, and the geometry of the jet bending is consistent with a 2° misalignment between the binary spin axis and the BH spin axis. (Additional Lense-Thirring precession is predicted in 1928+738 on time scale $> 10^3$ yr, unfortunately too long for an easy detection.)

2.6. *Double emission lines*

A class of quasars named after the prototype object 3C 390.3 exhibits broad emission lines with one or two peaks substantially Doppler shifted from the rest frame of the host galaxy. Gaskell (1983) first suggested that 3C 390.3 quasars are spectroscopic binaries consisting of two supermassive BHs, one or both of which give rise to the shifted emission line components. His analysis of the positions of the blue-shifted peak of Hβ line in 3C 390.3 showed that the line moved steadily closer to systemic velocity by more than 1500 km/s over the period of 20 years (Gaskell 1996). Little is known about the red-shifted line component, but its wavelength appears to have decreased by the equivalent of many hundred km/s (in ratio 1:2 to the blueshifts). The minimum period of radial velocity changes was estimated at 210 yr (the true value being likely of order 300 yr), and the masses and separation of the BHs at $2.2 \times 10^9 M_\odot$, $4.4 \times 10^9 M_\odot$, and ~ 0.3 pc. According to Gaskell (1996), the binary BH model is more attractive than several other alternative models of the origin of double peaks (e.g., biconical ejection, line emission from accretion disk ansae), because it accounts for the magnitude of velocity displacement, for the blue- and redshifted broad lines being equally common, and especially prevalent in systems not seen pole-on, for why some objects have one and some two peaks (in analogy with single and double lined spectroscopic binary stars), and for the variability of peaks and the continuum (e.g., non-correlated spectral variability of red and blue line components; Miller & Peterson 1990).

While 3C 390.3 is the best known example, the displaced broad lines are relatively common among quasars. In the case of a luminous quasar OX 169, this spectroscopic feature coexists with the morphological feature ("jet" of stars), both of which find a plausible explanation in the model involving galaxy merger and a BH pair (Stockton & Farnham, 1991). If the binary BH interpretation of the double lines is correct, this points to a large frequency of black hole pairs in AGN, since most binary BHs will be missed in observations, due to unfavorable inclination or orbital phase.

However, the spectroscopic-binary interpretation of double peaks was recently challenged by Newman et al. (1997) and Eracleous et al. (1997). The first paper discusses the variability properties of the double-peaked emission lines in Arp 102B, where an accretion disk origin is preferred over other models including the double system hypothesis. The second paper aims at disproving the binary BH hypothesis for this and additional two objects (3C 390.3, and 3C 332), based on spectroscopic monitoring spanning two decades. The most striking case is provided by the prototype, quasar 3C 390.3. A trend noticed by Gaskell in one of the Balmer-line peaks of 3C 390.3 before 1988 simply did not continue after that year. This immediately invalidates the 300 yr esimate of the orbital period, as well as formally requires a strangely large mass of the BH system, $> 10^{11} M_\odot$. Obviously, this value is hard to reconcile with the usually much smaller values derived from AGN luminosities. Eracleous et al. (1997) propose that transient but otherwise similar double-peaked emission lines in nearby active galactic nuclei (NGC 1097, M81, and Pictor A) are not likely to originate in binary broad-line regions, but may be produced by a single compact source in an as yet unspecified way. The current status of the "double-lined spectroscopic binary" interpretation for the 3C 390.3 systems is uncertain at present.

2.7. *Variability of emission - OJ 287*

The luminous BL Lac quasar (blazar) OJ 287 has high optical polarization and an extremely variable total and polarized flux, which has been extensively monitored (cf. Sillanpää et al. 1996 and references therein). Keeping relatively constant color indices, this quasar varies in B-band between B=12.5 and B=18.5 mag over time scales of years, with smaller fluctuations on short time scales (Gonzales-Perez *et al.*1996).

During the last hundred years, quasar OJ 287 has experienced several major outbursts as well as numerous smaller brightness fluctuations. The major outbursts occur at intervals of approximately 12 years (9 years in the rest frame of the quasar, which is at $z = 0.306$). Figure 2 illustrates the historical data. Stothers & Sillanpää (1997) reconfirmed recently that the near periodicity is real, based partly on the characteristic double-peak structure of each major outburst (seen in Fig. 2).

OJ 287 varies much more frequently with much higher amplitude of brightness variations than predicted by microlensing scenarios. The best existing models of the nearly-periodic variability all involve a binary BH interacting with disk(s) and creating jet(s) aligned nearly along the line of sight. The observed period is usually identified with the orbital period of the BH. Interaction-induced activity is particularly attractive in the case of OJ 287, because a binary BH model can account not only for the optical flux periodicity (Lehto & Valtonen 1996), but also the VLBI radio morphology, which can be explained by a precessing jet with a narrow opening angle, directed nearly toward us (Vicente et al. 1996). In addition, the imaging of the surrounding nebulosity (host galaxy) by Yanny et al (1997) support the notion of recent merger activity that might have led to a binary BH. The specific models of OJ 287 invoke gas disk-binary interaction, which we will describe briefly in section 4.2.

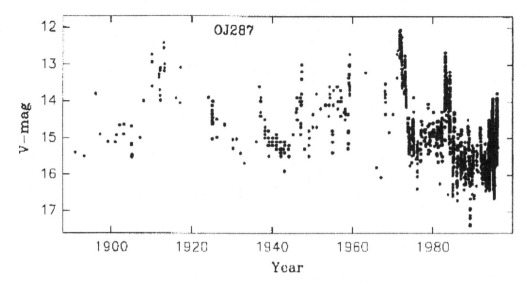

FIGURE 2. Fingerprints of a supermassive binary: Historic light curve showing nearly periodic outbursts of quasar OJ 287 (from: Silanpää et al. 1996)

3. Stellar dynamics around binary black holes

3.1. *The life of a massive binary in a galaxy*

Begelman, Blandford & Rees (1980; denoted BBR hereafter) divided the life of a pair of BHs in a merger remnant of two galaxies into several stages. First, the initial merger of two stellar systems (nuclei) should occur on a rapid time scale ($< 10^8$ yr) because the dynamical friction is very efficient. The second stage, lasting for a similar or only slightly longer time (inversely proportional to the secondary BH mass) is associated with the settling of the secondary toward the center of the merged nucleus, until the two black holes form a true binary with orbital speed exceeding the ambient velocity dispersion.

The subsequent binary hardening is controlled by the formation of the 'loss cone', a region in the phase space surrounding the binary containing highly elongated orbits, which gets depleted as a result of close encounters of stars with the binary. The loss cone is refilled on an exceedingly long time scale (of order $\sim$ solid angle of cone $\times$ two-body relaxation time > Hubble time). In addition, the semi-major axis will shrink significantly only after many loss-cone filling times.

The final stage of the evolution is the rapid binary hardening connected with energy losses due to emission of gravitational radiation. It only takes 10^8 yr for the BH to merge from the initial separation of a few times 0.01 pc, but the rate of evolution falls as the 4th power of distance, so that the gravitational radiation cannot be effective at $r \gg 0.01$ pc. At the end of the final stage of evolution, a brief physical merger of BHs should produce an observable burst of electromagnetic and gravitational waves (Thorne & Braginsky 1976, Fukushige et al. 1992a, Rajagopal & Romani 1995).

3.2. *The evolutionary bottleneck*

BBR estimated the time scales of the three evolutionary stages. The galaxy merger and the BH binary formation stages have an associated 10^6 yr time scale, at a binary separation scale of order 1–10 pc. The time scale then jumps due to loss cone problem to 10^{11} yr at separations below 1 pc, and stays constant until the separation decreases

Table 1. Do binary supermassive black holes typically merge in one Hubble time?

Year	Authors	Answer	$e =$?
1980	Begelman, Blandford & Rees	No/Yes	$e = 0$
1981	Roos	Yes/No	
1989	Valtaoja, Valtonen & Byrd	No	
1991	Ebisuzaki, Makino & Okamura	Yes(!)	$e \to 1$
1992	Fukushige, Ebisuzaki & Makino	Yes(!)	$e \to 1$
1992	Mikkola & Valtonen	Yes(?)	
1993	Kaastra & Roos	Yes	
1994	Governato, Colpi & Marashi	No/Yes	
1994	Polanrev & Rees	No(?)	$e \ll 1$
1996	Quinlan	Yes/No	$e \ll 1$
1997	Makino	No(?)	
1997	Quinlan & Hernquist	Yes/No	$e \ll 1$

to ~ 0.01–0.1 pc, after which the time scale begins to shorten catastrophically. Unless something dramatic happens (powerful collective effects in the galaxy, accretion of dwarf galaxies, or very large stochastic motion of the binary in the nucleus recently proposed by Quinlan & Hernquist 1997) the stellar-dynamical evolution will stall at the stage where the binary BH has a separation of ~ 0.1 to 1 pc. A separate mechanism for binary hardening seems necessary to decrease the orbital distance to that at which gravitational radiation emission takes over.

The predicted bottleneck of the stellar dynamical evolution, if real, would have significant observational ramifications. It would cause binary BH systems to be at least as widespread as single BHs, which does not seem to be observed. Moreover, if the entire evolution of a pair of BHs takes more time than the mean time between galactic mergers, then a triple or multiple system of BHs might form. The ultimate result of a few-body dynamics could be the tightening of the binary and the ejection into intergalactic space of a third BH (Valtonen *et al.*1994, Valtonen 1996).

In the following sections we discuss several possible ways in which the evolutionary stalemate can be avoided.

3.3. *The role of orbital eccentricity*

The orbital eccentricty of the binary may have a profound effect on its evolution. Low eccentricity assumed in the early work (BBR) is one of the culprits in creating the bottleneck of the stellar-dynamical evolution described above. If, however, eccentricity could grow secularly due to some factors, then even a slow decrease of semi-major axis would not hamper the rapid decrease of the pericenter distance. When that distance falls below a few times 10^{-2} pc, gravitational radiation takes over and rapidly shrinks the binary just as in the non-eccentric case. It is therefore essential to understand what can drive the eccentricity evolution of a binary BH. The research on that issue has an interesting history. Various works during the past 20 years reached diverging conclusions about the ability of the eccentricity to grow and to break the evolutionary deadlock. In Table 1 we summarize the main conclusion of some important papers, also listed in the references. Symbol $e \ll 1$ means 'non-extreme eccentricity', which includes low and intermediate e.

Similar to BBR, Roos (1981) estimated that two black holes initially embedded in much more massive nuclei will not be able to merge due to 3-body scattering of stars.

According to Roos, only BH of mass comparable with the mass of a surrounding stellar density cusp (and the original nucleus) have a chance to form a tight binary (< 0.1 pc). An even more evident difficulty is often encountered at a much larger separation $\gg 1$ pc, when the Chandrasekhar's dynamical friction formula used in numerical simulation of the orbital motion of the secondary black hole leads to a very slow settling, or no settling at all (Valtaoja et al. 1989). A similar conclusion was reached by Governato et al. (1994), who performed N-body merger simulations and found that a close BH binary does not form in a Hubble time when the progenitor galaxies have similar masses but different central densities, and the merger is not of an unlikely head-on type. Nevertheless, special cases or initial conditions always provide a loophole allowing the formation of a parsec-scale binary, although perhaps not as a predominant outcome.

This is where the issue of eccentricity becomes crucial. The argument originally given by BBR that "any eccentricity in the orbit will almost certainly be erased by frictional drag" does not hold either upon a closer analysis of the 3-body stellar dynamics, nor the dynamics of gas inflow (cf. below). As regards the stellar dynamics of two BHs plus a passing star, Roos (1981) noticed that a binary with $e = 0.6$ tends to grow more eccentric in his numerical calculations.

Papers published between 1991 and 1993 (see Table 1) developed the idea of a rapid eccentricity generation by the following simple mechanism: dynamical friction between a BH and its ambient stellar medium and hence the removal of angular momentum from the binary orbit are strongest when their relative velocities are smallest, i.e. at the apocenter of the binary orbit. This effect, unless countered by a sharp difference of stellar densities at apo/pericenter, leads, according to celestial mechanics, to a rapid eccentricity growth. However, in more detailed treatments of the system a distinction was made between the cusp and flat core regions in the merged nucleus, and the prediction of extreme eccentricities was generally not confirmed. Vecchio et al (1994) and Polnarev & Rees (1994) studied the motion of small black holes in a large, fixed stellar system and found only a moderate growth of eccentricity followed by eccentricity damping in the late stages. One reason for the changed conclusion was a different physical setup: small BH in a more massive stellar system vs. the opposite in earlier works, as well as a fuller recognition of the e-damping effect of gravitational radiation near periastron. Quinlan (1996) derived the binary semimajor axis and eccentricity changes from scattering experiments. The eccentricity growth was found to be moderate and qualitatively unimportant.

Rauch & Tremaine (1996) described the so-called resonant friction effect, which can strongly influence the orbital evolution of a binary BH. For sufficiently non-equal mass binaries, and provided that the stellar velocity ellipsoid is radially elongated, the eccentricity of the binary may grow, at least until suppressed by relativistic precession and effects of gravitational radiation. Quinlan & Hernquist (1997) continued their N-body calculation of mergers involving BHs, generally confirming the basic mechanism of the resonant friction theory. However, they concluded that conditions needed for eccentricity growth (esp. the radial bias in the velocity ellipsoid) are very unlikely to occur in real systems, especially merger remnants, and that therefore the eccentricity does not provide a resolution for the vexing problem of the time scale of the merger.

4. Gas dynamics around binary black holes

Gaseous interstellar medium, if available for accretion, i.e., abundant and having sufficiently low angular momentum, can feed a growing BH in the galaxy center much faster than the disrupted/accreted stars. Several mechanisms can supply the gas from large (kiloparsec) scales, down to the subparsec scales around the BH.

First, a galactic stellar bar or hierarchy of bars-in-bars, possibly mutually induced, forces shocks in the gas flow which efficiently remove the angular momentum from the gas, thus allowing accretion (Van Albada & Roberts 1981; Schlosman et al. 1989, 1990). Second, Heller & Schlosman (1994) modeled massive, gravitationally unstable disks, which can cause starbursts or AGN activity in the centers without the intervention of the stellar component. The large-scale instabilities and gas supply may be caused by merger activity or tidal interactions between galaxies.

As recognized by BBR, gas accretion opens an efficient route to overcoming the bottleneck in the stellar-system driven evolution of binary BH discussed in section 3. BBR proposed that gas accretion onto the more massive hole will cause the binary separation to decrease in accordance with the adiabatic constancy of the product Ma in a circular binary (a being the semi-major axis of the orbit). The gas was assumed to fall radially (with zero angular momentum) onto the dominant BH. They estimated that mass influx at a rate of 1 $M_\odot$/yr would shrink the binary in a characteristic time of 10^8 yr. The above assumptions are questionable in general, and are not supported by realistic modeling. Such a modeling has not yet been performed for the specific case of a massive binary BH. However, it was developed in a closely analogous theory of gas accretion onto the forming binary stars. Valuable lessons can be learned by at least qualitatively applying the latter to the massive BH binaries.

4.1. Dynamical effects of binarity: coplanar binaries

In the context of binary star formation, the same angular momentum problem is encountered as in BH evolution: the infalling gas can flatten into a disk but cannot flow onto the central object unless an internal viscosity mechanism operates to transfer the angular momentum outside. In galaxies, viscosity may be generated by gravitational instabilities, magnetic dynamo or, much less likely, convective or hydrodynamic instabilities.

Let us first describe the interaction of a disk with a coplanar binary. The coplanar geometry should be predominant, considering the process of resonant and hydrodynamic damping of inclination of a satellite (secondary) body interacting with a circumprimary disk. (Non-coplanar models will be discussed in section 4.2.)

Replacing an imaginary single central object by a binary (of the same mass) has a two-fold effect: firstly, the surrounding gas is strongly affected by gravitational torques from the binary, and secondly the binary orbit changes in response to the attraction of the non-axisymmetric gas distribution that the binary system causes. Given a sufficient secondary-to-primary mass ratio, the secondary object will cause a gap to open in the disk (Lin & Papaloizou 1993). It can be estimated that all secondaries of mass exceeding about $10^3 M_\odot$ open gaps in the AGN disks around $10^8 M_\odot$ primaries.

The gap in the case of a comparable-mass, eccentric, binary (mass ratio 3:7, disk semi-thickness $z/r \approx 0.1$, and α-viscosity with $\alpha \approx 0.1$) is illustrated in the top panel of Figure 3 with Smoothed Particle Hydrodynamics calculations by Artymowicz & Lubow (1996). This figure shows the prominent density waves in a disk, whose edge is at a radius predicted by theory (Artymowicz & Lubow 1994). Black circle indicates the 3:1 period commensurability (Outer Lindblad Resonance of the $m = 2$, $l = 1$ gravitational potential component, located at $r = 2.08a$).

The gap in the case of an equal-mass, circular binary surrounded by a disk with properties similar to that of the top panel (except for $\alpha \approx 0.02$) is smaller. The bottom panel of Figure 3 presents a snapshot from the Monotonic Transport code of W. Kley (1998, private comm.). The empty central circle between the two small accretion disks around the individual BHs is the area not covered by the computational grid. Both simulations assumed non-selfgravitating disks. Additional SPH simulations including selfgravity showed that this is justified as long as the Toomre stability parameter is larger than about 1.5.

The effects of a binary on a collisionless disk such as the stellar nuclear disk or a rapidly rotating core have not been studied in detail. We expect close qualitative similarities. All the structure (density waves etc.) in the gas disk is much more crispy than the corresponding density distribution of stars around a given binary BH, owing to a larger velocity dispersion in the latter. A central depletion (or "gap") will be created in the stellar, rapidly rotating nucleus, so that, after an appropriate projection onto the sky, the circumbinary disk (or ring) may better resemble the double nucleus of NGC 4468, shown in Fig. 1, than the model involving a single eccentric disk (cf. Kormendy *et al.*1997). (The galactic photocenter is in the middle between the two bright nuclei, and is not close to any one of them, as would be expected in the eccentric disk model). Clearly, much intesting modeling lies ahead.

In the standard theory, the opening of a disk gap was synonymous with the termination of the binary system growth by accretion ($\dot{M} = 0$). The circumprimary disk was considered effectively cut off from the supply of material from the circumbinary disk, in that case a *decretion* disk. This should lead to a relatively rapid drainage of the circum-BH disks onto their central BHs (triggering a brief period of AGN activity), after which the system would stop accreting gas and become dormant. The ensuing evolution of the orbit can be called *gravity-dominated*, a name reflecting the role of gravity as the only mechanism of angular momentum transport (and energy) from the orbit to the circumbinary disk, via density wave generation at Lindblad resonances in the disk. The result is a rapid shrinking of the binary. Its rate, or inverse time scale, can be written as (cf. Lubow & Artymowicz 1996)

$$\frac{\dot{a}}{a} \approx -\frac{j_d}{j_b} \frac{m_d}{M} \frac{1}{t_\nu}, \tag{4.3}$$

where j_d and j_b are the specific angular momenta at the disk edge and of the binary, m_d is the disk mass, and t_ν is the viscous time scale for gas accretion through AGN ($t_\nu \sim 10^6 - 10^7$ yr is required for generating appreciable $\dot{M} \sim 1 M_\odot/$yr in a disk around a $M = 10^8 M_\odot$ BH). We see from eq. (4.3) that the semi-major axis drops on a short time scale, essentially the gas inflow time t_ν.

However, some or all of the above results may be invalid if there are efficient non-gravitational ways of angular momentum and energy transfer, such as the advective transport. According to Artymowicz & Lubow (1996), gas can flow onto the binary system through the gap, in the form of usually two time-dependent streamers extending from the circumbinary disk. The most recent observations addressing the issue in the pre-main-sequence systems: Roddier *et al.*(1996), Mathieu *et al.*(1997), and Jensen & Mathieu (1997), all support the streaming model. In the cases named "efficient flow" by Artymowicz & Lubow (1996), contradicting the common assumption the gas flows preferentially onto the lighter component of the system, at a rate of the unimpeded mass flux in a disk around a single equivalent object. Taking the inspiration from advection-dominated accretion flows discussed elswhere in this volume, we may call this case of orbital coupling between the binary and the disk an *advection-affected* interaction. It will become clear from the following why we hesitate to call this process advection-dominanted.

The gas streams appear in two snapshots from the binary+disk simulations shown in Figure 3. In the top panel, the crosses denote the saddle points (Lagrange points) of the bar-like, slowly rotating effective potential ($m = 2$, $l = 1$). (For more detailed description of theory and simulations see Artymowicz & Lubow 1996, and Lubow &

Artymowicz 1996, 1997). Although the theory originally was not expected to apply to disks as thin as the AGN disks (having $z/r \ll 0.1$ at most radii), this restriction appears to have been an artifact of the insufficient numerical resolution of the SPH method (cf. different numerical technique of Różyczka & Laughlin 1997). Currently, using the Piecewise Parabolic Method and other high-resolution techniques, we are finding rarefied, fast, streaming flows in all types of binary-disk interaction. This suggests strongly that the inner disks associated with individual BH, and therefore the BH themselves, may be resupplied from, and as long-lived as, the external AGN disk. The BH activity due to gas accretion will therefore be preserved, if not enhanced, in a BH binary.

Numerical calculation allow case studies of orbital evolution without and with an efficient gap-crossing flow. In the first case, in addition to orbit contraction, the binary-disk interaction causes resonant eccentricity growth, first discussed by Artymowicz et al. (1991). The evolution consist of a rapid growth of e at small values of e ($e < 0.05$), followed by a very slow growth at high eccentricity ($e > 0.5$). Thus, in concrete applications, a moderate eccentricity is usually produced.

In the advection-affected flow the orbital evolution is complicated by the unknown ratio of the increments of orbital energy and orbital angular momentum, which in general must be computed in multidimensional hydrodynamical simulations. The theory is not yet ready to provide more than preliminary studies, but from the existing calculations it seems that the semi-major axis of BH system should decrease at a rate slower than that in the absence of advection. There is even some question if the advected angular momentum and energy are not sufficient to lead to a truly advection-dominated accretion flow, the one where the separation of BHs increases. The addition of mass always leads to the radial shrinking, while the addition of energy to expansion of the binary's orbit. It seems that in most cases the addition of mass onto the secondary component of the system tips the balance in favor of shrinking, but analytical estimates show that orbital expansion can accompany very large inflows of gas (mass flux through gap emptying the outer disk on a viscous time scale t_ν). If realizable under the specific conditions of AGN disks, the "unshrinking" effect could significantly slow down or at least temporarily reverse the tightening effect of the interaction of the binary BH with the stellar nucleus, with an uncertain long-term result depending on gas inflow history.

A much more robust prediction of the hydrodynamical models with eccentric binaries is a pronounced time modulation, with period equal to the orbital binary period, of the accretion rate onto the system components (BH and their circum-component disks). This suggests applications to OJ 287 and similar "outburst" systems with suspected massive BH pairs. The accretion rate onto BHs was found to depend very sensitively on the eccentricity of the binary. In principle, it is possible to emulate the shapes of the double outburst light curves in blazar OJ 287 by choosing $e \sim 0.5$ and non-equal mass ratio, so that the sharp peaks of $\dot{M}_1$ and $\dot{M}_2$, both appearing near the pericenter of the orbit, are time-lagged by appropriate amount. Detailed modeling has not been performed so far.

4.2. *Non-coplanar models: variability of OJ 287*

In this section we comment on several models proposed for the OJ 287 blazar, involving gas accretion onto a binary BH which orbits in the plane significantly inclined with respect to the accretion disk plane.

In the first model, the secondary BH passes almost vertically through the gaseous accretion disk around the more massive primary. The double structure of outbursts (sect. 2.7) is addressed by identifying the major superoutbursts occurring every ~ 12 years with "tidally induced mass flows towards the primary BH from the accretion disk surrounding it", while the secondary outbursts (superflares) are associated with the secondary crossing

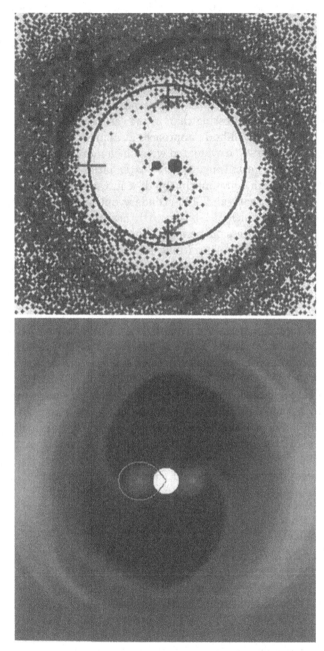

FIGURE 3. (*Top*) Eccentric binary black hole surrounded by a gaseous disk, supplying it with bursts of "fuel" via circumbinary gap (Artymowicz & Lubow 1996). (*Bottom*) A circular binary surrounded by a viscous disk producing a stationary inflow (W. Kley 1998, unpubl.)

the circumprimary disk (Sillanpää et al. 1988, Sundelius et al. 1997). The model has 7 free parameters plus several computationally motivated parameters (like the effective radii of the BHs, much larger than their Schwarzschild radii). It yields a good fit to the timing of the historical outbursts and very specific predictions regarding the next outburst (in year 2007). Unfortunately, the physical side of the modeling is weaker than its geometrical aspect. The numerical calculations were only performed with a restricted 3-body code (a collisionless disk of test particles), which cannot even approximately simulate the physics of BH passage, ejection of matter at the point of impact, or the tidal perturbation propagating through the disk. We have not been able to reproduce at all the main outbursts from this model, using an SPH code which simulates pressure and viscosity, and thus allows us, among others, to follow the propagation of density and bending waves in the disk (toward the primary). Much fewer SPH particles are ejected from the site of BH-disk crossing than in the case we switch off pressure and viscosity, in order to follow restricted 3-body approach. It appears that this type of model has severe problems which can only be evaluated with much more accurate and realistic simulations.

The second, more promising, type of explanations invokes two passages of the secondary BH through the circumprimary disk in rapid succession (due to a large e) as a cause of both the primary and the secondary outbursts. Lehto & Valtonen (1996) constructed a model in which the binary BH, has period 12.07 yr (redshifted), $e = 0.68$, and the relativistic precession period of the orbit equal to 130 yr. Masses of the BHs in the model are $1.7 \times 10^{10} M_\odot$ (unusually high, perhaps overestimated if we observe radiation of a Doppler-boosted jet?) and $10^8 M_\odot$. The authors of the model considered the process of the passage through the disk in some detail; however they concluded that hydrodynamical simulations would be desirable to understand the observable effects better.

The last category of models does not necessarily involve any passages through disks, just the creation of one (Katz 1997) or two (Villata 1998) jets by two BH+disk systems circling each other. The jets are subject to geodetic precession and, in the second case, the two jets are bent by the interaction of their magnetized plasma with the ambient medium. The combination of this bending with a long-term precession of the jet axes gives rise to a time-dependent orientation of the emitting outflows. The models are plausible, but having a smaller number of adjustable parameters have a tendency to predict too regular a spacing of outbursts, compared with observations. While the jury is out on which model is most believable, it is at least encouraging to be able to make specific model-based predictions and checking them against the future outbursts of OJ 287.

5. Conclusions

Binary supermassive black holes represent a relatively new area of research, despite the fact that the basic arguments for their existence were known already in the late 1970s (cf. the seminal paper by Belgelman et al. 1980). So far, the combined effects of their small spatial scale (they hide deep inside the nuclei of merged galaxies), mimicry (they can be easily mistaken for single BHs), and uncertain statistical frequency (their formation may or may not be common, and their lifetimes may or may not reach one Gyr), make them a very difficult trophy for observers.

We have summarized in section 2 the main lines of evidence for the existence of binary BHs such as: correlation of core radius with galaxy mass (which arises in a natural way if binary BHs form in a hierarchy of mergers); binary BHs as a possible reason for the distinction between radio-loud and radio-quiet AGNs; direct observations of double nuclei in galaxies (some very suggestive of a presence of a double massive objects, cf. Fig. 1); jet precession and wiggles, double broad emission lines from spectroscopic double BHs (interpretation in a bit of trouble after a crown example, quasar 3C 390.3, started disobeying the predictions recently); and the almost periodic variability of blazar OJ 287. The existing evidence is thus substantial and manifold, if not as direct and allowing unique interpretation, as one might wish. Therefore, any future efforts directed toward more detailed photometry, studying the kinematics of stars in the vicinity of the candidate BH pairs, continued multi-wavelength campains to study the variability etc., will be welcome. The currently uncertain status of the "double-lined spectroscopic binary" interpretation for the 3C 390.3 quasars motivates further observations. Finally, we might learn about the statistics of supermassive binaries from the detection/nondetection of gravitational wave bursts accompanying their mergers.

On the theoretical front, we need to better understand the complex lives of the binary BHs interacting with stellar and gaseous components of a galactic nucleus. We would like to gain guidance in the observational search from the theoretical predictions of the frequency of occurrence and lifetimes of these objects. We seem to have made a circle from the early 1980s when the role of the binary eccentricity was thought to be small, to mid-1980s when it was thought to be crucial for breaking the loss-cone problem with its associated stagnation of inward spiraling of the two BHs immersed in a stellar system, back to the currently prevalent opinion that eccentricity does not grow to extreme values and does not break the deadlock. If not the stars, then the interstellar gas may help, in ways that we have yet to study in detail. Recent progress in research on star formation, showing how pre-main sequence binaries interact with accretion disks, motivates further work on binary BHs using similar numerical, multidimensional, simulations, as well as theory (resonant disk-binary coupling theory, in particular). Reliable calculations are needed of the energy/angular momentum transfer in both the gravity-dominated and advection-affected modes of interaction of binaries coplanar with the disks. Some models of the variability involving non-coplanar disk-binary systems seem promising, and should be developed to allow prediction of observable effects of the proposed passages of a massive BH through the AGN disk. Assessing the plausibility of such models in view of the generally very short predicted future lifetimes of close BH pairs would be beneficial.

This research was supported by grants from the Swedish Natural Science Research Council (NFR) and Anna-Greta and Holger Crafoord's Foundation. Willy Kley kindly supplied the material for Fig. 3 (bottom).

REFERENCES

ARTYMOWICZ, P., CLARKE, C. J., LUBOW, S. H. & PRINGLE, J. E. 1991 The effect of an external disk on the orbital elements of a central binary. *ApJ* **370**, L35-L38.

ARTYMOWICZ, P. & LUBOW, S. H. 1994, Dynamics of binary-disk interaction. 1: Resonances and disk gap sizes. *ApJ* **421**, 651-667.

ARTYMOWICZ, P. & LUBOW, S. H. 1996 Mass Flow through Gaps in Circumbinary Disks. *ApJ* **467**, L77-L80.

BARNES, J. & HERNQUIST, L. 1992 Dynamics of interacting galaxies. *Ann. Rev. Astron. Astroph.* **30**, 705-742.

BEGELMAN, M. C., BLANDFORD, R. D. & REES, M. J. 1980, Massive black hole binaries in active galactic nuclei. *Science* **287**, 307-309.

BROWN, R. L. 1982, Precessing jets in Sagittarius A - Gas dynamics in the central parsec of the galaxy. *Astroph. J.* **262**, 110-119.

EBISUZAKI, T. MAKINO, J. & OKAMURA S. K. 1991, Merging of two galaxies with central black holes. *Nature* **354**, 212-214.

EKERS, R. D., FANTI, R., LARI, C. & PARMA, P. 1978 NGC326 - A radio galaxy with a precessing beam. *Nature* **276**, 588-590.

EMSELLEM, E. & COMBES, F. 1997 N-body simulation of the nucleus of M31. *Astron. Astroph.* **323**, 674-684.

ERACLEOUS, M., HALPERN, J.P., GILBERT, A.M., NEWMAN, J.A. & FILIPPENKO, A.V. 1997, Rejection of the Binary Broad-Line Region Interpretation of Double-peaked Emission Lines in Three Active Galactic Nuclei. *Astroph. J.* **490**, 216-226.

FUKUSHIGE, T., EBISUZAKI, T. & MAKINO, J. 1992a Gravitational wave bursts produced by merging of central black holes in galaxies. *Astroph. J.* **396**, L61-L63.

FUKUSHIGE, T., EBISUZAKI, T. & MAKINO, J. 1992b Rapid orbital decay of a black hole binary in merging galaxies. *Publ. Astron. Soc. Japan* **44**, 281-289.

GASKELL, C.M. 1983 Quasars as supermassive binaries. In Proc. 24th Lige Int. Astrophys. Colloq., Quasars and Gravitational Lenses (Cointe-Ougree: Univ. Lige), 473-476.

GASKELL, M. 1996 Evidence for Binary Orbital Motion of a Quasar Broad-Line Region. *Astroph. J.* **464**, L107-L110.

GOVERNATO, F., COLPI, M. & MARASCHI, L. 1994 The fate of central black holes in merging galaxies. *Mon. Not. Royal Ast. Soc.* **271**, 317-322.

GONZALEZ-PEREZ, J. N., KIDGER, M. R. & DE DIEGO, J. A. 1996 The microvariability characteristics of OJ 287.. *Astron. Astroph.* **311**, 57-68.

HARDEE, P. E., COOPER, M. & CLARKE, D. 1994 On jet response to a driving frequency and the jets in 3C 449. *Astroph. J.* **424**, 126-137.

HELLER, C. & SCHLOSMAN, I. 1994 Fueling nuclear activity in disk galaxies: Starbursts and monsters . *Astroph. J.* **424**, 84-105.

HUNSTEAD, R. W., MURDOCH, H. S., CONDON, J. J. & PHILLIPS, M. M. 1984 A QSO with precessing jets - 2300. *Mon. Not. Royal Ast. Soc.* **207**, 55-71.

HUTCHINGS, J. B. & CRAVEN, S. E. 1988 Markarian 110 - A twin-nucleus galaxy with an optical jet?. *Astron. J.* **95**, 677-681.

JAEGERS, W. J. & DE GRIJP, M. H. 1985 The radio structure of 3C 130 interpreted with a dynamical model. *Astron. Astroph.* **143**, 176-181.

JENSEN, E. L. & MATHIEU, R. D. 1997 Evidence for cleared regions in the disks dround pre-main-sequence spectroscopic binaries. *Astron. J.* **114**, 301-316.

KAASTRA, J. S. & ROOS, N. 1992 Massive binary black holes and wiggling jets. *Astron. Astroph.* **254**, 96-98.

KATZ, J. I. 1997 A precessing disk in OJ 287? *Astroph. J.* **478**, 527-529.

KORMENDY, J., BENDER, R., MAGORRIAN, J., TREMAINE, S., *et al.* 1997, Spectroscopic Evidence for a Supermassive Black Hole in NCG 4486B. *Astroph. J.* **482**, L139-L142.

KORMENDY, J. & RICHSTONE, D. 1995 Inward Bound—The Search For Supermassive Black Holes In Galactic Nuclei. *Ann. Rev. Astron. Astroph.* **33**, 581-624.

LAUER, T., FABER, S., GROTH, E., SHAYA, E., ET AL. 1993 Planetary camera observations of the double nucleus of M31. *Astron. J.* **106**, 1436-1447.

LAUER, T., TREMAINE, S., AJHAR, E., BENDER, R., ET AL. 1996 Hubble Space Telescope Observations of the Double Nucleus of NGC 4486B, ApJ, 471, L79-L82. 1

LEHTO, H. J., VALTONEN, M. J. 1996 OJ 287 Outburst Structure and a Binary Black Hole Mode. *Astroph. J.* **460**, 207-213.

LIN, D. N. C. & PAPALOIZOU, J. 1993 On the tidal interaction between protostellar disks and companion, in Protostars and Planets III, Eds. E. H. Levy et al. (Tuscon: Univ. Arizona Press), 749-835

LIU, C. T. & KENNICUTT, R., C. 1995 Spectrophotometric Properties of Merging Galaxies. *Astroph. J.* **450**, 547-558.

LU, J. F. 1990 Accretion disk-driven jet precession in active galactic nuclei. *Astron. Astroph.* **229**, 424-426.

LUBOW, S. H. & ARTYMOWICZ, P. 1996 Young binary star/disk interactions. In *Evolutionary Processes in Binary Stars*, Eds. R. A. M. Wijers et al., (Kluwer: Dordrecht), 53-80.

LUBOW, S. H. & ARTYMOWICZ, P. 1997 Young binary star/disk interactions. In IAU Coll. 163 Accretion Phenomena and Related Outflows, ASP Conf. Ser. No. 121, Eds. D. Wickramasinghe et al, 505-514.

MAKINO, J. 1997 Merging of Galaxies with Central Black Holes. II. Evolution of the Black Hole Binary and the Structure of the Core. *Astroph. J.* **478**, 58-65.

MATHIEU, R. D., BASRI, G., STASSUN, K., JOHNS, C., ET AL. 1997 The Classical T Tauri Spectroscopic Binary DQ Tau. I. Orbital Elements and Light Curves. *Astron. J.* **113**, 1841-1855.

MIKKOLA, S., VALTONEN, M. J. 1992 Evolution of binaries in the field of light particles and the problem of two black holes. *Mon. Not. Royal Ast. Soc.* **259**, 115-120.

MILLER, J. S. & PETERSON, B. M. 1990 Do the broad emission lines in Arp 102B arise in a relativistic disk?. *Astroph. J.* **361**, 98-100.

MOORE, B., LAKE, G. & KATZ, N. 1998 Morphological Transformation from Galaxy Harassment. *Astroph. J.* **495**, 139-151.

NEWMAN, J.A., ERACLEOUS, M., FILIPPENKO, A.V. & HALPERN, J.P. 1997 Measurement of an Active Galactic Nucleus Central Mass on Centiparsec Scales: Results of Long-Term Optical Monitoring of Arp 102B. *Astroph. J.* **485**, 570-580 .

OWEN, F. N., O'DEA, C. P., INUE, M. & EILEK, J. A. 1985 VLA observations of the multiple jet galaxy 3C 75. *Astroph. J.* **294**, L85-L88.

POLNAREV, A. G. & REES, M. J. 1994 Binary black hole in a dense star cluster. *Astron. Astrophys.* **283**, 301-312.

QUINLAN, G. 1996 The dynamical evolution of massive black hole binaries I. Hardening in a fixed stellar background. *New Astron.* **1**, 35-56.

QUINLAN, G. & HERNQUIST, L. 1997 The dynamical evolution of massive black hole binaries - II. Self-consistent N-body integrations *New Astron.* **2**, 533-554.

RAJAGOPAL, M. & ROMANI, R. 1995 Ultra–Low-Frequency Gravitational Radiation from Massive Black Hole Binaries. *Astroph. J.* **446**, 543-549.

RAUCH, K. & TREMAINE, S. 1996 Resonant relaxation in stellar systems. *New Astron.* **1**, 149-170.

REES, M. 1990 Black holes in galactic centers. *Science* **263**, 56-66.

RODDIER, C., RODDIER, F., NORTHCOTT, M., GRAVES, J. & JIM, K. 1996 Adaptive Optics Imaging of GG Tauri: Optical Detection of the Circumbinary Ring. *Astroph. J.* **463**, 326-335.

ROOS, N. 1981, Galaxy mergers and Active Galactic Nuclei *Astron. Astroph.* **104**, 218-228.

ROOS, N., KAASTRA, J. S. & HUMMEL, C. A. 1993 A massive binary black hole in 1928+738?. *Astroph. J.* **409**, 130-133.

RÓŻYCZKA, M. & LAUGHLIN, G. 1997 Hydrodynamical evolution of circumbinary accretion disks. In IAU Coll. 163 Accretion Phenomena and Related Outflows, ASP Conf. Ser. No. 121, Eds. D. Wickramasinghe et al, 792-793.

SCHLOSMAN, I., FRANK, J. & BEGELMAN, M. 1989 Bars within bars - A mechanism for fuelling active galactic nuclei. *Nature* **338**, 45-47.

SCHLOSMAN, I., BEGELMAN, M. & FRANK, J. 1990 The fuelling of active galactic nuclei. *Nature* **345**, 679-686.

SCHWEIZER, F. 1986 Colliding and merging galaxies. *Science* **231**, 227-234.

SCHWEIZER, F., MILLER, B., WHITMORTE, B. & FALL, M. 1996 HST Observations of Candidate Young Globular Clusters and Stellar Associations in the Recent Merger Remnant NGC3921. *Astron. J.* **112**, 1839-1873.

SILLANPÄÄ, A., HAARLA, S., VALTONEN, M. J., SUNDELIUS, B., & BYRD, G. G. 1988 OJ 287 - Binary pair of supermassive black holes. *Astroph. J.* **325**, 628-634.

SILLANPÄÄ, A., TAKALO, L. O., PURSIMO, T., LEHTO, H. J. *et al.* 1996a Confirmation of the 12-year optical outburst cycle in blazar OJ 287. *Astron. Astroph.* **305**, L17-L20.

SILLANPÄÄ, A., TAKALO, L. O., PURSIMO, T., NILSSON, K. *et al.* 1996b Double-peak structure in the cyclic optical outbursts of blazar OJ 287. *Astron. Astroph.* **315**, L13-L16.

SMITH, B.F. & MILLER, R.H. 1994 Multiple Core Galaxies: Implications for M31. *Bull. Amer. Astr. Soc.* **185**, #107.08 (abstract).

STOCKTON, A. & FARNHAM, T. 1991 OX 169: Evidence for a recent merger. *Astroph. J.* **371**, 525-534.

STOTHERS, R. B. & SILLANPÄÄ , A. 1997 Test of Periodicity in the Quasar OJ 287. *Astroph. J.* **475**, L13-L15.

SUNDELIUS, B., WAHDE, M., LEHTO & VALTONEN, M. J. 1997 A Numerical Simulation of the Brightness Variations of OJ 287. *Astroph. J.* **484**, 180-185.

THORNE, K. S. & BRAGINSKII, V. B. 1976 Gravitational-wave bursts from the nuclei of distant galaxies and quasars. *Astroph. J.* **204**, L1-L6.

TREMAINE, S. 1995 An Eccentric-Disk Model for the Nucleus of M31. *Astron. J.* **110**, 628-633.

VALTAOJA, L., VALTONEN, M. J. & BYRD, G. G. 1989 Binary pairs of supermassive black holes - Formation in merging galaxies. *Astroph. J.* **343**, 47-53.

VALTONEN, M. J. 1996 Triple black hole systems formed in mergers of galaxies. *Mon. Not. Royal Ast. Soc.* **278**, 186-190.

VALTONEN, M. J., MIKKOLA, S., HEINAMAKI, P. & VALTONEN, H. 1994 Slingshot ejections from clusters of three and four black holes, *Astroph. J. Supp.* **95**, 69-86.

VAN ALBADA, G. D. & ROBERTS, W. W. 1981 A high-resolution study of the gas flow in barred spirals. *Astroph. J.* **246**, 740-750.

VICENTE, L., CHARLOT, P. & SOL, H. 1996 Monitoring of the VLBI radio structure of the BL Lacertae object OJ 287 from geodetic data. *Astron. Astroph.* **312**, 727-737.

VILLATA, M., RAITERI, C. M., SILLANPÄÄ, A. & TAKALO, L.O. 1998 A beaming model for the OJ 287 periodic optical outbursts. *Mon. Not. Royal Ast. Soc.* **293**, L13-L16.

VECCHIO A., COLPI, M. & POLNAREV A. 1994 Orbital evolution of a massive black hole pair by dynamical friction. *Astroph. J.* **433**, 733-745.

WILSON, A. & COLBERT, E. 1995 The difference between radio-loud and radio-quiet active galaxies. *Astroph. J.* **438**, 62-71.

YANNY, B., JANNUZI, B. T. & IMPEY, C. 1997 Hubble Space Telescope Imaging of the BL Lacertae Object OJ 287. *Astroph. J.* **484**, L113-L116.

ZEPF, S. E. & ASHMAN K. M. 1993 Globular cluster systems formed in galaxy mergers. *Mon. Not. Royal Ast. Soc.* **264**, 611-622.

Large scale perturbation of an accretion disc by a black hole binary companion

By JOHN C. B. PAPALOIZOU[1],
CAROLINE TERQUEM[2,3], and DOUG N. C. LIN[2]

[1]Astronomy Unit, School of Mathematical Sciences, Queen Mary and Westfield College, Mile End Rd., London, E14NS, UK

[2] Lick Observatory, University of California, Santa–Cruz, CA 95064, USA

[3]Laboratoire d'Astrophysique, Université J. Fourier/CNRS, BP 53, 38041 Grenoble Cedex 9, France

We consider the global interaction of an accretion disc with a binary companion in inclined circular orbit. Self–gravity, pressure and viscosity may all be important for determining the details of the disc response. If bending waves can propagate efficiently through the disc faster than the time required for precession induced by the companion to occur, a large scale warp of small magnitude may be produced.

By way of illustration, we explore the possibility that the recently observed warp in the disc in the active galaxy NGC 4258 is produced by such a binary companion. Results indicate that it may be produced by a companion with a mass comparable to or larger than that of the observed disc.

---

1. Introduction

In some theories of galaxy formation, mergers of satellite galaxies may be important, leading to the formation of binary cores (Roos 1981, 1985; see also Artymowicz this volume). Interaction with a surrounding star cluster through dynamical friction is then expected to cause the binary orbit to lose energy and angular momentum causing the separation to decrease. If the cores contain black holes in their centres, black hole binaries could ultimately be formed (Begelman, Blandford & Rees 1980). It is believed that frictional processes can bring a modest sized black hole into a galactic core in much less than a cosmological timescale (see Polnarev & Rees 1994 and references therein) where it will be in approximately circular orbit.

Black hole binary systems in which the companions can have comparable mass and separations on a scale of 10^{18} cm have been considered by Begelman, Blandford & Rees (1980), and postulated to account for jet behavior in extragalactic sources by Kaastra & Roos (1992) and Roos, Kaastra & Hummel (1993), although the dense cores of globular clusters may be an alternative candidate for such a companion.

A companion black hole orbiting a central black hole with much larger mass on the scale of a parsec will undergo orbital decay as it interacts with the field stars in the galactic nucleus. For illustrative purposes, we consider a model in which the stellar density in the galactic core varies like r^{-n}, r being the distance from the galactic centre. Assuming that the velocity of the companion remains Keplerian as it spirals inwards, the timescale t_d on which the angular momentum L of the companion changes can be estimated as (Binney & Tremaine 1987)

$$t_d = \frac{L}{dL/dt} \sim \frac{C}{4\pi} \frac{1}{\omega} \frac{M}{M_p} \frac{M}{\rho_0 D^3}.$$

Here the central black hole mass is M and the companion mass is M_p. The stellar density at a distance $r = D$ from the centre is ρ_0, D is the orbital separation, the orbital period is $2\pi/\omega$ and it has been assumed that the velocities of the interacting stars are also approximately Keplerian. C is a number of order unity which depends on various parameters of the galactic core (see Binney & Tremaine 1987 for more details), so we set $C = 1$. For illustrative purposes, we adopt $D \sim 0.3$ pc, and for ρ_0 the value estimated for the galaxy M32, which has been resolved down to subparsec scale. Then ρ_0 lies between 10^6 (Lauer *et al.* 1992) and 10^7 $M_\odot$ pc^{-3}. This gives t_d between 10^7 and 10^8 yr, but it could be somewhat longer if the stellar distribution function is affected by the interaction in a way which removes the resonant stars important for giving rise to dynamical friction. This indicates that orbiting companions may survive for times thought to be characteristic of the lifetime of AGN (e.g. Woltjer 1959).

However, note that the orbital evolution of the companion may be significantly affected by interaction with an accretion disc (see, for example, Lin & Papaloizou 1993 and references therein). An orbiting companion excites waves which carry angular momentum which is exchanged between the orbit and the disc thus producing effects on the disc structure. If the disc is thin and has a small viscosity such that the Reynolds' number $> 40(M/M_p)$, a gap is expected to form which truncates the disc (Lin & Papaloizou 1993). Resonant interactions which may produce potentially observable global responses such as the secular interaction due to the time averaged potential of a companion in an inclined orbit are also important. Note that in this context there is no reason for considering the disc and orbit to be initially coplanar.

Accretion discs may be important in AGN (Lynden-Bell 1969) and self–gravity may be important for their dynamics (Paczynski 1977). The most definitive evidence for such a gaseous accretion disc is provided by the discovery of megamasers (Claussen, Heiligman & Lo 1984) around the nucleus of the mildly active galaxy NGC 4258 (Claussen & Lo 1986). Confirmation of Keplerian rotation was obtained from the radio interferometer (Nakai, Inoue & Miyoshi 1993) and VLBI observations (Greenhill *et al.* 1995). Based on the correlation between the spatial locations and radial velocities of the masers, Miyoshi *et al.* (1995) deduced that the masers are located at $R \sim 0.13 - 0.26$ pc around a black hole with a mass 3.6×10^7 $M_\odot$ (Watson & Wallin 1994; Maoz 1995). Recently the inner part of the disc has been modelled with an advection–dominated flow (Lasota *et al.* 1996).

Random deviations from the Keplerian rotation curve at the location of the masers are ~ 3.5 km s^{-1}. The lack of systematic deviations provides an upper limit for the disc mass of 4×10^6 $M_\odot$. The disc scale height $H < 3 \times 10^{-4}$ pc is such that $H/R < 2.5 \times 10^{-3}$ and the mid–plane temperature $T_c < 10^3$ K (Moran *et al.* 1995).

In addition, the high–velocity maser sources have negligible acceleration and are not colinear with the low–velocity masers (Greenhill *et al.* 1995). The radial dependence of declination of the red–shifted (with respect to the systemic velocity of NGC 4258) high–velocity masers is antisymmetric to that of the blue–shifted high–velocity masers (Miyoshi *et al.* 1995). One scenario to account for these observed properties is that the position angle of the rotation axis varies continually with radius by up to 0.2 radians (Herrnstein, Greenhill & Moran 1996). This corresponds to introducing a small warp into the disc model. The inclination of the local orbital plane with respect to the orbital plane at the innermost region of the disc, g, being the ratio of the vertical displacement to the local radius, varies with the disc radius. In at least one set of models, small–amplitude variations in the angle between the line of sight and the rotation axis are also introduced. This corresponds to introducing a "twist" into the warped disc. The magnitude and direction of the twist are not yet well constrained.

Here we study the secular interaction of an accretion disc for which self–gravity is important with a companion in an inclined orbit and make an application of the results to the warped disc in NGC 4258.

In the general model, the disc is subject to forces due to self–gravity, pressure, viscosity, and also to the gravitational forces arising from the central object and the perturbing companion. Since we shall focus on the secular response of the disc (see § 2.2), the perturbation due to the companion is effectively the same as that produced by a mass distributed uniformly along an inclined ring. Such a perturbation exerts a torque on the disc which causes it to precess.

Hunter & Toomre (1969) showed that an isolated self–gravitating disc subject to a position dependent vertical displacement generally precesses differentially. But this differential precession is prevented by gravitational torques produced by the distorted disc itself if it has a sharp edge. This is also the case if the disc orbits in the external potential due to a companion or flattened halo whose equatorial plane is misaligned with the disc plane (Toomre 1983; Dekel & Shlosman 1983; Sparke 1984; Sparke & Casertano 1988), provided departure from spherical symmetry of the external potential is relatively small. Then the disc can settle into a discrete mode referred to as the modified tilt mode. In the limit of spherical symmetry this reduces to a trivial rigid tilt. In a non spherically symmetric potential, the disc has to bend to alter the precession frequency at each radius so that the rate is everywhere the same (Sparke & Casertano 1988; Hofner & Sparke 1994) and thus the mode is modified. This situation is expected when a near Keplerian self–gravitating disc dominated by a central mass is perturbed by the secular effects due to a smaller orbiting companion.

But note that the effects of pressure may be important (Papaloizou & Terquem 1995) and be comparable to those due to self–gravity in determining the properties of the bending modes important in determining the disc response (Papaloizou & Lin 1995; Masset & Tagger 1996). For near Keplerian stable self–gravitating discs, with very small viscosity, a mild warp is expected if the sound crossing time through the disc is much smaller than the characteristic precession period induced by the companion. Then bending waves, which propagate with a speed comparable to the sound speed, can propagate through the disc sufficiently fast so that the different parts of it are able to "communicate" and adjust their precession rate to a constant value. This also occurs when viscosity is present, but the communication becomes diffusive rather than wave–like when the Shakura & Sunyaev (1973) viscosity parameter α applicable to vertical shear significantly exceeds the ratio of disc semi–thickness to radius (see Papaloizou & Pringle 1983; Demianski & Ivanov 1997). The theoretical expectation of uniform precession of inclined discs under conditions of adequate physical communication has been confirmed by the numerical simulations of Larwood *et al.* (1996) and Larwood (1997).

When the above conditions are satisfied, the disc precesses uniformly in the external potential. The precession frequency can be calculated from the condition that the disc is stationary in an appropriate uniformly rotating frame. This condition implies that the net torque exerted by the external potential and the Coriolis force is zero (Kuijken 1991). Further, under conditions of efficient bending wave propagation, the disc is expected to be only slightly bent or warped on a long lenghth scale thus reducing the effectiveness of viscous processes which might cause the warp to decay with a consequent alignment of disc and orbital planes. In general the most long–lived situation is one where the warp is relatively slight and thus can be regarded as a small perturbation. The decay might then be expected to be on a global viscous timescale.

2. Disc response to a companion in an inclined orbit

2.1. *Basic equations*

The dynamics of the disc is described by the equation of motion

$$\frac{\partial \mathbf{v}}{\partial t} + (\mathbf{v} \cdot \nabla) \mathbf{v} = -\frac{1}{\rho} \nabla P - \nabla \Psi + \mathbf{f_v}, \tag{2.1}$$

and the equation of continuity

$$\frac{\partial \rho}{\partial t} + \nabla \cdot (\rho \mathbf{v}) = 0, \tag{2.2}$$

where P is the pressure, ρ the density, $\mathbf{v}$ the flow velocity and Ψ the total gravitational potential. We allow for the possible presence of a viscous force per unit mass $\mathbf{f_v}$ but we shall assume that it does not affect the undistorted axisymmetric disc so that it operates on the perturbed flow only.

We write $\Psi = \Psi_{ext} + \Psi_G$, where Ψ_{ext} is the contribution to potential due to the central black hole and the perturber and Ψ_G is the contribution to the potential arising from the disc self–gravity given by

$$\Psi_G = -G \int_V \frac{\rho\,(\mathbf{r}')\,d^3\mathbf{r}'}{|\mathbf{r} - \mathbf{r}'|}, \tag{2.3}$$

the integral being taken over the disc volume, with $\mathbf{r}$ and $\mathbf{r}'$ denoting position vectors and G being the gravitational constant.

For our calculations, we adopt a polytropic equation of state $P = K\rho^{1+1/n}$, K and n being the polytropic constant and index respectively and the sound speed is given by $c_s = \sqrt{dP/d\rho}$.

2.2. *Perturbing potential*

We suppose that the disc is perturbed by a point mass M_p orbiting as a binary companion to an effective point mass M located at the disc centre. The disc outer radius is denoted by R and the binary separation by D. The disc may also be supposed to have a small inner radius R_i, but results are insensitive to this.

We consider an initially non rotating Cartesian coordinate system (x, y, z) centered on the central mass M. The z–axis is chosen to be along the rotation axis of the unperturbed disc. The associated cylindrical polar coordinates are (r, φ, z). We take the orbit of the perturbing mass to be in a plane which has an inclination angle δ with respect to the (x, y) plane and the line of nodes to coincide with the x–axis. The position vector of the perturbing mass is $\mathbf{D}$ with $D \equiv |\mathbf{D}|$.

The potential Ψ_{ext} due to the central and the orbiting masses is given by

$$\Psi_{ext} = -\frac{GM}{|\mathbf{r}|} - \frac{GM_p}{|\mathbf{r} - \mathbf{D}|} + \frac{GM_p \mathbf{r} \cdot \mathbf{D}}{D^3}$$

The last indirect term accounts for the acceleration of the origin of the coordinate system. We study the global warping response of the disc which is produced by terms in the potential which are odd in z and which have azimuthal mode number $m = 1$. We consider only the secular (zero frequency) term in the potential because the disc response to this is global. This is because the near equality of circular orbit and vertical oscillation frequencies of material in near Keplerian orbits results in a response of a global resonant nature. However, note that non zero–frequency terms may be important in structuring the disc at a more local level, resulting in the production of gaps and tidal truncation.

For example, if the viscosity is small, this may occur near to the inner Lindblad (2:1) resonance with the companion (see, for example, Lin & Papaloizou 1993). We later adopt a model of this kind such that $D = 1.5R$. The problem of considering a companion that penetrates the disc is thus avoided. The disc is of course expected to be warped in that case also, but the calculations in this case are more problematic than those considered here.

The term in the Fourier expansion of the potential which is of the required form is given by

$$\Psi'_{ext} = \frac{\sin\varphi}{4\pi^2} \int_0^{2\pi} d(\omega t) \int_0^{2\pi} [\Psi_{ext}(r, \varphi', z, \omega t) - \Psi_{ext}(r, \varphi', -z, \omega t)] \sin\varphi' d\varphi'. \quad (2.4)$$

To eighth order in D^{-1} we obtain:

$$\Psi'_{ext} = -\frac{3}{4} \frac{GM_p}{A^{5/4}D^3} rz \sin(2\delta) \sin\varphi \left[1 + \frac{a_1}{A}\left(\frac{r}{D}\right)^2 + \frac{a_2}{A^2}\left(\frac{r}{D}\right)^4 + \frac{a_3}{A^3}\left(\frac{r}{D}\right)^6 + \frac{a_4}{A^4}\left(\frac{r}{D}\right)^8\right],$$
$$(2.5)$$

with $A = \left(1 + r^2/D^2\right)^2$ and

$$a_1 = 2.1875\left(2 - 1.5\sin^2\delta\right),$$
$$a_2 = 6.0156\left(3 - 4.5\sin^2\delta + 1.875\sin^4\delta\right),$$
$$a_3 = 18.3289\left(4 - 9\sin^2\delta + 7.5\sin^4\delta - 2.1875\sin^6\delta\right),$$
$$a_4 = 59.2022\left(5 - 15\sin^2\delta + 18.75\sin^4\delta - 10.9375\sin^6\delta + 2.4609\sin^8\delta\right).$$

Note that because we work in the thin disc limit, terms smaller by factors containing powers of z/r have been neglected. It is convenient to use the complex potential

$$\Psi'_{ext} = i\frac{3}{4}\frac{GM_p}{A^{5/4}D^3} rf(r)z \sin(2\delta)\, e^{i\varphi}, \quad (2.6)$$

defined such that its real part is equal to the physical potential. The function $f(r)$ is the term in brackets in equation (2.5).

2.3. Equilibrium structure of the disc

In the absence of the companion, the disc is axisymmetric so that in cylindrical coordinates the velocity $\mathbf{v} = (0, r\Omega, 0)$, Ω being the angular velocity, and we have hydrostatic equilibrium such that:

$$\frac{1}{\rho}\frac{\partial P}{\partial r} = -\frac{\partial\Psi}{\partial r} + r\Omega^2, \quad (2.7)$$

$$\frac{1}{\rho}\frac{\partial P}{\partial z} = -\frac{\partial\Psi}{\partial z}. \quad (2.8)$$

These together with the assumed barotropic equation of state imply that Ω is a function of r alone.

The external potential is $\Psi_{ext} = -GM/\sqrt{r^2 + z^2}$. The effect of self–gravity is included under the approximation that the radial scale over which physical parameters vary is very much larger than the disc vertical scale height. Then the disc's self–gravity may be evaluated as if the disc had zero thickness. Then when, as allowed for here, the Toomre parameter $Q \sim \Omega c_s/(\pi G\Sigma)$, Σ being the surface density and c_s being evaluated in the midplane, is of order unity, self–gravity is more important by a factor of order r/H than pressure in equation (2.7). The angular velocity is given by equation (2.7) as

$$\Omega^2 = \Omega_K^2 + \frac{1}{r\rho}\frac{\partial P}{\partial r} + \frac{1}{r}\frac{\partial \Psi_G}{\partial r}, \tag{2.9}$$

where $\Omega_K = \left(GM/r^3\right)^{1/2}$ is the Keplerian angular velocity. Typically, for $Q \sim 1$, the contribution of self–gravity gives $\Omega - \Omega_K = O(H\Omega_K r^{-1})$.

In principle, the surface density distribution, $\Sigma = \int_{-\infty}^{\infty} \rho dz$, is determined by the viscous evolution of the disc. In the absence of a deterministic prescription for the effective viscosity, for illustrative purposes we arbitrarily choose $\Sigma(r) = \Sigma_0 \left(R/r - 1\right)$, Σ_0 being a constant. Formally Σ should taper to zero at the disc inner boundary. However, results are independent of this detail. Alternative functional forms describing the way Σ tapers to zero at the outer boundary have also been considered below.

2.4. Linear perturbations

In the limit where the perturbing/forcing potential is small compared to that due to the central mass and the amplitude of any free bending modes is small, the disc response can be described by a linear analysis. We here follow the standard procedure of vertical averaging (see, for example, Hunter & Toomre 1969, Sparke 1984, Papaloizou & Lin 1995). This should be valid when the radial wavelength of the disturbance is significantly greater than the disc thickness as is the case for the perturbations considered here. For general linear warps, the dependence of the displacement vector on φ and t may be taken to be through a factor $\exp[i(m\varphi + \sigma t)]$. For global warps, $m = 1$, and the mode frequency, σ, is zero for a secular response viewed in an appropriate rotating reference frame and we shall consider this case below.

We denote the Lagrangian displacement as $\boldsymbol{\xi} \equiv (\xi_r, \xi_\varphi, \xi_z)$. Here we have not assumed at the outset that the horizontal components of the displacement are zero, as it was shown in Papaloizou & Lin (1995) that these can produce significant effects in a self–gravitating but near Keplerian disc of the type considered here.

The inclination (vertical displacement/radius) associated with the tilt of the disc is ξ_z/r. This is in general assumed to be a slowly varying function of z such that its variation with z may be neglected. The applicability of this approximation is supported by the numerical calculations of Papaloizou & Lin (1995) which allowed for the possibility of variation of the vertical displacement associated with modes of the type we consider with z.

We first assume that we are working in an inertial frame in which the unperturbed disc appears steady. However, it may subsequently be necessary to transfer to a uniformly rotating frame in order to remove the rigid body precession of the disc resulting from the net perturbing torques on it. When this is done, the effects due to the coriolis force need to be added. Also we initially neglect the effects of viscosity.

The perturbed form of the vertical component of the equation of motion applicable to a gaseous disc with a barotropic equation of state (see Papaloizou & Lin 1995)

$$\frac{D^2 \xi_z}{Dt^2} = -\frac{\partial \Psi_G'}{\partial z} - \frac{\partial (P'/\rho)}{\partial z} - \frac{\partial \Psi_{ext}'}{\partial z}, \tag{2.10}$$

where perturbations to quantities are denoted with a prime and the convective time derivative D/Dt is here equivalent to multiplication by $i\Omega$.

The potential perturbation, Ψ_{ext}', is taken to be the secular contribution due to the companion in an inclined circular orbit (see § 2.2).

For a polytropic equation of state we also have $P' = \rho' c_s^2$. Using the above, equation (2.10) may be written

$$\Omega^2 \xi_z = \frac{\partial \Psi'_G}{\partial z} + \frac{\partial \left(\rho' c_s^2 / \rho \right)}{\partial z} + \frac{\partial \Psi'_{ext}}{\partial z}. \tag{2.11}$$

The perturbation to the gravitational potential of the disc is given by the Poisson integral

$$\Psi'_G = -G \int_V \frac{\rho' \left(\mathbf{r}' \right) d^3 \mathbf{r}'}{|\mathbf{r} - \mathbf{r}'|}, \tag{2.12}$$

while the perturbed continuity equation gives

$$\rho' = -\nabla \cdot (\rho \boldsymbol{\xi}). \tag{2.13}$$

We then find after an integration by parts, assuming that the disc density vanishes at its boundaries, that

$$\Psi'_G = -G \int_V \frac{\rho \left(\mathbf{r}' \right) \boldsymbol{\xi} \left(\mathbf{r}' \right) \cdot \left(\mathbf{r} - \mathbf{r}' \right) d^3 \mathbf{r}'}{|\mathbf{r} - \mathbf{r}'|^3}. \tag{2.14}$$

2.4.1. *The tilt equation*

Under the assumption that ξ_z is independent of z, it is possible to solve the perturbed equations of motion to find the other components of $\boldsymbol{\xi}$, which then have a linear dependence on z. A vertical averaging procedure can then be used to derive a tilt equation for $g = i\left(\xi_z/r \right) \exp\left(-i\varphi \right)$. At zero frequency, its modulus is the ratio of vertical to azimuthal velocity and also the disc inclination. The details of the derivation are tedious and are to be published elsewhere (Papaloizou, Terquem & Lin 1998), so we shall omit them here, quoting only the result:

$$\mathcal{L}\left(g \right) + \frac{i\Sigma}{r\Omega_K^2} \frac{\partial \Psi'_{ext}}{\partial z} = 0, \tag{2.15}$$

where the operator $\mathcal{L}$ is defined through

$$\mathcal{L}\left(g \right) = \frac{G}{r^3 \Omega_K^2} \int_{R_i}^{R} \hat{K}\left(r, r' \right) \Sigma(r) \Sigma(r') \left[g(r) - g(r') \right] r'^2 r^2 dr' + \frac{\partial}{\partial r} \left(\frac{\mu \Omega_K^2}{\kappa^2 - \Omega^2} \frac{\partial g}{\partial r} \right). \tag{2.16}$$

Here the kernel $\hat{K}$ is given by

$$\hat{K}(r, r') = \int_0^{2\pi} \frac{\cos \Phi \, d\Phi}{[r^2 + r'^2 - 2rr' \cos \Phi]^{3/2}}. \tag{2.17}$$

The square of the epicylic oscillation frequency is

$$\kappa^2 = \frac{2\Omega}{r} \frac{d(r^2 \Omega)}{dr}, \tag{2.18}$$

and $\mu = \int_{-\infty}^{\infty} \rho z^2 dz$. Note also that a factor $\exp(im\varphi)$ has been removed from Ψ'_{ext}.

Note that in deriving equation (2.15), frequencies of order $H\Omega_K r^{-1}$, are retained while those of order $H^2 \Omega_K r^{-2}$ are neglected by comparison. Consistent with this we have made the replacement $\Omega^2 = \Omega_K^2$ everywhere apart from in the denominator $\kappa^2 - \Omega^2$ in equation (2.16).

We comment that the above equation describes the global response of a thin, inviscid, near Keplerian self–gravitating disc with Toomre parameter Q down to order unity.

There are two identifiable contributions to the operator $\mathcal{L}$ defined by equation (2.16). The first comes from self–gravity and leads to the description of warps given by Hunter & Toomre (1969), Sparke (1984), Sparke & Casertano (1988), Kuijken (1991) and others. The second term in $\mathcal{L}$ can be identified as arising from pressure (see Papaloizou & Terquem 1995) and it can lead to comparable effects to those due to self–gravity in a near Keplerian disc with small viscosity and $Q \sim 1$. This is in contrast to a general galactic disc, because of the near Lindblad resonance that arises in the former case because $\kappa - \Omega = O(H\Omega r^{-1})$. However, as the disc becomes more viscous, the response becomes increasingly dominated by self–gravity (see below).

A useful property noted by Sparke & Casertano (1988) in the pure self–gravity problem is that, for a finite isolated disc, $\mathcal{L}$ is self–adjoint. That is, for any pair (g_1, g_2) that satisfy appropriate regularity conditions, including zero derivative at a Lindblad resonance if the pressure term is included:

$$\int_{R_i}^{R} g_2^* \mathcal{L}(g_1)\, dr = \left(\int_{R_i}^{R} g_1^* \mathcal{L}(g_2)\, dr \right)^*. \tag{2.19}$$

Furthermore, when there is no forcing, equation (2.15) has the solution $g = \text{constant}$ which corresponds to a rigid tilt in a pure Keplerian potential.

2.4.2. *Disk precession*

The governing equation is equation (2.15), namely

$$\mathcal{L}(g) = -\frac{i\Sigma}{r\Omega_K^2} \frac{\partial \Psi'_{ext}}{\partial z}. \tag{2.20}$$

But because the unforced problem (with $\Psi'_{ext} = 0$,) has $g = \text{constant}$ as a solution, the forced problem will not in general have a solution. Physically, this is because there is an unbalanced torque due to the companion which produces disc precession. To deal with this, we suppose the disc appears steady in a frame precessing about the total angular momentum axis with frequency ω_p. At first, let us suppose this coincides with the companion orbital angular momentum vector. Then we must add the vertical component of the coriolis force as a perturbing force along with that due to the external potential. This amounts to an additional perturbing force per unit mass being the real part of $-2i\omega_p \sin(\delta) r\Omega_K \exp(i\varphi)$. Incorporating this into equation (2.20) (i.e. adding it to $-\partial \Psi'_{ext}/\partial z$ after having removed the factor $\exp(i\varphi)$) results in this now becoming

$$\mathcal{L}(g) = \frac{\Sigma}{r\Omega_K^2} \left(2\omega_p r\Omega_K \sin \delta - i\frac{\partial \Psi'_{ext}}{\partial z} \right). \tag{2.21}$$

The integrability condition that determines ω_p is equivalent to the condition of zero net torque, and this gives

$$\omega_p = \int_{R_i}^{R} \frac{i\Sigma}{2r\Omega_K^2} \left(\frac{\partial \Psi'_{ext}}{\partial z} \right) dr \left/ \int_{R_i}^{R} \frac{\Sigma \sin \delta}{\Omega_K}\, dr. \right. \tag{2.22}$$

This equation can be interpreted as a gyroscope equation giving the precesion frequency as the ratio of appropriate disc torque and angular momentum components. Note here that the perturbing external potential is complex (see eq. [2.6]), ensuring that the calculated precession frequency is real. The factor of two in equation (2.22) accounts for an azimuthal average.

After finding ω_p from equation (2.22), equation (2.21) may be integrated to give g, to within the addition of an arbitrary constant inclination, which may be eliminated by choosing the coordinate system so that $g = 0$ at the disc inner boundary. Then if g is small elsewhere, the disc approximately precesses like a rigid body.

From equation (2.5), we see that $\omega_p \propto \cos \delta$ when $D \gg R$ such that the tidal potential becomes quadrupolar. The same dependance is found for a disc subject to the potential of a misaligned flattened halo, δ being in that case the angle between the disc plane and the symmetry plane of the potential (Kuijken 1991).

Up to now we have assumed that the disc angular momentum content can be neglected in comparison to that of the orbiting companion. Then the disc precesses about the companion angular momentum axis. For finite disc angular momentum content, the disc and companion both precess about the conserved total angular momentum vector. This can be taken account of by noting that the precession frequency ω_p as calculated above should be replaced by $\omega_p J_{orb}/J$, where J_{orb} is the companion's orbital angular momentum and J is the total angular momentum.

2.4.3. *The effect of viscosity*

In principle, viscosity can act on the disc so as to change its inclination with respect to the orbital plane. As long as the disc is globally warped, the timescale on which the disc inclination is expected to change is likely to be related to the global viscous timescale, namely $\Omega^{-1}(R/H)^2/\alpha$, where α is the standard Shakura & Sunyaev (1973) parameter. Since $\alpha \leq 1$, evolution on this timescale is long and has been neglected in comparison to the other processes considered in the above analysis, where frequencies on the order of $H^2 \Omega r^{-2}$ have been neglected. Thus, in the ordering scheme adopted, changes in the disc inclination with respect to the orbital plane should be negligible.

In a near Keplerian disc, viscosity acting on the vertical shear in horizontal motions may be important (Papaloizou & Pringle 1983). If an appropriate α can be defined through

$$\alpha = \frac{-1}{(\rho \Omega z)} \frac{\partial (\rho \nu)}{\partial z}, \tag{2.23}$$

and taken to be constant, viscosity can be taken into account by changing the denominator $\kappa^2 - \Omega^2$ of the last term of the operator $\mathcal{L}$ (eq. [2.16]) to $\kappa^2 - \Omega^2(1 - i\alpha)^2$ (Papaloizou, Terquem & Lin 1998). We note that when $\alpha \gg H/r$, the pressure term in $\mathcal{L}$ is small compared to that due to self–gravity, so that the warp is mainly controlled by self–gravity. This is illustrated below in the case of a model for NGC 4258.

We comment that incorporation of viscosity as described above does not affect the discussion leading to the determination of the precession frequency through equation (2.22). This is consistent with the idea that as long as the disc is only mildly warped, the internal viscous forces do not enable a torque to be exerted on the disc which could modify the precession rate or change the disc inclination, except possibly on a long timescale.

3. Application to NGC 4258

As an illustrative example, Papaloizou, Terquem & Lin (1998) applied the above formalism to discuss the warped disc in NGC 4258. Such a model was originally proposed to account for the origin of the megamasers (Neufeld & Maloney 1995). In this scenario, the warp provides a favorable condition for parts of the disc to be illuminated by the

X–ray source within the nucleus. In the regions between $0.13 - 0.26$ pc, the surface layers of the disc are heated to temperatures in the range $300\ K < T < 8,000\ K$ at which water maser production may occur (Collison & Watson 1995).

3.1. *Disk Parameters*

For the total luminosity of $L = 4 \times 10^{41}$ ergs s^{-1} attributed to this object, an efficiency ϵ (here taken to be 0.1) of rest mass conversion implies a mass accretion rate through the system of

$$\dot{M} = \frac{L}{\epsilon c^2} = 6 \times 10^{-5}\ M_\odot/yr. \tag{3.24}$$

But note that the accretion rate may be a lot higher than this in the inner parts if an advection dominated flow occurs there (Lasota *et al.* 1996).

The disc semi–thickness H is given by

$$\frac{H}{r} \sim \frac{c_s}{r\Omega} \sim 1.5 \times 10^{-3} \left(\frac{T_{200}r_{18}}{M_{7.6}} \right)^{1/2}. \tag{3.25}$$

Here $M_{7.6} = M/(4 \times 10^7\ M_\odot)$, $T_{200} = T_c/200\ K$, T_c is the midplane temperature and r_{18} is the radius in units of 10^{18} cm. From the observed upper limit $H/R < 2.5 \times 10^{-3}$, we find $T_c \sim T_{eff}$ due to irradiation and $T_{200}r_{18}/M_{7.6} < 2.8$.

Applying the conventional α prescription for an effective turbulent viscosity (Shakura & Sunyaev 1973), the mass transfer rate in a steady–state disc is

$$\dot{M} = 3\pi\alpha H^2\Omega\Sigma = 0.59\alpha\frac{M_D}{M} \left(\frac{T_{200}^2 M_{7.6}}{r_{18}} \right)^{1/2}\ M_\odot/yr, \tag{3.26}$$

where we have replaced $\pi\Sigma r^2$, with Σ being the surface density, by M_D, the disc mass interior to radius r. For NGC 4258, we find from equations (3.26) and (3.24) that

$$\alpha \sim 10^{-4}\frac{M}{M_D} \left(\frac{T_{200}^2 M_{7.6}}{r_{18}} \right)^{-1/2}. \tag{3.27}$$

The importance of self–gravity is measured by the Toomre Q parameter, $Q \sim (HM)/(rM_D)$, such that $Q \geq 1$ is required for stability to axisymmetric ring modes. From equation (3.25), we find that the disc becomes gravitationally unstable in this way if $M_D/M \geq 1.5 \times 10^{-3}$. From equation (3.27), we find $\alpha \sim 0.1$ for $M_D/M = 10^{-3}$. A smaller value of α would imply larger Σ, a smaller Q value and then a more unstable disc.

Based on the above estimates, we consider a model which is marginally self–gravitating with $Q \sim 1$ and $\alpha \sim 0.1$ such that $M_D \sim 3 \times 10^4\ M_\odot$. We note that Maoz (1995) has suggested that the radial intervals between successive maser sources may be related to the wavelength of transient spiral structure in a marginally self–gravitating disc. The value of α assumed is comparable to that expected to occur in a marginally self gravitating disc as a result of non–axisymmetric waves (Laughlin & Różyczka 1996).

Pringle (1996) has indicated that forces due to radiation pressure may be important in producing warps in discs. Such forces, as well as energy and momentum exchange occuring through a disc wind, could in principle be included in the formalism presented here. However, we shall limit ourselves in this discussion to including the effects arising from a companion only.

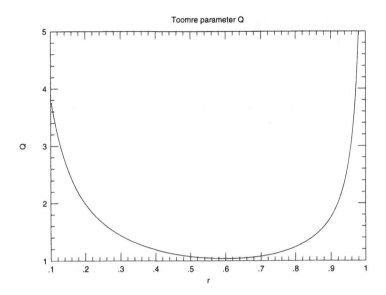

FIGURE 1. Toomre parameter $Q = \kappa c/(\pi G \Sigma)$ for $M_D(R) = 1.5.10^{-3}M$ and $(H/r)_{max} = 1.5.10^{-3}$. The other parameters are given in the text.

3.2. Numerical Results

For illustrative purposes, Papaloizou, Terquem & Lin (1998) solved equation (2.21) to show that the observed properties of the warped disc in NGC 4258 can be regulated by the perturbation on a marginally self–gravitating disc due to a low–mass companion on an inclined orbit.

For computational convenience, units are normalized such that $M = 1$, $R = 1$ and $\Omega_K(R) = 1$. The observations of Miyoshi *et al.* (1995) indicate that $M_D(R)$ of the disc is less than $10^{-2}M$. Marginally self–gravitating discs are then such that the disc semi-thickness is less than one percent of the radius. For illustrative purposes, we have chosen a model for which $\Sigma = \Sigma_0(R/r - 1)$, with a truncation at the inner edge which has no significant effect on the results. The constant Σ_0 was chosen such that $M_D(R) = 1.5 \times 10^{-3}M$, $\alpha = 0.1$, and the maximum value of H/r, $(H/r)_{max}$, was $\sim 10^{-3}$.

The equilibrium structure of the disc was calculated adopting a polytropic index 1.5. The Toomre parameter $Q = \kappa c_s/(\pi G \Sigma)$, with c_s being evaluated in the midplane, is plotted in Figure 1 for a model disc with $R_i = 0.1R$. The minimum value of Q is approximately unity, making the disc marginally self–gravitating.

For the companion providing the perturbation, we adopt an orbital inclination with respect to the disc of $\delta = \pi/4$. Results for other values of δ can be obtained by noting that, to a reasonable approximation, the precession frequency is proportional to $\cos \delta$ and g is proportional to $\sin 2\delta$. We also take the companion's mass to be such that $M_p = M_D(R)$ and the orbital radius to be $D = 1.5R$. If other parameters are fixed, $g \propto M_p$.

There is an approximate scaling relation such that g is essentially unaltered if M_D, M_p, and H are reduced by the same scaling factor Note too that Q is invariant under this scaling. However, for large $\alpha > H/r$, the small amount of twist induced by viscosity is reduced by the same scaling factor.

The induced precession frequency in this model is $\omega_p = -7.9 \times 10^{-5}(J/J_{orb})\Omega(R)$. A warp with a significantly higher amplitude than the perturbing to central mass ratio

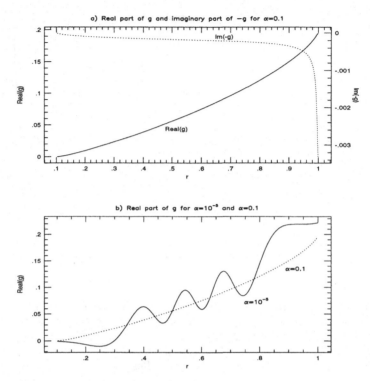

FIGURE 2. a) Real and imaginary parts of g and $-g$ respectively for a self–gravitating viscous disc with $\alpha = 0.1$. The other parameters are given in the text. b) Real part of g obtained for the cases when $\alpha = 10^{-5}$ and $\alpha = 0.1$.

is produced because of the resonant nature of the secular forcing. An approximate estimate for $|g|$ may be obtained by estimating the magnitudes of terms in equations (2.6) and (2.15) as

$$|g| \sim \frac{3M_pR^3}{8M_D(R)D^3}|\sin 2\delta|, \tag{3.28}$$

for significant inclination, this being the ratio of the time for a bending wave to propagate through the disc and the time for the disc to precess significantly. When $|g|$, and thus the warping, is small, wave propagation is effective at enabling the different parts of the disc to communicate so they can precess at the same rate.

In Figure 2.a, we plot both the real part of g and the imaginary part of $-g$ as functions of r. These quantities represent the inclination of the disc as seen edge on, viewed looking down the y and x–axis, where $\varphi = \pi/2$ and 0 respectively. Figure 2.b shows the real part of g obtained including both self–gravity and pressure terms but with very small $\alpha = 10^{-5}$. When $\alpha = 0$ the solution looks almost exactly the same. The case for $\alpha = 0.1$ is shown for comparison. We observe that when α is small, the response oscillates around that obtained with the larger α. This is due to the fact that when both self–gravity and pressure are present, both a slow and a fast bending wave can propagate in the disc (Papaloizou & Lin 1995). When α is large, the pressure term in equation (2.16) is small, so that only the fast wave corresponding to the purely self–gravitating case can propagate. In Figure 2.b, we see the superposition of the fast and slow waves, the former having a larger wavelength than the latter. The slow wave disappears once α becomes larger than H/r, as is the case here.

The presence of viscosity results in a non zero imaginary part for g. This results in a small twist in the warp pattern amounting to a few degrees in total. The magnitude of the total twist is regulated in turn by the magnitude of the viscosity.

4. Discussion

Secondary binary black hole companions may produce important effects when they interact with accretion discs. In addition to the general phenomenon of the excitation of waves which are able to cause an angular momentum exchange between the disc and orbit (Lin & Papaloizou 1993; Artymowicz 1994), possibly leading to gap formation and to orbital evolution most probably on the viscous timescale (see, for example, Lin & Papaloizou 1986; Syer, Clarke & Rees 1991), significant responses of a global character may also be produced if the binary orbit and disc plane are misaligned.

By way of illustration, we have explored the possibility that the recently observed warped disc in the active galaxy NGC 4258 is produced by a binary companion. Our results indicate that it can be produced by a companion with a comparable mass to that of the observed disc. For a disc mass of $10^{-3}M = 4 \times 10^4 \, M_\odot$, the required mass for the companion is comparable to that contained in the cores of the densest globular clusters. Either these cores or black holes with comparable masses can survive tidal disruption by the central black hole. The warp remains of modest magnitude as long as the companion induced precession time is everywhere significantly longer than the time required for a bending wave to propagate through the disc.

If orbital migration of a companion of the mass we have considered occurs as a result of tidal interaction with the disc, this is sufficiently thin that it is expected to occur on the viscous timescale $\sim 10^9$ yr.

The authors wish to thank L.J. Greenhill for useful conversations. This work is supported by PPARC through grant GR/H/09454 and by NASA through grant NAG 53059. C.T. acknowledges support for this work by the Center for Star Formation Studies at NASA/Ames Research Center and the University of California at Berkeley and Santa–Cruz, and by the EU.

REFERENCES

ARTYMOWICZ, P. 1994, ApJ, 423, 581

BEGELMAN, M.C., BLANDFORD, R.D., & REES, M.J. 1980, Nature, 287, 307

BINNEY, J., & TREMAINE, S. 1987, Galactic Dynamics (3d ed.; Princeton: Princeton Univ. Press)

CLAUSSEN, M.J., HEILIGMAN, G.M., & LO, K.Y. 1984, Nature, 310, 298

CLAUSSEN, M.J., & LO, K.Y. 1986, ApJ, 308, 592

COLLISON, A.J., & WATSON, W.D. 1995, ApJ, 452, L103

DEKEL, A., & SHLOSMAN, I. 1983, in IAU Symposium 100, Internal Kinematics and Dynamics of Galaxies, ed. E. Athanassoula (Dordrecht: Reidel), 187

DEMIANSKI, M., & IVANOV, P.B. 1997, A&A, 324, 829

GREENHILL, L.J., JIANG, D.R., MORAN, J.M., REID, M.J., LO, K.Y., & CLAUSSEN, M.J. 1995, ApJ, 440, 619

HERRNSTEIN, J.R., GREENHILL, L.J., & MORAN, J.M. 1996, ApJ, 468, L17

HOFNER, P., & SPARKE, L.S. 1994, ApJ, 428, 466

HUNTER, C., & TOOMRE, A. 1969, ApJ, 155, 747

KAASTRA, J.S., & ROOS, N. 1992, A&A, 254, 96

KUIJKEN, K. 1991, ApJ, 376, 467

LARWOOD, J.D. 1997, MNRAS, 290, 490

LARWOOD, J.D., NELSON, R.P., PAPALOIZOU, J.C.B., & TERQUEM, C. 1996, MNRAS, 282, 597

LASOTA, J.-P., ABRAMOWICZ, M.A., CHEN, X., KROLIK, J., NARAYAN, R., & YI, I. 1996, ApJ, 462, 142

LAUER, T.R., FABER, S.M., CURRIE, D.G., EWALD, S.P., GROTH, E.J., HESTER, J.J., HOLTZMAN, J.A., LIGHT, R.M., ONEIL, E.J., SHAYA, E.J., & WESTPHAL, J.A. 1992, AJ, 104, 552

LAUGHLIN, G., & RÓZYCZKA, M. 1996, ApJ, 456, 279

LIN, D.N.C., & PAPALOIZOU, J.C.B. 1986, ApJ, 309, 846

LIN, D.N.C., & PAPALOIZOU, J.C.B. 1993, in Protostars and Planets III, ed. E. H. Levy & J. Lunine (Tucson: Univ. Arizona Press), 749

LYNDEN-BELL, D. 1969, Nature, 233, 690

MAOZ, E. 1995, ApJ, 455, L131

MASSET, F., & TAGGER, M. 1996, A&A, 307, 21

MIYOSHI, M., MORAN, J., HERRNSTEIN, J., GREENHILL, L., NAKAL, N., DIAMOND, P., & INOUE, M. 1995, Nature, 373, 127

MORAN, J., GREENHILL, L., HERRNSTEIN, J., DIAMOND, P., MIYOSHI, M., NAKAI, N., & INOUE, M. 1995, Proc. Nat. Acad. Sc. U.S., 92, 11427

NAKAI, N., INOUE, M., & MIYOSHI, M. 1993, Nature, 361, 45

NEUFELD, D.A., MALONEY, P.R. 1995, ApJ, 447, L17

PACZYNSKI, B. 1977, ApJ, 216, 822

PAPALOIZOU, J.C.B., & LIN, D.N.C. 1995, ApJ, 438, 841

PAPALOIZOU, J.C.B., & PRINGLE, J.E. 1983, MNRAS, 202, 1181

PAPALOIZOU, J.C.B., & TERQUEM, C. 1995, MNRAS, 274, 987

PAPALOIZOU, J.C.B., TERQUEM, C., & LIN, D.N.C. 1998, ApJ, *in press*

POLNAREV, A.G., REES, M.J. 1994, A&A, 283, 301

PRINGLE, J.E. 1996, MNRAS, 281, 357

ROOS, N. 1981, A&A, 104, 218

ROOS, N. 1985, ApJ, 334, 95

ROOS, N., KAASTRA, J.S., & HUMMEL, C.A. 1993, ApJ, 409, 130

SHAKURA, N.I., & SUNYAEV, R.A. 1973, A&A, 24, 337

SPARKE, L.S. 1984, ApJ, 280, 117

SPARKE, L.S., & CASERTANO, S. 1988, MNRAS, 234, 873

SYER, D., CLARKE, C.J., & REES, M.J. 1991, MNRAS, 250, 505

TOOMRE, A. 1983, in IAU Symposium 100, Internal Kinematics and Dynamics of Galaxies, ed. E. Athanassoula (Dordrecht: Reidel), 177

WATSON, W.D., & WALLIN, B.K. 1994, ApJ, 432, L35

WOLTJER, L. 1959, ApJ, 130, 38

Stable oscillations of black hole accretion discs

By MICHAEL A. NOWAK[1] AND D. E. LEHR[2]

[1]JILA, Campus Box 440, Boulder, CO 80309-0440, USA

[2]Department of Physics, Stanford University, Stanford, CA 94305-0460, USA

The study of stable accretion disc oscillations relevant to black hole candidate (BHC) systems dates back over twenty years. Prior work has focused on both unstable and (potentially) stable disc oscillations. The former has often been suspected of being the underlying cause for the observed broad-band variability in BHC, whereas the latter has had little observational motivation until quite recently. In this article, we review both the observations and theory of (predominantly) stable oscillations in BHC systems. We discuss how variability, both broad-band and quasi-periodic, is characterized in BHC. We review previous claims of low frequency features in BHC, and we discuss the recent observational evidence for stable, high frequency oscillations in so-called 'galactic microquasars'. As a potential explanation for the latter observations, we concentrate on a class of theories– with a rich history of study– that we call 'diskoseismology'. We also discuss other recent alternative theories, namely Lense-Thirring precession of tilted rings near the disc inner edge. We discuss the advantages and disadvantages of each of these theories, and discuss possible future directions for study.

1. Overview of Black Hole States

Galactic X-ray sources are typically identified as black hole candidates (BHC) if they have measured mass functions indicating a compact object with $M \gtrsim 3\ M_\odot$, or if their high energy spectra (~ 1 keV $- 10$ MeV) and temporal variability ($\sim 10^{-3} - 10^2$ Hz) are similar to other BHC. A review of the general observations, a number of theoretical models, plus individual descriptions of approximately twenty galactic BHC can be found in Tanaka & Lewin (1995). A review of the timing analyses can be found in van der Klis (1994) and van der Klis (1995). (These latter reviews attempt to link the energy spectra and timing observations, and they draw analogies to similar observations of neutron stars in low mass X-ray binaries). A review of a subset of BHC with reasonably well-determined mass functions can be found in Nowak (1995).

The energy spectra of BHC have been historically labelled based upon observations of the soft X-ray band ($\sim 2 - 10$ keV). Intense, quasi-thermal flux is referred to as the "high" state. Non-thermal flux in this band, typically a power law with a photon index† of ~ 1.7, indicates that the BHC is in the "off" state (for extremely low intensity flux) or "low" state (for moderate intensity flux). The high state tends to have little variability, with a root mean square (rms) variability (cf. §2) of a few percent, whereas the low state tends to have an rms variability of several tens of percent. If a black hole candidate has a quasi-thermal soft X-ray component and significant high energy emission— occasionally modeled as a power law with a photon index ~ 2.5— it is said to be in the "very high" state [cf. Miyamoto, et al. (1991)]. The rms variability of the very high state tends to be greater than that of the high state, but less than that of the low state [cf. Miyamoto, et al. (1992), Miyamoto, et al. (1993)]. Note that the labels "off", "low", "high", and "very high" are purely qualitative in nature, but historically have been in popular use. No universally agreed upon quantitative definition for these states

---

† Photon index here and throughout shall refer to the photon count rate, such that photon index Γ implies # photons/keV/s/cm² $\propto E^{-\Gamma}$, where E is the photon energy.

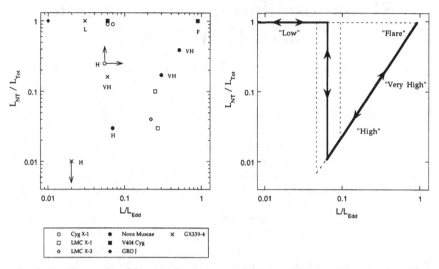

FIGURE 1. *Left:* The ratio of luminosity in a power law tail to total luminosity (L_{NT}/L_{Tot}) versus the ratio of total luminosity to Eddington luminosity (L/L_{Edd}) for black holes with well-determined mass functions. (Note: error bars are *not* plotted.) *Right:* Hypothetical evolutionary tracks for changes in BHC luminosity. [Figure adapted from Nowak (1995).]

exist. More recently, the terms "hard state" and "soft state" have begun to be used in place of "low state" and "high/very high state", respectively.

Figure 1 depicts this state behavior for several BHC with reasonably well-determined mass functions [cf. Nowak (1995)]. This figure shows the fraction of total luminosity in a power law-like tail as a function of the BHC's fractional Eddington luminosity. BHC tend to be dominated by a power law tail for luminosities below $\sim 10\%$ L_{Edd}, be completely dominated by a quasi-thermal, disc-like component near $\sim 10\%$ L_{Edd}, and then begin to show a power law-like tail for $L \gtrsim 10\%$ L_{Edd}. Above 10% L_{Edd}, the fraction of total luminosity in this power law tail (which typically has photon index $\Gamma \sim 2.5$) tends to increase with fractional Eddington luminosity. There are, of course, exceptions to the trends shown in Fig. 1; however, some transient BHC, such as Nova Muscae, have followed this trend very closely [cf. Miyamoto, et al. (1994)]. Historically, the soft flux component has been associated with a "classical" accretion disc, as discussed by Shakura & Sunyaev (1973).

2. Measuring Variability

The states of BHC have been further distinguished by their variability behavior, as discussed by van der Klis (1994), van der Klis (1995), Miyamoto, et al. (1992), and Miyamoto, et al. (1993). As shown in Figure 2, the greater the fraction of luminosity in a hard tail, the greater the amplitude of the observed variability. Miyamoto, et al. (1992) and Miyamoto, et al. (1993) further point out that states appear to have "canonical" variability behaviour not only in terms of amplitude, but also in terms of frequency distribution. This frequency distribution is typically described via Fourier transfrom techniques [cf. van der Klis (1994)], which we briefly describe below.

The key assumption in applying Fourier transform techniques is the assumption that the lightcurve represents a statistically stationary process [cf. Davenport & Root (1987)]. Effectively, this means that one assumes that individual data segments from the lightcurve are a good representation of an ensemble of *statistically* identical lightcurves. It is from

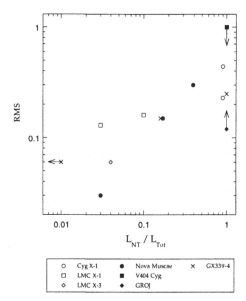

FIGURE 2. RMS variability versus ratio of luminosity in a power law tail to total luminosity (cf. Fig. 1). (Consistent energy bandpasses were not used for all depicted data points, although they predominantly represent a $\sim 2 - 35$ keV bandpass. Again, error bars are *not* plotted.) [Figure adapted from Nowak (1995).]

these individual lightcurve segments that one forms the power spectral density (PSD) [cf. van der Klis (1994)].

The PSD is calculated by dividing the lightcurve into segments of equal length and then taking a Fast Fourier Transform (FFT, i.e. discrete transform, cf. Press, et al. 1992) of each data segment. The squared amplitude of each individual FFT is then averaged together. Often one also averages over Fourier frequency bins. (Typically one chooses a scheme such that the number of frequency bins averaged over is $\propto f$, where f is the Fourier frequency.) This yields the resulting PSD for each lightcurve.

There are a number of different possible normalizations for the PSD; however, in this work we shall only refer to the normalization of Miyamoto, et al. (1992). For this normalization, the integral of the PSD over positive Fourier frequency yields the *square* of the total *root mean square (rms) amplitude* of the variability [Miyamoto, et al. (1992)]. That is, the PSD integrated over positive frequency yields $(\langle x^2 \rangle - \langle x \rangle^2)/\langle x \rangle^2$, where x is the photon count rate [cf. Miyamoto, et al. (1992), Miyamoto, et al. (1993)]. The rms can be calculated over specific Fourier frequency and energy spectral bandpasses. For a given narrow Fourier frequency interval, the rms is (to within a factor of $\sqrt{2}$) the fractional amount by which the lightcurve is sinusoidally modulated in that given Fourier frequency interval.

It is this measure of rms variability, typically calculated over frequency intervals of $\sim 10^{-3} - 10^2$ Hz in BHC, that leads to the characterization of power law tail-dominated sources (i.e. low and very high state sources) as being the most variable BHC. As a function of Fourier frequency, the PSD of all states can usually be described by (possibly multipley) broken power laws. However, occasionally superimposed on top of this broad band variability are narrow— compared to the the broad band PSD— features which are referred to as quasi-periodic oscillations (QPO). The properties of these features are the focus of the next section.

3. Quasi-periodic Variability observed in BHC

A quasi-periodic oscillation is usually taken to be *any* feature in the PSD that is well-fit by a Lorentzian of the form:

$$PSD(f) = \frac{R^2 \, Q \, f_{QPO}}{\pi \left[\, Q^2 (f_{QPO} - f)^2 + f_{QPO}^2 \, \right]} \, . \tag{3.1}$$

As written in eq. 3.1, this functional fit is relevant to the normalization of Miyamoto, et al. (1992). In the above, R is the fractional rms, f_{QPO} is the QPO frequency, and Q is the mode 'quality factor'. If Δf_{QPO} is the full-width half maximum of the QPO feature, then $Q \approx f_{QPO}/\Delta f_{QPO}$.

For $Q \lesssim 3$, it is difficult to attribute the QPO to a discrete feature, as opposed to being merely an extension of the broad-band variability. For $Q \gtrsim 10$, the QPO is usually quite apparent in the PSD, and may be attributable to a discrete process in the BHC system. The question then arises of whether or not such a QPO feature represents a quasi-stable mode in an accretion disc. If the QPO feature is attributable to a mode or set of modes, then there are many possible mechanisms for generating a finite feature width in the PSD [cf. van der Klis (1994)]. Among these possibilities are: multiple modes distributed over frequencies $f_{QPO} \pm \Delta f_{QPO}$, a driven-damped mode with damping time scale $\tau_D \approx Q f_{QPO}^{-1}$, a mode that executes a random walk in phase of $\sim 2\pi$ on time scales of order τ_D, etc. Rarely does one have the statistics to determine which of these processes is responsible for the Q of the QPO.

Prior to the launch of the *Rossi X-ray Timing Explorer (RXTE)*, observations of QPO in BHC were made with *Ginga*, *BATSE*, and *Granat/SIGMA*. All of these observed QPO were at frequencies $\lesssim 10$ Hz, although the above instruments were for the most part limited to $\lesssim 60$ Hz. In addition, the rms variability of these features were all $\lesssim 10\%$, with $Q \lesssim \mathcal{O}(10)$. Below, we review some of the historical observations of black hole QPO.

3.1. *Historical Observations*

Reviews of some of the observed QPO features in BHC can be found in Nowak (1995) and Tanaka & Lewin (1995). Here we consider separately low state and high/very high state observations of BHC .

In the low state, Cygnus X–1 has shown QPO with rms as high as 15% [cf. Kouveliotou, et al. (1992a), Ubertini, et al. (1994), Vikhlinin, et al. (1994)]. The features were seen to have frequencies ~ 0.04 and 0.07 Hz. However, they appeared more as broad peaks in the otherwise broken power law PSD, and did not appear as discrete, narrow features. Similar features, with frequencies 0.04 and 0.2 Hz were seen in the hard state of the X-ray transient GRO J0422+32 (a.k.a. Nova Persei) [Kouveliotou, et al. (1992b), Sunyaev, et al. (1992), Sunyaev, et al. (1993)].

The source GX 339–4, which has been observed to transit through soft and hard X-ray states, has shown a relatively narrow $[Q \sim \mathcal{O}(10)]$ QPO at 0.8 Hz [Grebenev, et al. (1991)] during its hard state. On the other hand, in its very high state GX 339–4 has also shown a narrow QPO at 6 Hz [Miyamoto, et al. (1991)]. Similar to this feature, the X-ray transient Nova Muscae has shown QPO in the range $3 - 8$ Hz during its very high state [Kitamoto, et al. (1992)].

In the soft state of LMC X-1 (which is the only state that has been observed in this source), there has been a claim of a 0.08 Hz feature with rms $\sim 4\%$ [Ebisawa, Mitsuda, & Inoue (1989)]. However, the count rates for these observations were extremely low, and it is unclear whether or not the putative variability detections described by Ebisawa, Mitsuda, & Inoue (1989) were below the effective Poisson noise limit. More recently,

Cui et al. (1997) has found evidence (based upon *RXTE* observations) of a $3 - 12$ Hz feature in the soft state of Cygnus X–1. We note, however, that this feature is very broad and is difficult to associate with a discrete feature distinct from the observed broad-band variability.

Prior to the recent *RXTE* observations, a fairly clean division of QPO frequencies occurs if one ignores the 0.08 Hz QPO attributed to LMC X–1. Low frequency QPO ($\lesssim 1$ Hz) are seen in low/hard states, while high frequency QPO (~ 6 Hz) are seen in very high/soft states. Again, two questions arise. First, are these QPO associated with modes distinct from the broad-band variability? Second, if the QPO are distinct modes, are they associated with an accretion disc or with another component (such as a corona) in the BHC system? In the next subsection, we discuss a set of observed features in two BHC that may indeed be attributable to an accretion disc.

3.2. QPO in 'Microquasars'

Recently, two rather unusual and dramatic X-ray transient galactic BHC have been observed: GRS 1915+105 [Mirabel & Rodriguez (1994), Morgan, Remillard, & Greiner (1997), Chen, Swank, & Taam (1997), Taam, Chen, & Swank (1997)] and GRO J1655-40 [Hjellming & Rupen (1995), Remillard et al. (1997)]. These sources were unusual in that they showed powerful, highly relativistic radio jets [Mirabel & Rodriguez (1994), Hjellming & Rupen (1995)]; hence, they have been dubbed 'microquasars'. Furthermore, both have shown dramatic X-ray variability.

As observed by *RXTE*, GRS 1915+105 has shown X-ray count rates up to 10^5 cts/sec, which, given a distance of $\gtrsim 12$ kpc [Mirabel & Rodriguez (1994)], indicates a peak luminosity of over 10^{40} erg/s [Morgan, Remillard, & Greiner (1997), Chen, Taam, & Swank (1997)]. As discussed in Morgan, Remillard, & Greiner (1997), GRS 1915+105 has also shown intense variability patterns, with the luminosity changing by factors of several in only a few seconds. Amidst this spectral behavior, a host of different QPO features have been observed, ranging in frequencies from $\sim 0.1 - 10$ Hz [Morgan, Remillard, & Greiner (1987), Chen, Swank, & Taam (1997), Taam, Chen, & Swank (1997)]. Many of these QPOs have multiple harmonics, and both correlations and anti-correlations with source luminosity have been observed.

However, among the various QPO features seen, a high frequency QPO at 67 Hz has stood out because it apparently does not appreciably vary in frequency [Morgan, Remillard, & Greiner (1997)]. During the first epoch that this QPO feature was observed, it was relatively narrow ($Q \sim 20$), weak (rms variability $\sim 0.3 - 1.6\%$), and varied in frequency by $< 3\%$, despite factors of ~ 2 variations in source luminosity. When viewed in restricted bandpasses the rms variability was seen to increase with energy, with a maximum rms variability $\sim 6\%$ in the highest energy bandpass ($\sim 10 - 20$ keV). During subsequent observations similar features were observed, always at 67 ± 2 Hz [even on occasions when no soft, excess flux above a power law tail was observed in the spectrum; Morgan 1997, private communication].

A high frequency QPO feature has also been observed in GRO J1655-40, with frequency 300 Hz [Remillard, et al. (1997)]. However, this QPO was only detected once, and then only by summing several, individual observations. It was therefore difficult to place limits on the stability of this feature's frequency. Like the 67 Hz feature in GRS 1915+105, this feature was weak (rms variability $\sim 0.8\%$). Furthermore, it was detectable only in the "hardest" spectra ($\sim 10 - 20$ keV). The 300 Hz feature in GRO J1655-40 was somewhat broader than the 67 Hz feature in GRS 1915+105. It is also notable in that a very accurate mass determination, $7 \pm 0.2 \, M_\odot$, has been made for the compact object in GRO J1655-40 [Orosz & Bailyn (1997)].

As we will discuss in §5, a number of properties displayed by these features, especially the 67 Hz feature in GRS 1915+105, are what one expects for stable, global oscillations of accretion discs. We will concentrate on theories of the oscillation of the innermost disc regions as potential explanations of these observations.

3.3. *QPO in Active Galactic Nuclei*

As discussed in §4, the characteristic time scales are one hundred thousand to one hundred million times longer in Active Galactic Nuclei (AGN) as compared to galactic BHC. Thus, if a 0.1 s QPO period is typical of a galactic black hole, then we might expect a 3 hour to year-long period to be typical for AGN. Such long time scales are very difficult for current missions to detect, as the shorter period is on the order of the *total* duration of a typical observation, and the longer period requires many dedicated observations. However, one AGN, IRAS18325–5926, has shown strong evidence for a 5.5×10^4 s periodicity that was observed for 9 cycles [Iwasawa, et al. (1998)]. Over these few cycles, there was no evidence for any strong change in phase, and the amplitude of the modulations was on the order of 10%. This is a somewhat stronger amplitude than observed for the 67 Hz and and 300 Hz features described above, and therefore are difficult to explain with the models described in §5. Further (longer) observations are required to determine to what extent this AGN QPO is or is not similar to the QPO seen in the microquasars.

4. Characteristic Accretion Disc Frequencies and Timescales

The fact that the 67 Hz QPO feature seen in GRS 1915+105 does not appear to vary in frequency as a function of luminosity suggests that this feature is tied to a fundamental frequency or time scale in the BHC system. As this frequency is apparently independent of accretion rate (i.e. luminosity), the relevant time scales are likely to be gravitational ones. The two most important parameters for the gravitational time scales of BHC are the black hole mass, M, and the angular momentum, J. We shall usually normalize the black hole mass to a solar mass. We will characterize the black hole angular momentum by the dimensionless parameter $a \equiv cJ/GM^2$, where c is the speed of light and G is the gravitational constant. For a Schwarzschild black hole, $a = 0$, while for a Kerr hole $a < 1$. Below we discuss the frequencies most relevant to orbits near the equatorial plane of a (possibly spinning) black hole. These frequencies are those most relevant to a thin accretion disc†.

In what follows, we shall set $c = G = 1$. Furthermore, physical length scales will be normalized to GM/c^2 (≈ 15 km for $M = 10\ M_\odot$), and frequencies will be normalized to c^3/GM ($\approx 3.2 \times 10^3$ Hz for $M = 10\ M_\odot$). In general, characteristic gravitational length scales increase linearly with M, whereas characteristic time scales decrease linearly with M. Therefore, whereas 1 ms may be a relevant time scale for a galactic black hole, days or months might be the relevant time scale for a massive black hole at the center of an AGN. There will be four main frequencies of concern to us, each a function of radius r: the Keplerian (i.e. orbital) frequency, the radial epicyclic frequency, the vertical epicyclic frequency, and the Lense-Thirring precession frequency. We describe each of these in turn below.

The Keplerian frequency is the frequency with which a free particle azimuthally orbits

† We define the equatorial plane to be the plane perpendicular to the black hole angular momentum axis that passes through the center of the black hole. We shall consider a cylindrical coordinate system (r, ϕ, z) set up on this plane, with $z = 0$ being in the plane.

the black hole. This frequency, to a viewer observing at infinity, is given by:

$$\Omega = \left(r^{3/2} + a \right)^{-1} \tag{4.2}$$

For a 10 $M_\odot$ Schwarzschild black hole, this is approximately 220 Hz at $r = 6$.

The radial epicyclic frequency is the frequency at which a free particle oscillates about its original circular orbit if it is given a radial perturbation. In classical mechanics, the square of the radial epicyclic frequency is given by $\kappa^2 = 4\Omega^2 + r\partial\Omega^2/\partial r$, and is exactly equal to Ω^2 for Keplerian orbits in an r^{-1} potential† [cf. Binney & Tremaine (1987)]. In General Relativity, there is a minimum radius for which an orbit is stable to radial perturbations. The innermost marginally stable orbit is at $r = 6$ for $a = 0$, and moves inward as a increases. The epicyclic frequency is given by

$$\kappa^2 = \Omega^2 \left(1 - \frac{6}{r} + \frac{8a}{r^{3/2}} - \frac{3a^2}{r^2} \right) \tag{4.3}$$

and is zero at the marginally stable orbit. For a Schwarzschild black hole, κ reaches a maximum at $r = 8$ and is approximately 71 Hz for $M = 10$ $M_\odot$. Note that for a fixed M, the maximum κ for $a \sim 1$ is approximately equal to Ω at the marginally stable orbit for $a \sim 0$. This coincidence of time scales has led to alternative suggestions that the 300 Hz feature in GRO J1655-40 can be attributed to either the Keplerian frequency at the marginally stable orbit or to the maximum epicyclic frequency, depending upon whether one adopts $a \sim 0$ or $a \sim 1$, respectively, for this source.

Just as the radial epicyclic frequency differs from the Keplerian orbital frequency in General Relativity, so does the vertical epicyclic frequency for a spinning black hole [Kato (1990), Kato (1993), Perez, et al. (1997)]. The vertical epicyclic frequency, $\Omega_\perp$, is the frequency at which a free particle oscillates about its original circular orbit if it is given a vertical perturbation. It is equal to the Keplerian orbital frequency for both Newtonian gravity and Schwarzschild black holes. In general, for non-zero black hole angular momentum it is given by:

$$\Omega_\perp^2 = \Omega^2 \left(1 - \frac{4a}{r^{3/2}} + \frac{3a^2}{r^2} \right) \tag{4.4}$$

As we will discuss in §5.2.3, this frequency is relevant in determining the frequencies of global "corrugation modes" in accretion discs [cf. Kato (1990), Kato (1993), Perez (1993), Ipser (1996)].

The final gravitational frequency that we need to consider is the Lense-Thirring precession frequency [Lense & Thirring (1918)]. If an azimuthally orbiting ring of matter is tilted out of the equatorial plane, it will begin to precess due to frame dragging effects if the black hole has non-zero angular momentum. This precession frequency, Ω_{LT}, is given by the expression:

$$\Omega_{LT} = \frac{2a}{r^3} \ . \tag{4.5}$$

Note that for $a \sim 1$ it is on the order of Ω, for radii near the marginally stable orbit. Furthermore, Ω_{LT} has a strong radial dependence (cf. §5.2.6).

In Figure 3 we plot these various frequencies for three angular momentum parameters, $a = 0$, $a = 0.5$, $a = 0.998$. In the inner regions of accretion discs around black holes, these frequencies typically range from $\sim 100 - 1000$ Hz, for $M = 10$ $M_\odot$, with the highest frequencies being achieved for $a \sim 1$.

---

† This is just the statement that a radial perturbation sends a circular orbit into an elliptical one, with an identical orbital period, for Newtonian free-particle orbits in an r^{-1} potential.

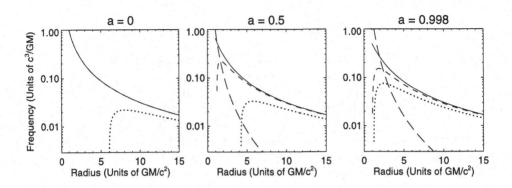

FIGURE 3. Characteristic frequencies in an accretion disc around a black hole, normalized to $c^3/GM \approx 3.2 \times 10^3$ Hz for $M = 10\ M_\odot$. The angular momentum parameter, $a = cJ/GM^2$, where J is the angular momentum of the hole. Solid line is the Keplerian frequency, Ω, dotted line is the radial epicyclic frequency, κ, short dashed line is the vertical epicyclic frequency, $\Omega_\perp$, and the long dashed line is the Lense-Thirring precession frequency, Ω_{LT}.

Note that, at least for the Schwarzchild case, the Keplerian and epicyclic frequencies can be reasonably approximated in Newtonian gravity via the use of a "pseudo-Newtonian" potential. Two such potentials that have been used in the literature are given by:

$$\Phi_{NW} = -r^{-1}\left[1 - \frac{3}{r} + \frac{12}{r^2} \right] \quad , \quad \Phi_{PW} = -r^{-1}\left[1 - \frac{2}{r} \right]^{-1} \quad , \qquad (4.6)$$

[cf. Nowak & Wagoner (1991), Pazczynski & Witta (1980)]. Both of these potentials have $\kappa = 0$ at $r = 6$, and the Φ_{NW} potential has been used for Newtonian calculations of modes in relativistic discs about Schwarzchild black holes [cf. Nowak & Wagoner (1991), Nowak & Wagoner (1992), Nowak & Wagoner (1993), Nowak, et al. (1997).]

All of the above frequencies have strong radial dependences, so it is unlikely that any one of them can lead to a narrow feature at a discrete frequency without some mechanism that couples a range of radii†. Such coupling is most likely achieved via hydrodynamic effects, i.e. pressure, viscosity, etc. To this end, the most relevant *hydrodynamical* frequencies are given by the frequency of sound (for a given wavelength), ω_{cs}, and the vertical Brunt-Väisälä (i.e. buoyancy) frequency, N_z. In terms of the pressure, P, the density, ρ, and the radial and vertical perturbation wavelengths, λ_r and λ_z, these frequencies are given by:

$$\omega_{cs}^2 = c_s^2 k^2 = c_s^2\,(2\pi)^2\,\left(\lambda_r^{-2} + \lambda_z^{-2} \right) \quad , \qquad (4.7)$$

$$N_z^2 = \left[\rho^{-2}\frac{\partial \rho}{\partial z} - (\gamma\rho P)^{-1}\frac{\partial P}{\partial z} \right] \frac{\partial P}{\partial z} \quad . \qquad (4.8)$$

For a thin disc, the sound speed is of order $h\Omega$, where h is the disc half-thickness [cf. Shakura & Sunyaev (1973)]. For radial wavelengths of $\mathcal{O}(r)$, the frequency of sound waves, $\sim (h/\lambda_r)\Omega$, is typically much less than the Keplerian frequency. The buoyancy frequency, as it deals with gradients on scales of $\mathcal{O}(h)$, can be comparable to the vertical epicyclic frequency [cf. Kato (1993)]. (Due to symmetry, however, the buoyancy

---

† Unless, of course, there is a "special" radius, such as the marginally stable orbit radius, which picks out a discrete frequency. However, it then may become very difficult to achieve a large luminosity modulation.

frequency vanishes at the disc mid-plane.) Both of these frequencies play a part in the theories discussed in §5.2.

5. Theoretical Models of Disc Oscillations

5.1. *Historical Review*

The study of oscillations relevant to accretion discs around black holes has a history that dates back more than 30 years. It was quickly realized that the α−disc model of Shakura & Sunyaev (1973) shows both viscous and thermal instabilities for high luminosities above $\mathcal{O}(10\% \, L_{\rm Edd})$ [Shakura & Sunyaev (1976)]. These instabilities have often been invoked to explain the intense, broad band variability seen in BHC low/hard states†.

As discussed in §3.1, until recently there has been little observational evidence for discrete (possibly stable) periodic modes in accretion discs about black holes. However, as was first realized by Kato & Fukue (1980), not only can accretion discs support discrete oscillatory modes, but also the effects of General Relativity modify the mode spectrum and determine the regions in which modes can be trapped. By considering adiabatic perturbations of an isothermal disc, Kato & Fukue (1980) showed that the roll-over of κ to zero at the marginally stable orbit leads to a cavity that effectively traps acoustic modes (i.e. p−modes) at the disc inner edge (cf. §5.2.1).

Later, Okazaki, et al. (1987) considered isothermal perturbations of isothermal discs and showed that it was also possible to trap what are essentially internal gravity modes (i.e. g−modes) near the epicyclic frequency maximum (cf. §5.2.2). Vishniac & Diamond (1989) considered travelling wave versions of g−modes that had an azimuthal dependence $\propto \exp[im\phi]$, with $m = 1$. These modes were invoked as a possible mechanism for angular momentum transport in an accretion disc. [Note, however, that the strictly Newtonian modes considered by Vishniac & Diamond (1989) were not trapped modes, and furthermore they had very small vertical wavelengths, i.e. $\lambda_z \ll h$.]

In a series of papers Nowak & Wagoner (1991,1992,1993) adopted a Lagrangian perturbation approach [cf. Lynden-Bell & Ostriker (1967), Freidman & Schutz (1978a,b)] to the study of disc oscillations. They showed that the acoustic modes of Kato & Fukue (1980) and Okazaki, et al. (1987) are high and low frequency limits, respectively, of the same dispersion relationship, which itself is the strong-rotation limit of the simplest form of the "helioseismology" dispersion relationship [cf. Hines (1960)]. Hence, they gave these modes the name "diskoseismology". These works used a pseudo-Newtonian potential (eq. 4.6) to mimic the effects of General Relativity; however, they also generalized the calculations to adiabatic perturbations of non-isothermal discs. As discussed by Freidman & Schutz (1978a,b), by adopting a Lagrangian perturbation approach one can calculate a conserved (in the absence of dissipative forces) "canonical energy" for the modes‡. Accurately determining the canonical energy allows one to determine the effects of various dissipative effects, such as viscosity [Nowak & Wagoner (1992)], and allows one to consider various excitation mechanisms for the mode [Nowak & Wagoner (1993), Nowak, et al. (1997)]. Nowak & Wagoner (1993,1997) also considered ways in which the modes might lead to actual luminosity modulations (cf. §5.2.5).

Ipser & Lindblom (1992) developed a scalar potential formalism for calculating modes

---

† It should be noted, however, that as shown in Figure 1 and discussed in Nowak (1995), low/hard state luminosities are nominally below the instability limits of Shakura & Sunyaev (1976).

‡ As Freidman & Schutz (1978a,b) point out, assigning an energy to a mode in a rotating medium is an extremely subtle issue, and many prior works were incorrect in determining this quantity.

of rotating systems in full General Relativity. Gradients of the scalar potential are related to the Lagrangian perturbation vector in both the pseudo-Newtonian formalism [Nowak & Wagoner (1992)] and the fully General Relativistic formalism [Perez, et al. (1997)]. Perez, et al. (1997) used this formalism to consider the fully relativistic version of $g-$modes. In addition, a third class of modes, the $c-$modes (cf. §5.2.3) recently have been identified and described with fully relativistic formalisms [Kato (1990), Kato (1993), Perez (1993), Ipser (1996).]

5.2. *'Diskoseismology'*

The main distinguishing feature of most of the theoretical modes described in the works cited above is that the mode frequencies predominantly depend upon fundamental gravitational frequencies and are not strongly effected by hydrodynamic processes if $h \ll r$. That is, for a thin disc, changes in density and/or pressure (which we assume to be correlated with luminosity) may effect the physical extent of a mode or its amplitude, but they do not greatly effect the mode's frequency. Furthermore, the expected mode frequencies, $\gtrsim 100$ Hz, are comparable to the gravitational frequencies in the innermost regions of the accretion disc.

From these standpoints, the QPO features discussed in §3.1 are *not* good candidates for 'diskoseismic' modes. They are of very low frequency, and furthermore several, such as the $3 - 8$ Hz QPO in Nova Muscae [Miyamoto, et al. (1994)], have been seen to vary in frequency by fractionally large amounts. Likewise, the low frequency ($\sim 0.1 - 10$ Hz) QPO seen in both GRS 1915+105 and GRO J1655-40 have been observed to be very highly variable in frequency [Morgan, Remillard, & Greiner (1997), Remillard, et al. (1997)].

However, the 67 Hz feature observed in GRS 1915+105 has been *consistently* observed at 67 ± 2 Hz, despite the fact that this source's luminosity has varied by factors of two or more over the epochs during which this feature has been observed. Although there has been only one detection of the 300 Hz feature in GRO J1655-40, it was consistent with being at a steady frequency. The 300 Hz, if attributable to a disc time scale, indicates a phenomenon occurring very close to the disc inner edge. For these reasons, we identify both of these features as candidate 'diskoseismic' modes.

For pseudo-Newtonian potential calculations of diskoseismic modes, one defines a Lagrangian perturbation vector that describes the displacement of a fluid vector *relative to its unperturbed path* [cf. Freidman & Schutz (1978a)]. All perturbation quantities compare the *displaced*, perturbed fluid element to the (moving) fluid element in the unperturbed flow. If we take $\vec{\xi}$ to be the displacement vector, then the Lagrangian variation of the velocity, $\Delta\vec{v}$, is given by $\partial\vec{\xi}/\partial t$ [Freidman & Schutz (1978a)]. [As in Nowak & Wagoner (1991), we use Δ to denote a Lagrangian perturbation.] The Lagrangian perturbation of the density is derived from mass conservation, and is given by $\Delta\rho = -\rho\nabla\cdot\vec{\xi}$. For adiabatic oscillations, Lagrangian perturbations of the pressure are related to Lagrangian variations of the density by $\Delta P/P = \gamma\Delta\rho/\rho$, where γ is the adiabatic index [cf. Freidman & Schutz (1978a), Nowak & Wagoner (1991)].

Calculationally, it is somewhat easier to determine the mode structure by defining a scalar potential [Ipser & Lindblom (1992), Nowak & Wagoner (1992)], $\delta V(r, z) \equiv \delta P/\rho$, where here δ refers to an Eulerian perturbation†. The potential δV is seen to be the Eulerian perturbation of the enthalpy, and furthermore the Lagrangian perturbation is related to gradients of this potential [Nowak & Wagoner (1992)]. If one looks for modes

---

† An Eulerian perturbation is one that compares the perturbed fluid to the unperturbed fluid at a *fixed* coordinate [Freidman & Schutz (1978a)].

$\propto \exp[-i(m\phi + \sigma t)]$, where t is time and σ is the inertial mode frequency, then in a WKB approximation the modes satisfy a dispersion relationship of the form

$$\left[\omega^2 - \Upsilon(r)\Omega^2\right]\left(\omega^2 - \kappa^2\right) = \omega^2 c_s^2 k_r^2 \ , \tag{5.9}$$

where $\omega \equiv \sigma + m\Omega$ is the corotating frequency, $\Upsilon(r)$ is a slowly varying separation function (which comes from separating the basic fluid perturbation equation into radial and vertical components), Ω and κ are the Keplerian rotation frequency and epicyclic frequency, respectively, c_s is the speed of sound, and $k_r \equiv 2\pi/\lambda_r$ is the radial wavenumber.

This dispersion relation is essentially the same as the simplest form of the helioseismology dispersion relation, except that κ takes the role of the buoyancy frequency, and $\sqrt{\Upsilon(r)}\Omega$ takes the role of the 'acoustic cutoff' frequency [cf. Hines (1960)]. Nowak & Wagoner (1991, 1992) showed that there are two general classes of solutions to this dispersion relation: high and low frequency. We identify the high frequency ($\omega^2 \gtrsim \kappa^2$) modes with acoustic p−modes [cf. Kato (1980), Nowak (1991), Perez (1993)], and the low frequency ($\omega^2 \lesssim \kappa^2$) modes with internal gravity g−modes [cf. Okazaki et al. (1987), Vishniac & Diamond (1989), Nowak & Wagoner (1992), Perez, et al. (1997)]. Very qualitatively, gravitational effects set certain fundamental oscillation frequencies, which are then modified by pressure effects. If the pressure forces 'assist' the gravitational forces, then a high frequency is achieved. If the pressure forces 'retard' the gravitational forces, then a low frequency is achieved. (We will return to this qualitative notion below when we consider mode excitation, cf. §5.2.4.)

The fully General Relativistic version of the diskoseismology equations leads to a WKB dispersion relationship of a very similar form. As was shown in Perez, et al. (1997), for the relativistic equations a WKB analysis still allows approximate separation of the governing equation into radial and vertical dependences. The radial component of the fluid perturbations satisfies the WKB relation

$$\frac{d^2 W}{dr^2} + \alpha^2 \left[\Psi\left(\frac{\Omega_\perp}{\omega}\right)^2 - 1\right](\kappa^2 - \omega^2)W = 0 \ , \tag{5.10}$$

where $\alpha(r)$ is inversely proportional to the speed of sound at the mid-plane and $\Psi(r)$ is a slowly-varying separation function. The eigenfunction $W(r)$ is proportional to a radial derivative of the potential δV (again defined as $\delta P/\rho$) and to the radial component of the Lagrangian fluid displacement [Perez, et al. (1997)].

From the radial WKB equation (5.10) one can identify three classes of modes that are trapped. The p−modes, defined by $\Psi(\Omega_\perp/\omega)^2 < 1$, are trapped where $\omega^2 > \kappa^2$; g−modes, defined by $\Psi(\Omega_\perp/\omega)^2 > 1$, are trapped where $\omega^2 < \kappa^2$. The third class of modes (unique to the fully relativistic treatment), the c−modes, are defined by $\Psi(\Omega_\perp/\omega)^2 \cong 1$ and may in principle be trapped in either region. [The separated equation governing the vertical component involves a more complicated second-order linear operator which contains the vertical buoyancy (Brunt-Väisälä) frequency, N_z.]

The frequencies of all classes of modes are proportional to $1/M$, but their dependences on the angular momentum of the black hole are quite different. In principle this would allow one to measure the angular momentum of the black hole if more than one type of mode were to be detected in the same source. Alternatively, if one could infer the mass of the black hole through the motion of a companion which feeds the disc, such as has been done for GRO J1655-40 [Orosz & Bailyn (1997)], one could determine the angular momentum of the black hole by a single mode observation (given a correct identification for the class of mode observed). Below we briefly describe each class of mode, and we present representative examples in Figures 4–6.

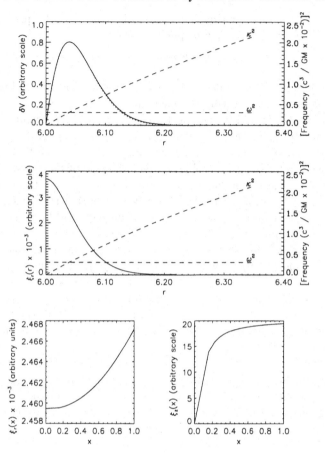

FIGURE 4. A representative $m = 0$ p−mode. Here we show the Eulerian potential, δV, as well as the radial component of the Lagrangian perturbation vector as a function of radius r, and normalized vertical coordinate, x. [Taken from Perez (1993).]

5.2.1. *P-modes*

As discussed by Kato (1980) and Nowak & Wagoner (1991), p−modes are trapped in regions where $\omega^2 > \kappa^2$. For $m = 0$, this leads to a narrow trapping region ($\Delta r_{\mathrm{mode}} \ll 1$) at the disc inner edge where κ^2 goes to zero†. These modes have frequencies a factor of a few less than the radial epicyclic maximum, and tend to have $\xi_r \gtrsim \xi_z$ (i.e. radial perturbations greater than the vertical perturbations) [Nowak & Wagoner (1991)]. Due to their narrow confinement very near the disc inner edge [where the 'no-torque' disc boundary condition leads to very little luminosity modulation; Shakura & Sunyaev (1973)], we do not expect these modes to have significant luminosity modulation. We present an example of a p−mode in Fig. 4.

5.2.2. *G-modes*

Internal (gravity) modes are trapped where $\omega^2 < \kappa^2$, in the region where κ achieves its maximum value and where the disc is hottest. The lowest modes can have significant vertical displacements ($\xi_z \gtrsim \xi_r$) and relatively large radial extents $\Delta r \approx GM/c^2$. We believe that these modes will produce the greatest luminosity modulations in the disc;

† Trapped p−modes might also exist in the large, outer region of the disc; Silbergleit 1997, private communication.

thus, they may be the most observable class of modes [Nowak & Wagoner (1992), Perez, et al. (1997)].

To analytically approximate the eigenfunctions and eigenfrequencies of the lowest $g-$modes, WKB solutions to the separated radial and vertical equations of fluid perturbations can be obtained. From the symmetry of the governing equations, it is sufficient to consider eigenfrequencies $\sigma < 0$ and axial mode integers $m \geq 0$. The resulting frequencies, $f = -\sigma/2\pi$, of the lowest radial ($m = 0$) $g-$modes are given by

$$f = 714\,(1 - \epsilon_{nj})(M_\odot/M)\,F(a)\ \text{Hz},$$

$$\epsilon_{nj} \approx \left(\frac{n + \frac{1}{2}}{j + \delta}\right)\frac{h}{r}. \tag{5.11}$$

Here $F(a)$ is a known, monotonically increasing function of the black hole angular momentum parameter a, with $F(a = 0) = 1$ and $F(a = 0.998) = 3.44$. The properties of the disc enter only through the small correction term ϵ_{nj}, which involves the disc thickness $2h(r)$ and the radial (n) and vertical (j) mode numbers, with $\delta \sim 1$. Typically $h/r \sim 0.1\,L/L_{Edd}$ for a radiation-pressure dominated optically thick disc region, where L/L_{Edd} is the ratio of the luminosity to the Eddington (limiting) luminosity†. [Higher axial $g-$modes with $m > 0$ have a somewhat different dependence on a than the radial modes; Perez, et al. (1997)].

The 67 Hz feature observed in GRS 1915+105 has been associated with a diskoseismic $g-$mode [Nowak, et al. (1997)]. If this association is correct, then equation (5.11) predicts a black hole mass of 10.6 $M_\odot$, if the hole is non-rotating, up to 36.3 $M_\odot$, if the hole is maximally rotating. (Further aspects of this identification are discussed in §5.2.4, §5.2.5 below.) If we identify the 300-Hz feature observed in GRO J1655-40 with the fundamental $g-$mode oscillating in an accretion disc surrounding a 7.0 $M_\odot$ black hole [as determined from spectra of the companion star; Orosz & Bailyn (1997)], then equation 5.11 implies that its angular momentum is 93% of maximum.

5.2.3. *C-modes*

To analytically obtain the approximate eigenfrequencies for the $c-$modes, one can solve the separated WKB perturbation equations in the regime $\Psi(\Omega_\perp/\omega)^2 \cong 1$ [Lehr, Wagoner, & Silbergleit (1997), private communication]. Here we consider $c-$modes that are non-radial ($m \geq 1$), nearly incompressible oscillations trapped in the very innermost region of the disc with eigenfrequencies $|\sigma| \sim m\Omega(r_c) - \Omega_\perp(r_c)$, where r_c is the upper radial bound of the mode. For $m = 1$, eigenfrequency of the fundamental $c-$mode is approximately the Lense-Thirring frequency evaluated at the radius r_c:

$$|\sigma| \cong \frac{2a}{r_c^3}. \tag{5.12}$$

Note that for the $m = 1$ mode, r_c is typically no greater than 10% larger than the marginally stable orbit radius. The physical structure of this mode resembles a tilted inner disc which slowly precesses about the black-hole spin axis (cf. Fig 6). Since the mode is nearly incompressible, there is little temperature or pressure fluctuation in the disc, and hence a $c-$mode may produce very little *intrinsic* luminosity modulation (cf. 5.2.5). The mode may be observable, however, since the projected area of the inner

---

† Although the frequencies of the modes do not have a strong dependence upon L/L_{Edd}, the modes are no longer effectively trapped at large radii if $L/L_{Edd} \gtrsim 0.3$. For such high luminosities, the outer evanescent region for the $g-$modes becomes narrow in radius, thereby allowing $g-$modes to effectively couple to travelling waves in the outer regions of the disc. The modes thus "leak" energy to larger radius.

Disk Oscillating About Black Hole

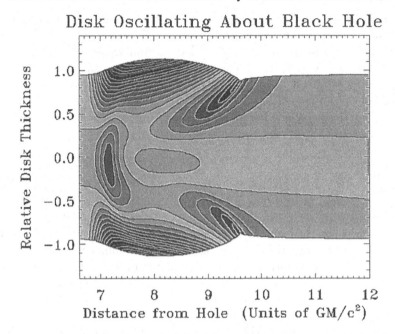

FIGURE 5. A cross section of a disc with an $m = 0$ $g-$mode. The disc thickness, as a function of radius, has been normalized to 1. Shaded contours correspond to the Eulerian pressure perturbations normalized to the *local* unperturbed pressure values. (The apparent asymmetry about the disc mid-plane is a numerical artifact, due to inadequate resolution of the contouring routine.)

disc changes with time. This allows the possibility of coronal emission being Compton reflected and modulated in the disc inner regions. Calculations to determine the extent of the resulting modulation are currently being undertaken [Nowak & Reynolds (1997), private communication].

Note that GRO J1655-40 does have a high inclination, $\sim 70°$, to our line of sight [Orosz & Bailyn (1997)] making this a potentially promising mechanism for producing the observed 300 Hz feature. (Also, Compton reflection features would peak near ~ 30 keV, which is also consistent with the observations.) If we associate the 300 Hz feature with a $c-$mode in a disc about a 7 $M_\odot$ black hole, then $a \approx 0.8$ for this source. Note that this is different from both the prediction made from assuming a $g-$mode in this source, as well as the prediction of Cui, et al. (1998) who associated this feature with Lense-Thirring precession at a single radius (cf. 5.2.6). In Fig. 6 we show the $c-$mode frequency as a function of the angular momentum parameter a.

5.2.4. *Excitation and Damping Mechanisms*

It is possible to use a parameterized stress tensor to estimate the effects of turbulent viscosity on the $g-$modes [Nowak & Wagoner (1992,1993)]. The canonical energy of a radial mode is $E_c \sim \sigma^2 \rho (\xi_z^2 + \xi_r^2)\ dV$, where here dV is the volume occupied by the mode. Isotropic turbulence produces a rate of change $dE_c/dt \equiv -E_c/\tau$, with $\tau \sim |\ \alpha\sigma\ [h^2/\lambda_r^2 + h^2/\lambda_z^2]\ |^{-1}$. Here λ_r and λ_z, respectively, are the radial and vertical mode wavelengths, and α is the standard Shakura-Sunyaev $\alpha-$parameter [Shakura & Sunyaev (1973)]. The corresponding quality factor is given by

$$Q_{jn}^{-1} = (|\sigma|\tau)^{-1} \sim \left[j^2 + (h/r)n^2\right]\alpha\,, \tag{5.13}$$

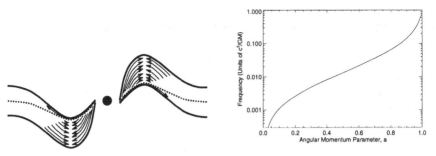

FIGURE 6. *Left:* Schematic drawing of a c–mode in an accretion disc (arbitrary amplitude). *Right:* Frequencies of the fundamental $m = 1$ c–mode frequency as a function of the black hole angular momentum parameter $a \equiv cJ/GM^2$. Frequencies are normalized to $c^3/GM \approx 3.2 \times 10^3$ Hz for $M = 10~M_\odot$.

as $\lambda_z \sim h/j$ and $\lambda_r \sim \sqrt{hr}/n$, where j and n are of order of the number of vertical and radial nodes in any particular eigenfunction. Thus, for $\alpha \ll 1$, we can have high mode Q [cf. Nowak & Wagoner (1993); Nowak, et al. (1997)].

The above estimates are for *isotropic* viscosity. If the turbulence does not efficiently couple to the vertical gradients of the modes, then the mode Q value is increased by a factor $\sim (j\lambda_r/h)^2$ [cf. Nowak & Wagoner (1993)]. Aside from damping modes, turbulence can also potentially excite modes. Velocity perturbations in the disc, $\delta\vec{v}$, are made up of a mode component, $\delta\vec{v}_M$, and a turbulent component, $\delta\vec{v}_T$. Viscous damping arises from terms of the form $\delta v_{Mi}\delta v_{Tj}$, while mode excitation arises from terms of the form $\delta v_{Ti}\delta v_{Tj}$ [Nowak & Wagoner (1993)]. It is possible to make simple estimates of the magnitude of the turbulent excitation, and balance this against the turbulent damping [Nowak & Wagoner (1993)]. The modes are excited to an amplitude of $|\xi^z| \sim \alpha(h/\lambda_r)^{3/2}~h$, and $\sim \alpha\sqrt{\lambda_r/h}~h$, for isotropic and anisotropic viscosity, respectively [Nowak & Wagoner (1993)]. If turbulence is playing the dominant role in damping and exciting the modes, then we have the following constraints. For isotropic turbulence, the Q value is large only for $\alpha \ll 1$; however, this implies a correspondingly small amplitude. For anisotropic viscosity, a larger mode amplitude and a higher Q is achieved for a given α. However, in either case it is likely very difficult to achieve amplitudes as large as required to agree with the observations of GRS 1915+105 and GRO J1655-40 (cf. §5.2.5).

Another potential excitation mechanism for g–modes is 'negative radiation' damping [Nowak, et al. (1997)]. As a first approximation, the g–modes are taken to be adiabatic. In reality, one expects there to be small entropy changes due to various effects, the most notable one being radiative losses. If one has a radiation pressure dominated atmosphere, as is likely in high-luminosity discs, one properly should use

$$\Delta P = \gamma\frac{P}{\rho}\Delta\rho + \gamma\frac{P}{s}\Delta s ~, \tag{5.14}$$

where s is the specific entropy. The effects of this term for a radiation pressure dominated atmosphere can be approximated by writing an 'effective' adiabatic index, γ', to be substituted into the relationship between the Lagrangian variations of the pressure and density. Nowak, et al. (1997) showed this term to be

$$\gamma' \approx \gamma\left(1 + i\frac{\gamma c}{4\omega\tau_{es}h}\right) \equiv \gamma(1 + i\alpha') ~, \tag{5.15}$$

where τ_{es} is the scattering depth of the disc and h is the disc half thickness. In the above, our ignorance of the disc's true vertical structure is subsumed into the parameter α', which one expects to be of $\mathcal{O}(\alpha)$ for a radiation pressure-dominated disc.

It is the imaginary component of the effective adiabatic index that leads to radiative damping for $p-$modes but conversely leads to excitation of the $g-$modes. One can see this by perturbing the approximate dispersion relationship $\omega^2 \approx \kappa^2 \pm c_s^2 k_r^2$, where the $+$ is for $p-$modes and the $-$ is for $g-$modes. Perturbing the frequency and the sound speed together, we have:

$$\delta\omega\omega \approx \pm \delta c_s c_s k_r^2 \ , \qquad (5.16)$$

where $\delta c_s^2 \sim i\alpha' c_s^2$. Thus, the imaginary component of γ' leads to damping of $p-$modes and excitation of $g-$modes. Qualitatively, as the $p-$mode compresses gas, radiation leaks out, thereby providing less restoring force from the compressed region, and therefore leads to damping. As discussed above, the $g-$mode is in some sense opposite as the pressure works *against* the gravity. Hence, radiative leaking leads to less effective pressure restoring forces and thereby to $g-$mode growth.

5.2.5. *Luminosity Modulation*

For most simple α-models, the energy generation rate per unit volume in the accretion disc is approximated by $\alpha P(r,z)\Omega(r)$. The local pressure is modulated by $p-$ and $g-$modes (and to a much lesser extent by $c-$modes); thus, there is the possibility that these modes can lead to an observable luminosity modulation. Furthermore, the modes not only perturb the pressure, but they also perturb the locations of the disc boundaries. As discussed in Nowak, et al. (1997), the integrated variation of the energy generation rate is therefore

$$\delta L \sim 2\pi \left[\int_{r_I + \xi_r(r_I)}^{r_O + \xi_r(r_O)} r' dr' \int_{-z_0 + \xi_z(-z_0)}^{z_0 + \xi_z(z_0)} \alpha\Omega P'(r', z') dz' \right.$$
$$\left. - \int_{r_I}^{r_O} r\,dr \int_{-z_0}^{z_0} \alpha\Omega P(r,z)\, dz \right], \qquad (5.17)$$

where $P'(r', z') \equiv P(r,z) + \Delta P(r,z)$, $r' \equiv r + \xi_r(r)$, $z' = z + \xi_z(z)$, and r_I, r_O, z_0 are the disc boundaries (r_O can be taken to go to ∞ without loss of generality).

The g-modes have the most promising combination of pressure modulation and physical extent, which makes them the most likely candidates to be observed by direct luminosity modulation of this sort. However, for standard disc models, only a few percent of the total luminosity is generated in the region of the disc where the $g-$modes can exist. Thus, we expect only of $\mathcal{O}(1\%)$ rms modulation due to the $g-$modes. This is consistent with the observations of GRS 1915+105 and GRO J1655-40; however, as shown in Nowak, et al. (1997), even this modest luminosity modulation requires large mode amplitudes. Specifically, $\xi_z \sim h$ is required†.

Qualitatively, as the $g-$modes exist in the inner, hotter disc regions, one expects the modulated luminosity to be harder than the average *disc* luminosity. However, both the 67 Hz and 300 Hz QPO seem to be strongest in the *power law tail*, which is problematic for the $g-$mode interpretation of these modes if one associates the power law tail with a corona distinct from the accretion disc.

On the other hand, although one expects the $c-$modes to have relatively little luminosity modulation due to pressure fluctuations, these modes may be effective at modulating

---

† Note, however, that Nowak, et al. (1997) only considered Schwarzschild black holes. As a disc about a Kerr black hole releases a fractionally greater amount of its energy in the $g-$mode region, one may only require a mode amplitude a factor of several smaller.

reflected emission. If the power law tail is due to a corona above a cold accretion disc, then part of this power law will be Compton reflected [Magdiarz & Zdziarski (1995)], and hence may be modulated by the warping associated with the $c-$modes. Furthermore, one expects the reflected emission to peak in the $20-30$ keV range [Magdiarz & Zdziarski (1995)], which is consistent with the observations of both GRS 1915+105 and GRO J1655-40.

5.2.6. *Alternative Models*

The diskoseismology $g-$mode interpretation of the 67 Hz and 300 Hz has several points in its favor. First, it reproduces the correct time scales. Second, the $g-$mode frequencies, like the observed 67 Hz frequency, are expected to be relatively insensitive to luminosity fluctuations. Third, mode Q of $\mathcal{O}(20)$ is achievable with a reasonable $\alpha-$parameter. Fourth, there is an identified mechanism (negative radiation damping) for exciting the modes. Fifth, the $g-$modes are capable of modulating the luminosity of the disc, possibly as much as the observed $\sim 1\%$. Furthermore, we qualitatively expect the modes to be harder than the mean disc spectrum.

The are two major arguments against the $g-$mode interpretation. First, one requires a large mode amplitude (although this might be reduced slightly if one considers near maximal Kerr holes). Second, the observed QPO features appear strongest in the power law tails, which are not necessarily associated with a disc component. Aside from the $c-$modes discussed above, there are several other alternative hypotheses to the $g-$modes.

One possibility is fluctuations in the location of the sonic point [Honma, Matsumoto, & Kato (1992)], which can lie just inside of the marginally stable orbit, where disc orbits are going from circular orbits to near free-fall trajectories. As discussed by Honma, Matsumoto, & Kato (1992), the time scales associated with this 'pulsational instability' of the sonic point are fairly rapid. Again, one needs to show that sufficient luminosity modulation is possible with such a mechanism.

Milsom & Taam (1997) have numerically shown that for some accretion disc models, one finds acoustic $p-$modes that, although not trapped, peak at the location of the epicyclic frequency maximum†. Furthermore, their frequency is that of the epicyclic frequency maximum. Many predictions of the $p-$mode hypothesis, such as the mass and angular momentum of the central object, are identical to those made by the $g-$mode hypothesis. As for the $g-$modes, it is also difficult to achieve $\mathcal{O}(1\%)$ luminosity modulation, and likewise it may also be difficult to produce features as hard as those observed in GRS 1915+105 and GRO J1655-40.

More recently, Cui, Zhang, & Chen (1998) have proposed that the observed QPO features are associated with Lense-Thirring precession. Specifically, these authors choose the QPO frequency to be the Lense-Thirring frequency at the radius at which the accretion disc effective temperature is maximum, and thus derive $a \sim 1$ for both GRS 1915+105 and GRO J1655-40. There are a number of objections to this picture. First, choosing the radius to be that of the effective temperature maximum is somewhat arbitrary, and one might expect that the radius at which $r^2 F_\gamma$ is maximized, where F_γ is the disc photon flux over the observed energy bands, is a more natural choice. Second, neither F_γ nor $r^2 F_\gamma$ are sharply peaked functions, therefore it is extremely unlikely that one can generate a feature with $Q \sim 20$, as was observed for the 67 Hz feature in GRS 1915+105. [From this point of view, the $c-$modes discussed in §5.2.3 are the natural, 'global' features to be

---

† Qualitatively, one can understand this by considering the $p-$mode dispersion relationship, which is $(\omega^2 - \kappa^2) \approx c_s^2 k^2$. The $p-$mode group velocity is seen to go to zero at $\omega^2 = \kappa^2$. Thus $p-$modes with the maximum epicyclic frequency "stall" at the location of the epicyclic frequency maximum.

associated with Lense-Thirring precession.] Third, due to frame dragging effects in the presence of viscous forces, one expects that the inner region of the disc will be flattened and constrained to the equatorial plane on a time scale of $\mathcal{O}(1s)$ [Bardeen & Petterson (1975)]‡. Fourth, Cui, Zhang, & Chen (1998) do not identify any luminosity modulation mechanism. Finally, Cui, Zhang, & Chen (1998) do not demonstrate that there is a viable excitation mechanism for the modes [although they suggest that the radiative warping mechanism of Pringle (1996) may be at work]. However, as discussed above, the $c-$modes, which are qualitatively similar to the Lense-Thirring precession suggested by Cui, Zhang, & Chen (1998), may yet provide a viable explanation for the observed high frequency QPO features.

6. Summary

As was discussed in §5.1, the study of stable oscillations in accretion discs has a long and rich history. However, it was not until the advent of *Ginga* in the late 1980's and of *RXTE* in only the past few years that this field has also become an observational one.

Prior to the launch of *RXTE*, there were only a few observations of QPO features in BHC. Most of these features were of low frequency $\lesssim 10$ Hz (this was likely mainly due to instrumental limitations), were not seen during more than one epoch, and were often broad ($Q \lesssim 10$) and often variable in frequency. Few theories have been put forth that adequately describe these observations.

In the past two years since the launch of *RXTE*, a wealth of new observational information has been obtained. Again, a number of low-frequency, often broad and highly variable features have been observed. Among the wealth of features seen with *RXTE*, two high frequency features stand out: the 67 Hz feature in GRS 1915+105 and the 300 Hz feature in GRO J1655-40. The former is notable for its steady frequency, whereas the latter is notable for its high frequency which suggests very strongly that it comes from very close to the probable 7 $M_{\odot}$ black hole in this system.

The high frequency of both features and the stability of the 67 Hz feature suggests that these QPO might be related to stable oscillations in the inner regions of accretion discs. We have described a class of theories, which we refer to as 'diskoseismology', that might offer an explanation for these observations. The main motivations for attributing these features to diskoseismic modes are that these modes: 1) are related to a "natural" frequency in the disc (i.e. the maximum epicyclic frequency); 2) their spectra are expected to be characteristic of the inner, hottest regions of the disc; 3) their frequencies are relatively insensitive to changes in luminosity; and 4) they have low rms variability.

This latter feature, although in agreement with the observations, is the strongest constraint. These modes *cannot* be applied to systems that show $\gtrsim 10\%$ rms variability over a wide range of energy bands. Furthermore, it may be difficult to explain the observed spectral hardness of the QPO. To these ends, one needs to begin to consider more detailed *dynamical* disc models that address the production of the hard radiation. Also, $g-$modes have been the major focus of recent study; however, $c-$modes, which are related to Lense-Thirring precession of warped accretion discs, may offer a better expla-

---

‡ Cui, Zhang, & Chen (1998) mistakenly claimed that this time scale was very long; however, they were quoting the time scale for the black hole's angular momentum to align with the binary orbital angular momentum (which is of order millions of years). The time scale for the inner region of the disc to flatten into the orbital plane is significantly shorter. Note that the $c-$modes should also be susceptible to damping by the 'Bardeen-Petterson' effect; however, as they are global modes we expect the damping time scale to be slightly longer than that suggested by Bardeen & Petterson (1975).

nation of some of the data. Finally, in all of the above work magnetic fields have been neglected. As shown by Balbus & Hawley (1992), even weak magnetic fields can play an important dynamical role. The incorporation of magnetic fields into the diskoseismology calculations is one of the next major steps that needs to addressed.

Even if the features seen in GRS 1915+105 and GRO J1655-40 do not turn out to be diskoseismic modes, they point out two important lessons. First, BHC systems can produce relatively stable, high-frequency features. Second, the *Rossi X-ray Timing Explorer* is capable of detecting and characterizing these features despite their weak variability. In a very real sense, despite nearly twenty years of research, the study of stable oscillations in black hole accretion systems has just begun.

The authors would like to acknowledge useful conversations with Robert Wagoner, Mitchell Begelman, Brian Vaughan, Chris Perez, Ron Remillard, and Ed Morgan. M.A.N. was supported in part by an LTSA grant from NASA (NAG 5-3225). D.E.L. was supported in part by a NASA GSRP Training Grant NGT5-50044.

REFERENCES

BALBUS, S. A., & HAWLEY, J. F. 1992 A powerful local shear instability in weakly magnetized disks. I - Linear analysis. II - Nonlinear evolution. *Ap. J.* **376** 214–233.

BARDEEN, J. M., & PETTERSON, J. A. 1975 *Ap. J.* **195** L65.

BINNEY, J., & TREMAINE, S. 1987 *Galactic Dynamics*. Princeton Press.

CHEN, X., SWANK, J. H., TAAM, R. E. 1997 The pattern of correlated X-ray timing and spectral behavior in GRS 1915+105. *Ap. J.* **477** L41–L44.

CUI, W., ZHANG, S. N., FOCKE, W., & SWANK, J. H. 1997 Temporal properties of Cygnus X-1 during the spectral transitions. *Ap. J.* **484** 383–393.

CUI, W., ZHANG, S. N., & CHEN, W. 1998 Evidence for frame dragging around spinning black holes in X-ray binaries. *Ap. J.* **492** L53–L57.

DAVENPORT, W. B., JR., & ROOT, W. L. 1987 *An Introduction to the Theory of Random Signals and Noise*. IEEE Press.

EBISAWA, K., MITSUDA, K., & INOUE, H. 1989 Discovery of 0.08-Hz quasi-periodic oscillations from the black hole candidate LMC X-1. *P. A. S. J.* **41**, 519–530.

FRIEDMAN, J. L., & SCHUTZ, B. F. 1978 Lagrangian perturbation theory of nonrelativistic fluids. *Ap. J.* **221** 937–957.

FRIEDMAN, J. L., & SCHUTZ, B. F. 1978 Secular Instability of rotating newtonian stars. *Ap. J.* **222** 281–296.

GREBENEV, S. A., SYUNYAEV, R. A., PAVLINSKII, M. N., & DEKHANOV, I. A. 1991 Detection of quasiperiodic oscillations of X-rays from the black-Hole candidate GX 339-4. *Sov. Astron. Lett.* **17**, 413–415.

HINES, C. O. 1960 *Can. J. Phys.* **38** 1441.

HJELLMING, R. M., & RUPEN, M. P. 1995 Episodic ejection of jets by the X-ray transient GRO:J1655-40. *Nature* **375** 464.

HONMA, F., MATSUMOTO, R., & KATO, S. 1992 Pulsational instability of relativistic accretion disks and its connection to the periodic X-ray time variability of NGC 6814. 1992 *P. A. S. J.* **44** 529–535.

IPSER, J. R., & LINDBLOM, L. 1992 On the pulsations of relativistic accretion disks and rotating stars - the Cowling approximation. *Ap. J.* **389** 392–399.

IPSER, J. R. 1996 Relativistic accretion disks: low-frequency modes and frame dragging. *Ap. J.* **458** 508–513.

IWASAWA, K., FABIAN, A. C., BRANDT, W. N., KUNIEDA, K., MISAKI, K., REYNOLDS, C. S., & TERASHIMA, Y. 1998 Detection of an X-ray periodicity in the Seyfert galaxy IRAS18325-5926. *M. N. R. A. S.*, in Press.

KATO, S., & FUKUE, J. 1980 Trapped radial oscillations of gaseous disks around a black hole. *P. A. S. J.* **32** 377–388.

KATO, S. 1990 Trapped one-armed corrugation waves and QPOs. *P. A. S. J.* **42** 99–113.

KATO, S. 1993 Amplification of one-armed corrugation waves in geometrically thin relativistic accretion disks. *P. A. S. J.* **45** 219–231.

KITAMOTO, S., TSUNEMI, H., MIYAMOTO, S., & HAYASHIDA, K. 1992 Discovery and X-ray properties of GS 1124–683 (=Nova Muscae). *Ap. J.* **394**, 609–614.

KOUVELIOTOU, C., FINGER, M. H., FISHMAN, G. J., MEEGAN, C. A., WILSON, R. B., PACIESAS, W. S. 1992a *I. A. U. Circ.* 5576.

———. 1992b *I. A. U. Circ.* 5592.

LENSE, J. & THIRRING, H. 1918   *Phys. Z.* **19** 156.

LYNDEN-BELL, D. & OSTRIKER, J. P. 1967 On the stability of differentially rotating bodies. *M. N. R. A. S.* **136** 293–310.

MAGDZIARZ, P. AND ZDZIARSKI, A. 1995 Angle-Dependent Compton reflection of X-rays and gamma-rays. *M. N. R. A. S.* **273** 837–848.

MILSOM, J. A., & TAAM, R. E. 1997 Two-dimensional studies of inertial-acoustic oscillations in black hole accretion discs. *M. N. R. A. S.* **286** 358–368.

MIRABEL, & RODRIGUEZ, 1994 A superluminal source in the galaxy. *Nature* **371** 46.

MIYAMOTO, S., KIMURA, K., KITAMOTO, S., DOTANI, T., & EBISAWA, K. 1991 X-ray variability of GX 339-4 in its very high state. *Ap. J.* **383** 784–807.

MIYAMOTO, S., KITAMOTO, S., IGA, S., NEGORO, H., & TERADA, K. 1992 Canonical time variations of X-rays from black hole candidates in the low-intensity state. *Ap. J.* **391** L21–L24.

MIYAMOTO, S., IGA, S., KITAMOTO, S., & KAMADO, Y. 1993 Another canonical time variation of X-rays from black hole candidates in the very high flare state? *Ap. J.* **403** L39–L42.

MIYAMOTO, S., KITAMOTO, S., IGA, S., & HAYASHIDA, K. 1994 Normalized power spectral densities of two X-ray components from GS 1124-683. *Ap. J.* **435** 398–406.

MORGAN, E. H., REMILLARD, R. H., & GEINER, J. 1997 RXTE observations of QPOs in the black hole candidate GRS 1915+105. *Ap. J.* **482** 993–1010.

NOWAK, M. A. 1995 Towards a unified view of black hole high energy states. *P. A. S. P.* **718** 1207–1216.

NOWAK, M. A., & WAGONER, R. V. 1991 Diskoseismology: probing accretion disks I. Trapped adiabatic oscillations. *Ap. J.* **378** 656–664.

NOWAK, M. A., & WAGONER, R. V. 1992 Diskoseismology: probing accretion disks II. G-modes, gravitational radiation reaction, and viscosity. *Ap. J.* **393** 697–707.

NOWAK, M. A., & WAGONER, R. V. 1993 Turbulent generation of trapped oscillations in black hole accretion disks. *Ap. J.* **418** 187–201.

NOWAK, M. A., WAGONER, R. V., BEGELMAN, M. C., & LEHR, D. E. 1997 The 67 Hz feature in the black hole candidate GRS 1915+105 as a possible "diskoseismic" mode. *Ap. J.* **477** L91–L94.

OKAZAKI, A., KATO, S., & FUKUE, J. 1987 Global trapped oscillations of relativistic accretion disks. *P. A. S. J.* **39** 457–473.

OROSZ, J. A., & BAILYN, C. D. 1997 Optical observations of GRO J1655-40 in quiescence. I. A precise mass for the black hole primary. *Ap. J.* **477** 876–896.

PAZCZYNSKI, B. & WITTA, P. 1980 *A. & A.* **88** 23.

PEREZ, C. A. 1993 *Ph.D. Thesis, Dept. of Physics, Stanford University.*

PEREZ, C. A., SILBERGLEIT, A. S., WAGONER, R. V., & LEHR, D. E. 1997 Relativistic diskoseismology. I. Analytical results for "gravity modes". **476** 589–604.

PRESS, W. H., TEUKOLSKY, S. A., VETTERLING, W. T., & FLANNERY, B. P. 1992 *Numerical Recipes in FORTRAN. The Art of Scientific Computing, 2nd Edition.* Cambridge Press.

PRINGLE, J. E. 1996 *M. N. R. A. S.* **281** 357–361.

REMILLARD, R. A., MORGAN, E. H., MCCLINTOCK, J. E., BAILYN, C. D., OROSZ, J. A., & GREINER, J. 1997 In *Proceedings of the 18th Texas Symposium on Relativistic Astrophysics* (eds. A. Olinto, J. Frieman, & D. Schramm) Univ. Chicago Press.

SHAKURA, N. I., & SUNYAEV, R. A. 1973 Black holes in binary systems. observational appearance. *A. & A.* **24** 337–355.

SHAKURA, N. I., & SUNYAEV, R. A. 1976 A theory of the instability of disk accretion on to black holes and the variability of binary X-ray sources, galactic nuclei, and quasars. *M. N. R. A. S.* **175** 613–632.

SUNYAEV, R. A., CHURAZOV, E., GILFANOV, M., NOVIKOV, B., GOLDWURM, A., PAUL, J., MANDROU, P., & TECHINE, P. 1992 *I. A. U. Circ.* 5593.

SUNYAEV, R. A., ET AL. 1993 Broad-band X-ray observations of the GRO J0422+32 X-ray nova by the "Mir-Kivant" observatory. *A. & A.* **280** L1–L4.

TAAM, R. E., CHEN, X., SWANK, J. H. 1997 Rapid bursts from GRS 1915+105 with RXTE. *Ap. J.* **485** L83–L86.

TANAKA, Y., AND LEWIN, W. H. G. 1995, Black hole binaries. In *X-ray Binaries* (eds. W. H. G. Lewin, J. van Paradijs, & E. P. J. van den Heuvel), Cambridge University Press.

UBERTINI, P., BAZZANO, A., COCCHI, M., LA PADULA, C., POLCARO, V. F., STAUBERT, R., KENDZIORRA, E. 1994 Hard X-ray timing observation of the Crab pulsar and Cygnus X-1. *Ap. J.* **421** 269–275.

VAN DER KLIS, M. 1994 Similarities in neutron star and black hole accretion. *Ap. J. S.* **92** 511–519.

————. 1995 Rapid aperiodic variability in X-ray binaries. In *X-ray Binaries* (eds. W. H. G. Lewin, J. van Paradijs, & E. P. J. van den Heuvel), Cambridge University Press.

VIKHLININ, A., ET AL. 1994 Discovery of a low-frequency broad quasi-periodic oscillation peak in the power density spectrum of Cygnus X-1 with Granat/SIGMA. *Ap. J.* **424** 395–400.

VISHNIAC, E. T., & DIAMOND, P. 1989 A self-consistent model of mass and angular momentum transport in accretion disks. *Ap. J.* **347** 435–447.

Spotted disks

By A. BRACCO[1], A. PROVENZALE[1,2],
E. A. SPIEGEL[3] AND P. YECKO[4]

[1]Istituto di Cosmogeofisica, Corso Fiume 4, I-10133 Torino, Italy

[2] JILA, University of Colorado, Box 440, Boulder, CO 80309-0440, USA

[3] Dept. of Astronomy, Columbia University, New York, NY 10027, USA

[4] Dept. of Physics, University of Florida, Gainesville, FL, USA

Rotating, turbulent cosmic fluids are generally pervaded by coherent structures such as vortices and magnetic flux tubes. The formation of such structures is a robust property of rotating turbulence as has been confirmed in computer simulations and laboratory experiments. We defend here the notion that accretion disks share this feature of rotating cosmic bodies. In particular, we show that the intense shears of Keplerian flows do not inhibit the formation of vortices. Given suitable initial disturbances and high enough Reynolds numbers, long-lived vortices form in Keplerian shear flows and analogous magnetic structures form in magnetized disks. The formation of the structures reported here should have significant consequences for the transport properties of disks and for the observed properties of hot disks.

1. Introduction

Whenever it has been possible to observe rotating, turbulent fluids with good resolution, it has been seen that individual, intense vortices form (Bengston & Lighthill 1982; Hopfinger *et al.* 1982; Dowling and Spiegel 1990). In each case where the medium is a good conductor, as in the solar convection zone, the vortices are instead magnetic flux tubes but the guiding principle remains the same: vector fields, whether vorticity or magnetic fields, are amplifed and concentrated by the combined effects of rotation and turbulence.

It seems natural that one should apply this principle to the theory of accretion disks (Dowling & Spiegel 1990). However, as critics of this idea have objected, accretion disks are exceptional among rotating turbulent objects in the strong shears that these bodies are believed to possess and this shear might lead to rapid destruction of any structures that tend to form. This objection is here overruled by a large number of simulations. Naturally, this does not prove that coherent structures must form on disks, but it does strengthen the argument that disks are likely to follow the norm of rotating, turbulent bodies. We may point also to numerical results suggesting that vortices form on disks (Hunter & Horak 1983) as well as experimental work (Nezlin & Snezhkin 1993).

Even if it is granted that vortices and magnetic flux tubes can form on accretion disks, one must naturally ask whether this matters for processes such as transport rates or for observational properties. For both these issues, we suggest that there can indeed be an important role of vortices. As we shall explain in the fluid dynamical review given below, there is an important property of rapidly rotating fluids called the potential vorticity that is approximately conserved following the motion of the fluid (Ghil & Childress 1987). When a vortex is perturbed (for example, by interaction with other vortices) it will be set to oscillating and will radiate fluid dynamical waves, especially Rossby waves (Llewellyn Smith 1996). This radiation represents a loss of specific angular momentum and, to conserve potential vorticity, the vortex must move to a location with a different mean angular velocity. This results in radial angular momentum transport. Moreover, coherent vortices trap passive tracers for long times (Elhmaidi *et al.* 1993; Babiano *et al.*

254

1994) and are responsible for much of the transport of passive and active constituents in rotation-dominated flows, see Provenzale *et al.* (1997) for a general review and Tanga *et al.* (1996) for an application to the early solar nebula.

It may be true that the fluid dynamics of disks, particularly of the turbulence in them, is so poorly understood that one should not worry about such effects, unless they threaten to become dominant. On this issue, unfortunately, we remain uncertain. But the possible observational effects of vortices (and flux tubes) do seem to us to have a real importance.

The centrifugal force or magnetic pressure produced by vortices produce local extrema in the total pressure (gas plus radiation). Hence, these structures can act either like holes in the disks or thickenings according to whether the vortices are cyclonic (having the sense of the local rotation) or anticyclonic. The holes may even go all the way through the disk in some cases, allowing the rapid escape of hard radiation from within the disk. Such radiation, coming from within an optically thick hot disk, can produce spots on disks that may affect the observed spectra of AGNs (Abramowicz *et al.* 1991).

In this paper, we shall not enter into the discussion of these motivating topics, but rather shall stick to the immediate fluid dynamical issues. We shall review some of the fluid equations of the disk problem and report on simulations that suggest that vortices form in the presence of intense shears. We realize that this suggestion has not gained general acceptance among specialists in disk theory, but are encouraged by the recent softening of their attitude toward the issue of vortex formation on disks. Thus, several years ago, when we suggested at the cataclysmic variable meeting in Eilat that disks would have spots, with the consequence that phenomena associated with the solar cycle might be expected around disks, including hot coronas and long term variations, the summarizer of that meeting disparaged these suggestions. At the meeting in Iceland, at which the present version was reported, the summarizer merely ignored our remarks and did not mention them at all. Given this apparent warming of the climate of opinion, we are encouraged in this case to report our findings here. These strongly suggest that disks are not so different from other rotating cosmic bodies and that they produce coherent structures that may play a role in their large scale properties.

2. Thin-Layer Theory

Our discussion focuses on fluid dynamical issues with little attention paid to the thermal aspects of accretion flows. The Navier-Stokes and continuity equations then govern the fluid fields, velocity $\mathbf{v} = (u, v, w)$ and density ρ, which depend on position and time. These equations are

$$\frac{D\mathbf{v}}{Dt} = -\frac{1}{\rho}\nabla p - \nabla \Phi + \mathbf{D_v} \tag{2.1}$$

and

$$\frac{D\rho}{Dt} + \rho \nabla \cdot \mathbf{v} = 0 \tag{2.2}$$

where $D/Dt = \partial/\partial t + \mathbf{v} \cdot \nabla$ is the material derivative, p is pressure and Φ is the potential of the (imposed) gravitational field. We shall be concerned with flow at some distance from the central object and will assume that $\Phi = -GM/R$, where R is the distance from that object and M is its mass. The last term in (2.1) is the viscous force per unit mass, which, for constant viscosity, is

$$\mathbf{D_v} = \nu[\nabla^2 \mathbf{v} + \frac{1}{3}\nabla(\nabla \cdot \mathbf{v})] \ . \tag{2.3}$$

In discussions of rotating and turbulent fluids the vorticity field, $\boldsymbol{\omega} = \nabla \times \mathbf{v}$, proves to be a very important quantity. We obtain an equation for it by taking the curl of eq.(2.1); this is

$$\frac{D\boldsymbol{\omega}}{Dt} + \boldsymbol{\omega}(\nabla \cdot \mathbf{v}) = \boldsymbol{\omega} \cdot \nabla \mathbf{v} + \frac{\nabla\rho \times \nabla p}{\rho^2} + \mathbf{D_\omega} \tag{2.4}$$

where $\mathbf{D_\omega} = \nabla \times \mathbf{D_v}$. We see that if the surfaces of constant p and ρ are not coincident, vorticity is generated by the so-called baroclinic term. We shall not include this effect here. Instead, we assume that the fluid is barotropic, which means that the pressure is a function only of the density. A familiar example of such a relation is the adiabatic gas law

$$p = K\rho^\gamma \tag{2.5}$$

where K and γ are constants. This holds for a perfect gas when the specific entropy, $S = c_v log(p/\rho^\gamma)$, is constant in space and time.

In a thin layer of fluid, there is not much horizontal vorticity and we are concerned mainly with the component of $\boldsymbol{\omega}$ parallel to the rotation axis, which, in the case of a disk with rough cylindrical symmetry, we associate with the z-axis, or vertical direction. We call this component of vorticity, the vertical component, ζ. When $\boldsymbol{\omega} \approx (0, 0, \zeta)$, the fluid velocity is approximately horizontal. This is the thin-layer approximation that is used in the study of shallow layers in geophysical and planetary fluid dynamics. The idea is that if the vertical velocity were not small the layer would not be thin. Then we may approximate (2.4) as

$$\frac{D\zeta}{Dt} + \zeta\nabla \cdot \mathbf{u} = D_\zeta \tag{2.6}$$

where $\mathbf{u} = (u, v)$ is the horizontal velocity (in the plane of the disk) and D_ζ is the z-component of $\mathbf{D_\omega}$.

There are two ways to derive thin-layer equations in a more formal and systematic way other than the appeal to physical intuition that we make here. One is by integrating over z, with certain assumptions on the vertical structures of the fields and the other is to make expansions in terms of small thickness (Ghil & Childress 1987; Qian *et al.* 1991). These approaches work as well on compressible as on incompressible fluids, as long as the motions remain reasonably subsonic. The resulting equations are those of a compressible fluid, but it is the surface density that is varying strongly, whether or not the real density varies. We see this in the integrated continuity equation.

To reduce (2.2) to an equation for the surface density we consider here only the case where the layer is symmetric under the transformation $z \to -z$. We assume that the disk has a surface on which the density goes to zero and that this is given by a height function h as $z = \pm h(x, y, t)$. The surface density is then defined as

$$\sigma(x, y, t) = \int_{-h}^{h} \rho(x, y, z, t)dz. \tag{2.7}$$

For barotropic, axisymmetric stationary disks, it may be shown that the velocity is independent of z. Naturally, this will not be exactly right for turbulent, time-dependent disks without axisymmetry, but if the violations are mild, we may use this picture in first approximation. Then, if we integrate the continuity equation over the depth of the disk, we obtain

$$\frac{D\sigma}{Dt} + \sigma\nabla \cdot \mathbf{u} = 0. \tag{2.8}$$

We see from the structure of this equation that the dynamical description of a thin disk is that of a compressible, two-dimensional fluid with density σ. But σ is really a surface density and its variations can reflect either real density variations or thickness variations. Since the thin-layer description is the same for gases and liquids, apart from coefficients in the equations, the term density wave has therefore to be used with caution since the real density variations associated with variations of σ need not be large. In particular, the waves associated with gravitational instability in thin layers are gravity waves and not sound waves (Qian & Spiegel 1994).

Combining equations (2.6) and (2.8) we obtain

$$\frac{Dq}{Dt} = D_q \qquad (2.9)$$

where

$$q(x, y, t) = \frac{\zeta(x, y, t)}{\sigma(x, y, t)} \qquad (2.10)$$

is the potential vorticity in the shallow-layer approximation and the dissipation term is $D_q = D_\zeta / \sigma$.

In the absence of dissipation ($D_q = 0$) in (2.9) the potential vorticity becomes a material invariant, that is, $Dq/Dt = 0$. This condition is central to many discussions of planetary and geophysical fluid dynamics (Pedlosky 1987). The approach in those subjects is to use an approximation to provide the velocity in the D/Dt of this equation. The approximation usually used is geostrophy, in which the Coriolis force is exactly balanced by the pressure gradient. In the presence of the intense shear of accretion disks, this approximation is less useful than in those other subjects and we shall, in this section, adopt another approach.

In order to concentrate on the behaviour of vorticity and the effects of shear, we shall omit the variations in surface density, both globally and locally. Then, by requiring that σ be constant we find that $\nabla \cdot \mathbf{u} = 0$ where $\mathbf{u} \equiv (u, v)$ is the (horizontal) velocity in the disk and $\mathbf{r} = (x, y)$ are the horizontal coordinates. Evidently, the resulting description is formally equivalent to two-dimensional incompressible hydrodynamics and we may introduce a stream function ψ such that

$$(u, v) = \left(-\frac{\partial \psi}{\partial y}, \frac{\partial \psi}{\partial x} \right) . \qquad (2.11)$$

Then the (total) vertical vorticity,

$$\zeta = \nabla^2 \psi , \qquad (2.12)$$

is the potential vorticity and we use equations (2.6), which can be written in the form

$$\frac{\partial \zeta}{\partial t} + [\psi, \zeta] = D_\zeta \qquad (2.13)$$

where $[\psi, \zeta] = (\partial_x)\psi \partial_y \zeta - (\partial_x \zeta)\partial_y \psi$.

For a solenoidal velocity, the viscosity term simplifies and we have followed standard practice in fluid dynamics by considering a variety of representations of it, all of which may be written as

$$D_\zeta = (-1)^{p-1} \nu_p \nabla^{2p} \zeta \qquad (2.14)$$

with ν_p taken constant. The case $p = 1$ corresponds to the conventional Newtonian viscosity but values of $p > 1$, the so-called hyperviscous cases, have also been used. Hyperviscosity is a device used in turbulence simulations to represent the effects of eddy viscosity, see e.g. McWilliams (1984). Its advantage is that it cuts off the viscous tail of the spectrum rapidly so that high spatial resolution is more easily achieved in a given computational time.

With this specification of D_ζ, we first use (2.12)- (2.13) to study the formation of coherent structures. This is physically reasonable when (a) the local vorticity of the rotation of the disk is larger than the vorticity of perturbations of the basic state and (b) the dynamics takes place on a scale smaller than that over which the disk thickness varies appreciably. Since vorticity has the dimensions of inverse time, (a) is valid when the dynamics of the perturbed flow is slower than the typical rotation time scale of the whole disk at the given radius. If we indicate by $\tilde{\zeta}$ a typical vorticity perturbation, and by Ω the mean angular velocity at the given radius, we expect that (2.13) is valid in the limit $\tilde{\zeta}/f \ll 1$ where $f = 2\Omega$ is the local Coriolis frequency.

Since the typical order of magnitude of $\tilde{\zeta}$ may be written as $[\tilde{\zeta}] = U/L$ where U and L are a typical velocity and length scale of the perturbation, one can express the above constraint by the requirement $Ro = U/(fL) \ll 1$, where Ro is called the Rossby number. The smallness of the Rossby number is a requirement for approximating the shallow-layer equation (2.9) by the quasi-geostrophic vorticity equation (Pedlosky 1987). Since $\Omega = \Omega(|\mathbf{r}|)$, the Rossby number depends on the radial coordinate of the disk.

In the absence of dissipation ($D_\zeta = 0$), eq.(2.13) admits an infinite number of conserved quantities, two of which are quadratic invariants. These are the kinetic energy $E = 1/2 \int (\nabla \psi)^2 dx dy$ and the enstrophy $Z = 1/2 \int (\nabla^2 \psi)^2 dx dy$. The conservation of these two quantities induces a direct cascade of enstrophy from large to small scales and an inverse (from small to large scales) cascade of kinetic energy (Charney 1971; Rhines 1979; Kraichnan & Montgomery 1980). This is different from the more familiar direct cascade of energy in three-dimensional turbulence, where the energy flows from large to small scales.

For freely decaying ($D_\zeta \neq 0$) barotropic turbulence in shear-free environments, intense long-lived vorticity concentrations are observed to form after an energy dissipation time. These coherent objects are characterized by a broad distribution of size and circulation and they contain most of the energy and the enstrophy of the system (McWilliams 1984; McWilliams 1990). The question addressed here is whether the strong Keplerian shears of accretion disks modify this well-documented behaviour.

3. Simulations with Keplerian Shear

In the case of the Keplerian disk, where the shear and differential rotation are quite strong, the question of vortex formation and survival remains open and we address it in this section. By using the same formalism as in previous simulations of vortices, we may isolate the effect of the shear, which favours the propagation of a class of dispersive waves, called Rossby waves (Pedlosky 1987), that do not occur in plane layers with rigid background rotation. Rossby waves transport energy away from a source of perturbation and so can inhibit the formation of large-scale vortices. Analytical studies and numerical simulations have shown that, in the so-called β-plane — a slab with linear differential rotation, $2\Omega = f(y) = f_0 + \beta y$ — coherent vortices form and dominate the dynamics on small scales, while Rossby waves inhibit the formation of vortices at large scales. The larger is β, the smaller is the scale below which vortices can survive (Cho & Polvani 1996a; 1996b).

Here, we report simulations based on a standard pseudo-spectral code with 2/3 dealiasing and resolution 512×512 grid points in a square box with periodic boundary conditions. After checking that the value of p had no significant effect on the outcome, if sufficient running time was allowed, we have used the hyperviscosity with $p = 2$ for most of this work.

We start each calculation with an initial condition corresponding to a perturbed Keplerian disk, namely

$$\zeta(\mathbf{r}, t = 0) = \Omega(r) + \tilde{\zeta}(\mathbf{r}, 0) \tag{3.15}$$

where $r = |\mathbf{r}|$, $\Omega(r) = Kr^{-3/2}$ is the Keplerian vorticity profile, K is a constant depending on the mass of the central object, and $\tilde{\zeta}(\mathbf{r}, 0)$ is the initial vorticity perturbation which is superposed on the Keplerian profile. The center of the disk is placed at the center of the simulation box at $\mathbf{r} = (0, 0)$ and the initial Keplerian profile has been smoothed at the center to eliminate the vorticity singularity in $\mathbf{r} = 0$.

In this section, we use dimensionless variables. The space scale is fixed by defining the simulation box to have size 2π. The time scale is fixed by the imposed values of the energy and of the enstrophy. In the simulations discussed below, we have chosen the initial kinetic energy of the unperturbed Keplerian disk to be $E_{kep} = 2.0$. This gives a Keplerian initial enstrophy $Z_{kep} \approx 6$. At $r = \pi/4$, the local Keplerian vorticity is $\Omega(\pi/4) \approx 6$, which gives a typical rotation time $T_{kep}(r = \pi/4) = 2\pi/\Omega \approx 1$. The simulations have in general been run up to a total time $T = 20$, which means ≈ 20 rotations at $r = \pi/4$. The initial energy of the perturbation, $\tilde{E}$, has then been fixed in the range $0 \leq \tilde{E}/E_{kep} \leq 0.1$. The eddy viscosity coefficient has been fixed at the value $\nu_2 = 5 \cdot 10^{-8}$.

Simulations starting with a pure Keplerian profile (i.e., $\tilde{E} = 0$) have indicated that this profile is stable to small disturbances. The perturbations involved in normal numerical procedures produce no instability and the disk shear slowly decays due to dissipation. The decay time is extremely long and the shear survives for hundreds of rotation times $T_{kep}(\pi/4)$.

In the course of this study we have considered three types of initial perturbations $\tilde{\zeta}(\mathbf{r}, 0)$: a single vortex structure, radially localized waves with various azimuthal wavenumbers, and a random initial perturbation field. All of these initial conditions produce analogous results. In the following, we present the evolution of the disk resulting from a random initial perturbation field $\tilde{\zeta}$, as this represents the most physically plausible type of perturbation. The initial perturbation has been generated as a narrow-band random vorticity field with energy spectrum

$$E(k) = E_0 \frac{k^n}{[(m/n)k_0 + k]^{m+n}} \tag{3.16}$$

where E_0 is a normalization factor; we used $k_0 = 10$, $n = 5$ and $m = 30$. The Fourier phases are randomly distributed between 0 and 2π. In order to avoid unrealistic periodicity effects, the perturbation has been set to zero for $r > 2\pi/3$, such that no perturbation is present at the edges of the simulation box.

For small perturbation energies, the vorticity perturbations are sheared away by the Keplerian flow, and the disk quickly returns to an unperturbed, slowly decaying Keplerian profile. This behavior is consistent with the linear stability of Keplerian profiles.

For large perturbation energies, the initial random vorticity field self-organizes into coherent vortices, similarly to what happens in non-shearing conditions. We refer to the sense of the basic Keplerian rotation as cyclonic (counterclockwise in the figures shown here). In the presence of the basic Keplerian shear cyclonic disturbances are rapidly sheared away, and are not able to organize themselves into individual vortices. By contrast, initial anticyclonic (negative vorticity) perturbations grow and form coherent vortices. Once formed, these anticyclonic vortices merge with each other, and generate larger vortices. This growth by merger is halted by the Keplerian shear, that, analogously to what happens for turbulence on the β-plane, sets an upper limit to the size of the surviving vortices. The anticyclonic vortices are coherent in the sense that they live much

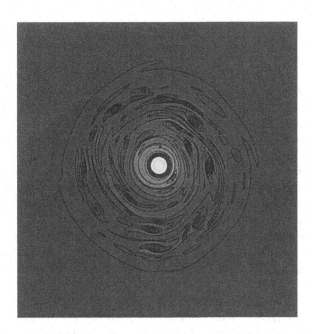

FIGURE 1. Vorticity field of a perturbed two–dimensional Keplerian disk, at time $t = 5$.

longer than the typical eddy turnover time of the perturbation, $\tilde{T} = \tilde{Z}^{-1/2}$ where $\tilde{Z}$ is the perturbation enstrophy.

Figures 1, 2 show the vorticity field for a simulation, in which the initial perturbation energy was normalized so that $\tilde{E}/E_{kep} \approx 0.005$. The vorticity is shown at the times $t = 5$ (figure 1) and $t = 10$ (figure 2). Figure 3 shows a section of the vorticity field at time $t = 0$ (the initial condition) and at time $t = 5$. The average vorticity profile remains Keplerian for the whole simulation and the rotation curves show bumps like those seen on rotation curves of galaxies.

These results show that long-lived, coherent anticyclonic vortices form in Keplerian shears, starting from an initially random perturbation field. Several other simulations, including ones with different energies in the Keplerian profile, different shapes of the initial perturbation, and different ratios between the energy of the perturbation and that of the unperturbed disk have given analogous results, provided that the perturbation energy $\tilde{E}$ was larger than a certain threshold. For the value of the hyperviscosity $\nu_2 = 5 \cdot 10^{-8}$ used in the previous simulation, the requirement is that $\tilde{E}/E_{kep}$ should be larger than about 0.001 for sustained vortices to form. The fate of nonlinear perturbations did not, as might have been supposed, lead to disorganized motions, but to coherent structures. These are anticyclonic vortices and they typically survive for tens or even hundreds of rotation times before being dissipated. Whether the final dissipation is a numerical artifact or a real phenomenon is not decidable as yet and this may be relevant to a discussion about whether the Keplerian profile should be considered as nonlinearly unstable or not (Balbus and Hawley 1996; Dubrulle and Zahn 1991). When we ran simulations with Newtonian viscosity, $p = 1$, and $\nu_1 = 2 \cdot 10^{-4}$ in eq.(2.14), we found that coherent vortices form in this case as well. The only difference is that the vortices obtained with $p = 1$ are slightly broader than those obtained with a $p = 2$ hyperviscosity, as commonly observed for turbulence simulations in a homogeneous background flow.

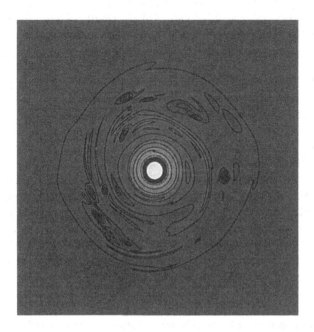

FIGURE 2. Vorticity field of a perturbed two–dimensional Keplerian disk, at time $t = 10$.

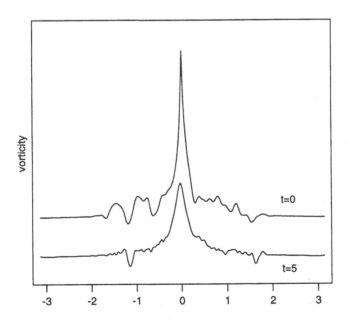

FIGURE 3. Section of the vorticity field of a perturbed two–dimensional Keplerian disk, at time $t = 0$ and $t = 5$.

The size of the viscosity coefficient plays an interesting role. The larger is the viscosity, the broader are the vortices. For large values of ν_2, small vortices cannot form and only a few of the largest vortices survive. Conversely, the smaller is the viscosity, the more numerous the small-scale vortices are. Simulations with $\nu_2 = 5 \cdot 10^{-9}$ show the presence of many small-scale anticyclonic vortices that penetrate well inside the central regions of the Keplerian disk. The smaller is the viscosity, the smaller is the minimum energy ratio $\tilde{E}/E_{kep}$ which is required in order to generate the vortices.

The roles of both the amplitude of the initial disturbance and of the viscosity emerge in the definition of two length scales that are used in discussions of geostrophic turbulence. In the present context, a length relating to the shear, $L_S = \sqrt{U/\Omega'}$ may be defined where $\Omega' = d\Omega/dr$ and U is a characteristic initial velocity disturbance. The shear length, L_S is the disk analogue of the Rhines scale of geostrophic turbulence theory (see Cho and Polvani 1996a; 1996b). In both situations, where the rotation rate varies in one direction, it is found that perturbations with scales greater than L_S become highly anisotropic. It is believed that such structures do not participate in the inverse cascade of energy and remain elongated. Their decay takes place through radiation of Rossby waves.

Distinct vortices can form only with sizes less than L_S. On the other hand, vortices that are too small will be rapidly destroyed by viscosity. The viscous decay time of a structure of size L is of the order of L^{2p}/ν_p. For a vortex to live several rotation periods, this time must be much larger than Ω^{-1}, which requires $L \gg L_\nu = (\nu_p/\Omega)^{1/2p}$. Thus, we expect that the copious formation of coherent vortices will be possible only if $L_\nu \ll L_S$. If the viscosity is too large, or the shear is too strong, L_ν and L_S become comparable and vortices cannot form. For $p = 1$ and with $\Omega/\Omega' = r$, the criterion is that $Ur/\nu \gg 1$. Though this is in a sense to be expected, the ingredients in the derivation are somewhat different than they are in the usual derivation of criteria involving the Reynolds number. Our Reynolds number is the ratio of the β term from the Coriolis force to the viscous term.

We have observed in these simulations that vortices form from general perturbations having initial sizes of the order of L_ν. The vortices then merge with each other, and the size of the surviving vortices grows with time. In a homogeneous background, the only limit to this growth is set by the size of the domain. In a shear flow, the scale L_S sets an upper limit to the vortex size. Thus, the older is a vortex, the closer its size is to L_S. At late evolutionary stages, most anticyclonic vortices have a size comparable with L_S. Since the strength of the local shear depends on the radius r, and $L_S = L_S(r)$, the range of radii in which vortices form will depend on the variation of ν, hence density, with r. In the present simulations with constant ν_p, we see larger vortices farther out.

4. Magnetic Effects

The inclusion of a magnetic field in the disk can produce linear instability and so change the fluid dynamics in an important way (Balbus and Hawley 1998). We study the effect of a purely horizontal magnetic field on the disk dynamics discussed in the previous section in the usual MHD approximation with displacement current neglected. The field is assumed to be not so strong as to modify the thin layer assumptions. We express it in terms of a scalar magnetic potential $a(\mathbf{r}, t)$ such that the horizontal field $\mathbf{B} = (B_x, B_y)$ is given by $B_x = -\partial a/\partial y$ and $B_y = \partial a/\partial x$. This magnetic field is horizontally non-divergent so any vertical field that is present must be independent of z.

The current is $\nabla \times \mathbf{B}$ and is in the vertical direction. This does not fit in well with the thin layer image and is more suited to the picture of two-dimensional incompressible flow. We nevertheless proceed formally to consider what happens when the Lorentz force

$\mathbf{j} \times \mathbf{B}$ is included in the dynamics of the previous section, where the current $\mathbf{j}$ is purely in the z-direction and its magnitude is given by

$$j = \nabla^2 a . \tag{4.17}$$

The equation of motion for the vertical vorticity becomes

$$\frac{\partial \zeta}{\partial t} + [\psi, \zeta] - [a, j] = D_\zeta \tag{4.18}$$

where the various quantities are as before. In addition, the induction equation for the magnetic field, when expressed in terms of the magnetic potential, is

$$\frac{\partial a}{\partial t} + [\psi, a] = D_a . \tag{4.19}$$

The Ohmic dissipation D_a term is given by

$$D_a = (-1)^{q-1} \eta_q \nabla^{2q} a . \tag{4.20}$$

In the absence of dissipation, eqs.(4.18) and (4.19) have three quadratic invariants: the total (kinetic plus magnetic) energy $E = 1/2 \int (\nabla \psi)^2 dx dy + 1/2 \int (\nabla a)^2 dx dy$, the integral of the squared magnetic potential, $A = \int a^2 dx dy$, and the cross helicity $H = \int \mathbf{u} \cdot \mathbf{B} dx dy$. The simultaneous conservation of these quantities induces a direct cascade of energy, from large to small scales, and an inverse cascade of magnetic potential, from small to large scales (Fyfe & Montgomery 1976; Fyfe *et al.* 1977). One thus expects to see coherent structures in the current field rather than in the vorticity. Since the three-dimensional MHD equations have three quadratic invariants as well, the fact that energy cascades to small scales, makes 2D MHD closer to the full three-dimensional MHD problem than 2D Navier-Stokes is to 3D hydrodynamics.

The 2D MHD system has a wide range of dynamical behaviours. In the limit of weak magnetic field, the magnetic potential behaves as a passive scalar, and the term $[a, j]$ in eq.(4.18) can be discarded. In this limit, one recovers the behavior of the 2D Navier-Stokes equations (2.13). An initially weak magnetic field does however grow with time, and in later stages the dynamics is significantly affected by magnetic forces.

An important diagnostic quantity is the global correlation between the velocity and the magnetic fields defined as $C = H/E$. If the velocity and magnetic fields are nearly aligned, the absolute value of the global correlation coefficient is close to one. Several simulations have shown that, if $|C|$ is larger than about 0.2 initially, then it grows with time toward the value $|C| = 1$. In this process, called dynamic alignment, Lorentz forces cause the magnetic and velocity fields to become parallel (or anti-parallel) to each other. On the other hand, if $C \approx 0$ initially, then the two fields remain uncorrelated.

Numerical simulations of the dissipative 2D MHD equations with $D_\zeta \neq 0$ and $D_a \neq 0$, in a homogeneous background reveal the formation of current and vorticity sheets (Biskamp & Welter 1989). Coherent magnetic vortices also form and dominate the current distribution and the dynamics of the system (Kinney *et al.* 1995) so that, at later times, the fluid vorticity plays a minor role, concentrating in sheets at the periphery of the magnetic vortices. In the individual magnetic vortices, there may be a non-zero correlation between vorticity and current. The sign of this correlation is different in different vortices, leading to an approximately zero global correlation between the magnetic and the velocity fields.

The question that we consider here is whether magnetic vortices can form in a shearing background. To address this issue, we have run a series of simulations on the behavior of a magnetized, barotropic Keplerian disk, by modifying the code described in the previous section to deal with (4.18)-(4.19). In the runs discussed below, we have chosen

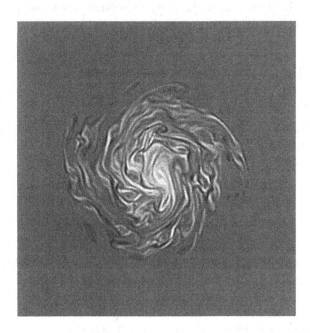

FIGURE 4. Vorticity at time $t = 10$ for a magnetized, two–dimensional initially Keplerian disk with no initial background current.

a hyperviscous dissipation with $p = q = 2$, $\nu_2 = 5 \cdot 10^{-8}$ and $\eta_2 = 2.5 \cdot 10^{-7}$. This gives a magnetic Prandtl number $Pr_{mag} = \nu_2/\eta_2 = 0.2$. Runs with $Pr_{mag} = 1$ have provided analogous results.

In simulating a magnetized disk, we have to specify the initial background current distribution. Though it seems likely that an initial field will quickly develop a strong toroidal presence, we have decided to leave the matter open, and so we show the results of two different simulations. In the first, we assume no background current; here the magnetic field is present only as a perturbation of a basically unmagnetized background state. In the second type of simulation, we assume the presence of a Keplerian background current, with the same amplitude as the background vorticity field at the outset. In both cases, the background vorticity is assumed to be Keplerian, as in Section 3.

In the case with no background current, the background Keplerian vorticity field has energy $E_{kep} = 0.5$. The typical rotation time at $r = \pi/4$ is $T_{kep}(\pi/4) \approx 2$ and the simulation has been run up to a time $T = 20$. In this simulation, we initially perturb the magnetic field away from zero and do not perturb the vorticity; the perturbation energy is $\tilde{E} \approx 0.004 E_{kep}$. The spectrum of the initial magnetic perturbation is given by eq.(3.16). This case is characterized by an almost zero correlation between the initial magnetic and velocity fields.

Figures 4 and 5 show the vorticity and the current at time $t = 10$ and they reveal that strong magnetic vortices form in the presence of an initial Keplerian shear. As already observed by previous studies without a background shear, the vorticity is dominated by magnetic effects, with fluid vorticity concentrating in sheets and in weaker structures associated with the magnetic vortices. In the evolved fields, there is no global correlation between vorticity and current.

A significant difference from the non-magnetic case is found in the evolution of the

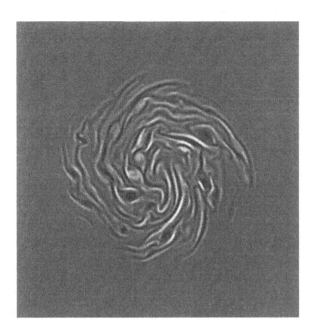

FIGURE 5. Current at time $t = 10$ for a magnetized, two–dimensional initially Keplerian disk with no initial background current.

mean vorticity profile. Without the magnetic field the Keplerian profile is linearly stable to perturbations and we find that the average radial profile remains Keplerian during the evolution. In the MHD simulation shown here, the presence of the magnetic perturbation leads to a modification of the background Keplerian profile, which becomes flatter as the evolution proceeds. This is not surprising, as the magnetic torques enter the dynamics significantly.

Figures 6 and 7 show the vorticity and the current obtained when the current has a background Keplerian component $J(r)$, equal in strength to the background vorticity profile $\Omega(r)$. No vortices form in this case, the two fields becoming rapidly fully correlated and frozen into alignment due to the presence of an initial correlation $|C| > 0.2$. The terms $[\psi, \zeta]$ and $[a, j]$ then cancel each other in eq.(4.18), and the field remains in its Keplerian state, slowly decaying due to dissipation.

These simulations have thus shown that, when there is no background Keplerian current, coherent magnetic vortices can be generated. In the magnetic field, the strong cyclone-anticyclone asymmetry observed for purely hydrodynamic disks is absent. The vorticity field is dominated by the magnetic effects, and weaker vorticity concentrations can be observed in the magnetic vortices. In the absence of a background Keplerian current, the vorticity profile does not remain Keplerian. These results are restricted to the unrealistic case of purely horizontal magnetic field and do not include the influence of an external gravitational force that might favor the maintenance of the Keplerian background flow. In the appendix we describe the inclusion of this effect and observe that it does not change the general character of the results.

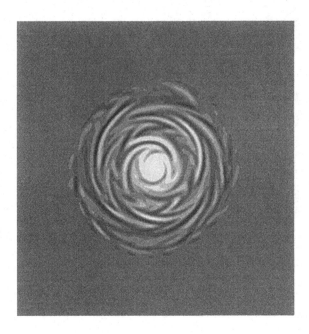

FIGURE 6. Vorticity at time $t = 10$ for a magnetized, two–dimensional initially Keplerian disk with an initially Keplerian background current.

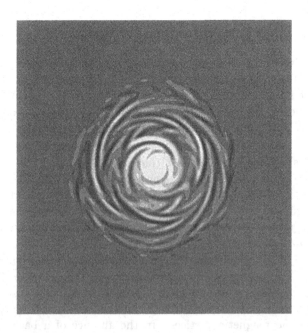

FIGURE 7. Current at time $t = 10$ for a magnetized, two–dimensional initially Keplerian disk with an initially Keplerian background current.

5. Disks With Variable Surface Density

The full thin-layer theory is like that of a two-dimensional compressible fluid, where density fluctuations are important, except that in the thin-layer approximation these fluctuations propagate as compressible gravity waves rather than acoustic waves (Qian & Spiegel 1994). We present here some first results from numerical simulations of the full thin-layer problem. This problem is computationally more demanding than the version we have just considered so that they involve even more modest parameter values than the incompressible case. We begin by recalling some of the theoretical background for the thin-layer simulations (Balmforth *et al.* 1992). To do that, we temporarily ignore the viscous term in order to keep the discussion simple; it is restored for the numerical computations.

In the full thin-layer theory of section 2, we solve the continuity equation (2.8) together with the vertically integrated counterpart of the momentum equation (2.1). Since vertical motions are negligible in thin layers, we have hydrostatic balance in the vertical direction. The simplified $\hat{z}$-component of the momentum equation then integrates to:

$$\int_{-h}^{h} \frac{1}{\rho} \frac{\partial p}{\partial z} dz = \int_{-h}^{h} \frac{\partial H}{\partial z} dz = -\int_{-h}^{h} \frac{\partial \Phi}{\partial z} dz, \qquad (5.21)$$

where we have introduced the specific enthalpy H according to the definition provided by the first equality in (5.21). For a thin layer, the thickness of the disk, from h above to $-h$ below the plane, is small compared to the radial coordinate, r, so this equation tells us that the enthalpy is well approximated by

$$H = \frac{GM}{2r^3} \left(h^2 - z^2\right) . \qquad (5.22)$$

Since the pressure, and therefore the enthalpy, vanishes at $z = \pm h$, (5.21) and (5.22) become

$$\eta(x, y, t) = H(x, y, 0, t) = \frac{GMh^2}{2r^3}, \qquad (5.23)$$

where η is the specific enthalpy to be used in the thin layer. Then, an effective equation of state for the layer can be found by substituting the polytropic formula for enthalpy, $H = K\frac{\gamma}{\gamma-1}\rho^{\gamma-1}$, into (5.21), allowing the density to be expressed as a function of the disk thickness; we get

$$\rho = \left[\frac{\gamma - 1}{\gamma K} \frac{GM}{2r^3} \left(h^2 - z^2\right)\right]^{1/\gamma-1} . \qquad (5.24)$$

This relation, when inserted into the definition of the surface density, equation (2.7), yields the vertically integrated equation of state,

$$\sigma = \sigma_0 \left[\eta^{\frac{\gamma+1}{\gamma-1}} r^3\right]^{1/2}, \qquad (5.25)$$

where σ_0 is a constant. Because of the dependence on the local vertical gravity, the equation of state depends on r.

The horizontal momentum equation is now

$$\frac{D\mathbf{u}}{Dt} = -\nabla (\eta + \phi) + \mathbf{D_u} \qquad (5.26)$$

where u is the two-dimensional velocity $\mathbf{u} = (u, v)$, ϕ is the gravitational potential in the plane $(\phi(x, y) = \Phi(x, y, 0))$ and the viscous term has been restored. The expression for the potential vorticity of the disk has been given in equation (2.10).

We have solved equations (2.8), (5.25) and (5.26) in finite-difference form, on grids of 256×256 and 512×512 using a modified form of the Miami Isopycnal Coordinate Model (MICOM), a second order algorithm which incorporates a flux-corrected transport of the surface density. This code has been used successfully for a wide variety of oceanic flows; in particular, it has been used to study the ocean thermocline, another thin layer whose equations are like the ones studied here. The special feature of this code, not contained in typical shallow fluid codes, is that it allows for a vanishing layer thickness (Sun *et al.* 1993).

For the initial conditions, we need a background disk. For this we used steady axisymmetric models with angular velocity profile given by

$$\Omega = \left(\frac{1}{r} \frac{d\bar{\eta}}{dr} + \frac{1}{r} \frac{d\phi}{dr} \right)^{1/2}, \tag{5.27}$$

where $\bar{\eta}$ is the time-independent enthalpy. When (as is usual in the mean state) the enthalpy variations are negligible, we recover from (5.27) the Keplerian velocity $\Omega_K = r^{-3/2}$ for the potential $\phi = -1/r$.

Since enthalpy variations are not negligible, the variations of the background surface density lead to both a departure from Keplerian flow, according to (5.27), and to a modulation of the background potential vorticity, according to (2.10) in a way that is analogous to topographic effects frequently encountered in geophysical problems.

To simplify the application of boundary conditions, we have taken the case of an annulus with enthalpy distribution

$$\bar{\eta}(r) = \eta_0 (r - r_i)^{n_i} (r_o - r)^{n_o}, \tag{5.28}$$

where the subscripts i and o mean inner and outer. This $\bar{\eta}$ vanishes at two radii ($r = r_i$ and $r = r_o$), always chosen to be well within the computational domain. Since we apply boundary conditions only on the edges of the domain, this makes for an insensitivity to the boundary conditions.

The computational domain extends from -2 to 2 and we have used typical values for the inner and outer disk radii of $1/4$ and $7/4$, respectively. The computations were performed in a frame co-rotating with the radius $r = 1$. The angular velocities do not depart significantly from Keplerian values; at $r = 1/3$ the orbital period is ≈ 1.

In figure 8 we display the enthalpy for three steady disks, having $n_i = \frac{3}{4}$ and $n_o = \frac{3}{4}, \frac{4}{4}, \frac{5}{4}$. The corresponding potential vorticity is also plotted (middle), but since q is not well defined where σ vanishes, we plot only those regions where the surface density exceeds some minimum value: $\sigma > \sigma_0$. The singular nature of q indicates the presence of a boundary layer at the disk edges. At present we are interested only in the behavior far from such boundary layers, so we can choose to represent the inverse of q, or what is called the *potential thickness* (figure 8, bottom), rather than the potential vorticity.

In shallow fluids, gravitational forces will act to restore perturbations made to the layer thickness. But in the presence of rotation, this process excites local shears in the flow. Large perturbations will be deformed in this process since there is a scale above which rotation dominates vertical stability. In geophysics this critical scale is called the (Rossby) deformation radius, L_D (Pedlosky 1987), given by $L_D^2 = 2\bar{\eta}r/f^2 h$. Vortices larger than the Rossby radius have quiet centers encircled by rings of rapidly varying potential vorticity. This manifestation of L_D has been revealed by numerous laboratory and computational results, which show how L_D constrains the extent of the perturbing shear layers within a vortex.

To study the action of potential vorticity and the behavior of potential vortices in disks,

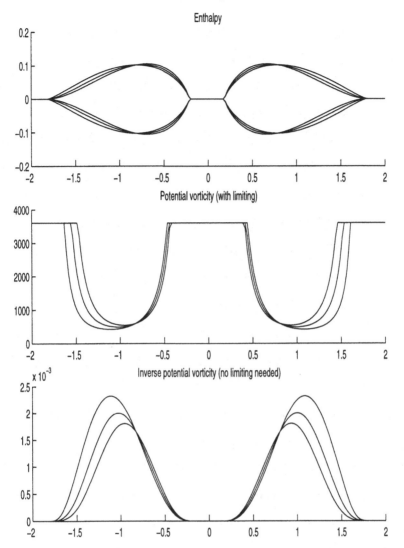

FIGURE 8. Enthalpy (top), potential vorticity (middle), and potential thickness (bottom) for three thin disks having respectively $n_o = 3/4$, $4/4$ and $5/4$.

we perturb the steady solutions given above. One way to introduce small disturbances of q is to simply perturb the thickness. To leading order, the perturbed potential vorticity can be expressed as

$$q = \bar{q} - \frac{\sigma'}{\bar{\sigma}}\bar{q} + \frac{\zeta'}{\bar{\sigma}}, \qquad (5.29)$$

where the last term is absent if we do not perturb the absolute vorticity. Using this expression, we build a random distribution of perturbed potential vorticity, whose energy peaks at a wavelength $\lambda \sim R_{disk}/10$, and superpose this onto the steady disk.

The evolution of the initial disk given by (5.27) and (5.28) leads to an approximately steady disk with some negligible migration of surface density toward $r = 0$ as a result of viscous stress. When the initial condition includes a perturbed potential vorticity as described above and in (5.29), the evolution is dominated by the interaction of compress-

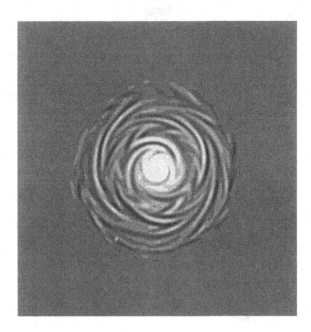

FIGURE 9. Potential thickness (inverse potential vorticity) at time $t = 2$ for a perturbed thin disk with $n_i = 3/4$ and $n_o = 5/4$.

ible vortices with each other and with the background shear. The disk at time $t = 2$ is shown in figure 9.

The most obvious features found in the potential vorticity are: (1) vortices which are too small are quickly eliminated by the combined actions of the shear and dissipation; (2) small potential vortices, if they are above this scale, rapidly coalesce to form larger, more coherent features; (3) cyclonic fluctuations tend to be easily absorbed into the background shear of the disk; (4) the more robust anticyclones can persist for many disk rotations.

6. Conclusions

At first sight it might seem that strong shears provide an abundant source of angular momentum that would feed vortex production and that would tightly wind local magnetic anomalies. However, shears also work against the conventional vortex formation process by making the environment highly anisotropic. In the context of the Jovian atmosphere, shears also occur and the work on that situation suggests that vortices cannot form if they are too large (Cho and Polvani 1996a; 1996b). Instead, large structures give rise to bands in the jovian atmosphere. We may expect a similar behavior in disks and some of our shallow water calculations have revealed large scale vortical structures with banded, spiral character (Yecko 1995). However, the calculations reported here show that when the local disturbances are not so large, vortices form in disks and that similar coherent structures form in mildly magnetized disks.

In order to form these coherent structures, we need to have initial perturbations of finite amplitude. There is no doubt that these are available in magnetized disks since they are linearly unstable. The case of nonmagnetized disks is apparently in dispute.

The systems we have studied are not unstable within the numerical schemes we have been using. Rather, they are excitable. This means that a finite amplitude disturbance with amplitude above a given threshold goes through an interesting evolution (in fact formation of vortices) before the system returns to its initial Keplerian state. If there were no magnetic fields, we would need to provide a source of perturbations to produce vortices. These could be present even in nonturbulent disks around black holes since they may be buffeted continually by incoming objects. On the other hand, magnetized disks will certainly have a continual source of perturbations.

In any case, our numerical simulations of two-dimensional, barotropic disks and of thin-layer, compressible disks have shown that sufficiently energetic vorticity perturbations evolve into long-lived, coherent vortices that may heavily affect the disk dynamics in several ways before they decay. In the presence of magnetic effects, coherent magnetic vortices are formed instead of purely kinetic vorticity structures. These will likely give rise to a good deal of variability and other phenomena seen on stars with similar activity.

We are grateful to Neil Balmforth and Steve Meacham for discussion and comments. AP is grateful to JILA for hospitality and support during much of his work on this report.

Appendix A. Forcing Keplerian Flows

In the previous sections we have discussed the evolution of a free Keplerian disk, i.e., of a freely-decaying, initially perturbed Keplerian profile. In the pure hydrodynamic case, the average profile remains Keplerian, except for the fact that it is slowly dissipated by the small hyperviscosity. In the MHD case, the average vorticity profile rapidly becomes non-Keplerian.

On the other hand, we may suppose that external forces, which are not described by the simple evolution equations used above, try to mantain a background Keplerian rotation. These could be, for example, the gravity of the central mass and the slow infall of material at the edge of the disk. In this section, we study the evolution of a perturbation of a Keplerian disk, whose average profile is forced to remain Keplerian.

Suppose that the vorticity field is the sum of two terms, a constant Keplerian background $\Omega(r)$, imposed from the outside, and a perturbation $\tilde{\zeta}(\mathbf{r}, t)$ which is free to evolve. Dissipation is assumed to act only on the perturbation field $\tilde{\zeta}$. For a purely hydrodynamic disk, the equations of motion become in this case

$$\frac{\partial \tilde{\zeta}}{\partial t} + [\tilde{\psi}, \tilde{\zeta}] = \tilde{D}_\zeta \ - [\Psi, \tilde{\zeta}] - [\tilde{\psi}, \Omega] \qquad (A\,30)$$

where $\Omega(r)$ is the constant Keplerian profile, $\Psi(r)$ is the corresponding stream-function, $\tilde{\zeta}$ and $\tilde{\psi}$ are the vorticity perturbation and perturbation stream-function, and $\tilde{D}_\zeta = (-1)^{p-1}\nu_p\nabla^{2p}\tilde{\zeta}$ is the dissipation. The Keplerian profile is a solution to the equation $[\Psi, \Omega] = 0$, this term is thus absent from eq.(A 30).

The term $F = -[\Psi, \tilde{\zeta}] - [\tilde{\psi}, \Omega]$ represents an external forcing on the evolution of the perturbation. If the perturbation were capable of extracting energy from the overall rotation of the disk, then it could grow and eventually stabilize, without being necessarily dissipated. From another point of view, this approach allows for studying the stability of a Keplerian disk, by following the evolution of an initial perturbation. This is just a special case of the general study of the stability of (barotropic) shear flows. Numerical simulation of eq.(A 30) has shown that anticyclonic vortices form in this case as well, notwithstanding the continuous forcing. The perturbation does, however, decay with time and simulations run with other values of the disk parameters — viscosity, ratio of the perturbation energy to the energy of the background Keplerian flow, *etc.* — have led to analogous results.

In the MHD case, the magnetic field acts as a catalyst that allows the vorticity perturbation to extract energy from the Keplerian shear. For a forced MHD disk the equations are

$$\frac{\partial \tilde{\zeta}}{\partial t} + [\tilde{\psi}, \tilde{\zeta}] - [\tilde{a}, \tilde{j}] = \tilde{D}_\zeta - [\Psi, \tilde{\zeta}] - [\tilde{\psi}, \Omega] \qquad (A\,31)$$

$$\frac{\partial \tilde{a}}{\partial t} + [\tilde{\psi}, a] = \tilde{D}_j \qquad (A\,32)$$

where $\tilde{a}$ and $\tilde{j} = \nabla^2 \tilde{a}$ are the perturbation on the magnetic potential and the current and $\tilde{D}_j = (-1)^{q-1} \eta_q \nabla^{2q} j$. We have supposed no background Keplerian current and the forcing is acting only in the vorticity equation.

The numerical simulation of eqs.(A 31, A 32) has shown that magnetic vortices do form also in the presence of a continuous forcing. In this case, however, the perturbation energy grows with time, and the Keplerian MHD disk is unstable. The presence of the magnetic fields acts as a trigger for the instability, and the kinetic energy of the perturbation grows with time. Notwithstanding the Keplerian forcing on vorticity, also in this case the background vorticity profile does not remain Keplerian.

REFERENCES

ABRAMOWICZ, M. A., LANZA, A., SPIEGEL, E. A. & SZUSZKIEWICZ, E. 1991 *Nature* **356**, 41–43.

BABIANO, A., BOFFETTA, G., PROVENZALE, A. & VULPIANI, A. 1994 *Phys. Fluids* **6**, 2465–2474.

BALBUS, S. A. & HAWLEY, J. F. 1996 *Astrophys. J.* **467**, 76–86.

BALBUS, S. A. & HAWLEY, J. F. 1998 *Rev. Mod. Phys.* , in press.

BALMFORTH, N. J., MEACHAM, S. P., SPIEGEL, E. A. & YOUNG, W. R. 1992 *Ann. N.Y. Acad. Sci.* **675**, 53–64.

BENGSTON, L. & LIGHTHILL, J., EDS. 1982 *Intense Atmospheric Vortices*, Springer-Verlag.

BISKAMP, D. & WELTER H. 1989 *Phys. Fluids B* **1**, 1964–1979.

CHARNEY, J. G. 1971 *J. Atmos. Sci.* **28**, 1087–1094.

CHO, J. Y.-K. & POLVANI, L. M. 1996a *Science* **273**, 335–337.

CHO, J. Y.-K. & POLVANI, L. M. 1996b *Phys. Fluids* **8**, 1531–1552.

DOWLING, T. E. & SPIEGEL, E. A. 1990 *Ann. N.Y. Acad. Sci.* **617**, 190–216.

DUBRULLE, B. & ZAHN, J.-P. 1991 *J. Fluid Mech.* **231**, 561–573.

ELHMAIDI, D., PROVENZALE, A. & BABIANO, A. 1993 *J. Fluid Mech.* **257**, 533–558.

FYFE, D. & MONTGOMERY, D. 1976 *J. Plasma Phys.* **16**, 181.

FYFE, D., JOYCE, G. & MONTGOMERY, D. 1977 *J. Plasma Phys.* **17**, 317.

GHIL, M. & CHILDRESS, S. 1987 *Topics in Geophysical Fluid Dynamics* Springer-Verlag.

HOPFINGER, E. J., BROWAND, F. K. & GAGNE, Y. 1982 *J. Fluid Mech.* **125**, 505–534.

HUNTER, J. H. & HORAK, T. 1983 *Astrophys. J.* **265**, 402–416.

KINNEY, R., MCWILLIAMS, J. C. & TAJIMA, T. 1995 *Phys. Plasmas* **2**, 3623–3639.

KRAICHNAN, R. H. & MONTGOMERY, D. 1980 *Reports Progr. Phys.* **43**, 547–619.

LLEWELLYIN SMITH, S. G. 1996 *Vortices and Rossby-wave radiation on the beta-plane* , Thesis, DAMTP, Cambridge University.

MC WILLIAMS, J. C. 1984 *J. Fluid Mech.* **146**, 21–43.

MC WILLIAMS, J. C. 1990 *J. Fluid Mech.* **219**, 361–385.

NEZLIN, M. V. & SNEZHKIN, E. N. 1991 *Rossby Vortices, Spiral Structure and Solitons*, Springer-Verlag.

PEDLOSKY, J. 1987 *Geophysical Fluid Dynamics*, Springer-Verlag.

PROVENZALE, A., BABIANO, A. & ZANELLA, A. 1997 in *Mixing: Chaos and Turbulence*, NATO-ASI, Cargese, France; Plenum, in press.

QIAN, Z. S., SPIEGEL, E. A. & PROCTOR, M. R. E. 1991 *Stab. App. An. Cont. Med.* **1**, 73–93.

QIAN, Z. S. & SPIEGEL, E. A. 1994 *Geophys. Astr. Fluid Dyn.* **74**, 225–243.

RHINES, P. B. 1979 *Ann. Rev. Fluid Mech.* **11**, 401–411.

SUN, S., BLECK, R. & CHASSIGNET, E.P. *J. Phys. Ocean.* **23**, 1877–1884.

TANGA, P., BABIANO, A., DUBRULLE, B. & PROVENZALE, A. *ICARUS* **121**, 158–170.

YECKO, P. A. 1995 *Ann. N.Y. Acad. Sci.* **773**, 95–110.

Self-organized criticality in accretion disks

By PAUL J. WIITA AND YING XIONG

Department of Physics and Astronomy, Georgia State University, Atlanta, GA 30303, USA

The concept of self-organized criticality has gained substantial favor as a way of understanding many disparate physical systems containing fluctuations which lack characteristic length and time scales. A brief review of some astrophysical applications of this idea is given. The possibility that coherent structures and observable variability are produced in accretion disks through a self-organized critical state is considered. New simulations indicate that while this model does produce light-curves and power-spectra for the variability which fit observations, the universality of self-organized criticality is not achieved. This is because the character of the variations depend quite substantially on the ratio of the inner to outer radii of the unstable disk region.

1. Introduction

Coherent structures in accretion disks can be produced through the growth of specific hydrodynamical mechanisms. Vortices (e.g. Abramowicz *et al.* 1992; Bracco *et al.*, this volume) and shocks induced by companion objects (e.g. Sawada, Matsuda, & Hachisu 1986; Chakrabarti & Wiita 1993) provide such physical mechanisms that can yield significant long-lived (multi-orbit) perturbations on disks.

Such coherent structures can produce "bright-spots" on the disk surfaces which, when coupled with the disk's differential rotation and gravitational lensing of photons passing close to the central black hole, can reproduce many aspects of the variations seen in galactic black hole-candidates and active galactic nuclei (AGN). The X-ray light curves and power spectra (Abramowicz *et al.* 1989, 1991; Zhang & Bao 1991), optical/UV light curves and power spectra (Wiita *et al.* 1991; Mangalam & Wiita 1993) can both be understood within this bright spot model. The anti-correlation between X-ray variability and luminosity in AGN confirmed by Lawrence & Papadakis (1993) can also naturally be explained using this model (Bao & Abramowicz 1996). Finally, it has recently been demonstrated that the surprisingly strong coherence in the temporal variability seen between different X-ray energies for galactic black hole binaries (Vaughn & Nowak 1997) is also capable of being understood within the bright-spot framework (Abramowicz *et al.* 1997). This scenario also predicts energy dependent polarization variations in the X-ray bands (Bao, Wiita & Hadrava 1996; Bao *et al.* 1997) which could be detected by future X-ray satellites with polarimetric capabilities.

Another way to produce extended coherent structures in disks is available if accretion disks exhibit self-organized criticality (SOC). This concept arose from the realization that a huge range of physical systems display power-law or scale-invariant correlations over decades in time or space. "Flicker noise" or "1/f noise" is seen in a huge range of physical systems, ranging from stock market prices, through the height of the Nile river, to quasar light curves (e.g. Press 1978). Bak, Tang & Wiesenfeld (1987, 1988) showed extended systems with many metastable states can naturally evolve into a critical state with no fundamental length- or time-scales.

In § 2 we first discuss characteristics of the SOC state, and then we describe astrophysical applications. In particular, in § 3 we indicate how this might work in accretion disks, as first proposed by Mineshige, Ouchi & Nishimori (1994). In § 4 we display additional computations of modified SOC models for disks which display a goodly range of light curves and variability power spectra. We find that the slopes of the power spectra are quite sensitive to the ratio of the inner and outer disk radii.

2. Characteristics of self-organized criticality

The standard toy model for self-organized criticality is a sandpile, and the original papers by Bak *et al.* (1987, 1988) performed numerical simulations relevant to this physical model, as reduced to a cellular automata problem. The idea is to drop grains of sand randomly onto a pile until it builds up to a critical slope or "angle of repose". Once the pile is at SOC, the addition of a single extra grain of sand anywhere causes the local slope to be too steep. This leads to a small avalanche which readjusts the local shape to just below critical. But now the moved sand will steepen the neighboring slope, perhaps making it, in turn, too steep. Avalanches continue until the local slope everywhere adjusts to be at, or just below, the critical value.

The system becomes stationary when perturbations can just propagate the length of the system, and a full-fledged SOC state of any physical system has the following characteristics:

- a distribution of minimally stable regions of all sizes;
- small perturbations can yield avalanches of all sizes;
- the lack of any characteristic length scale produces a featureless power-law spectrum of avalanche sizes;
- the critical state is insensitive to initial conditions.

However, different physical systems have different power-law indices for the relation between the number of events (avalanches) that occur with different sizes or strengths. These indices depend on (Bak *et al.* 1988; O'Brien *et al.* 1991): the number of spatial dimensions; the symmetry of the system; and the attractor of the dynamics of the system.

The mathematics of the one-dimensional sandpile model can be expressed very simply (Bak *et al.* 1988). This formulation can be easily generalized to two and three dimensions and to more complex phenomena, so it is worthwhile summarizing here how such a simple simulation is performed.

Let $h(x)$ be the integer height at a lattice site, x, and the slope,

$$z(x) \equiv h(x) - h(x+1). \tag{2.1}$$

Add a grain at x and then: $\quad z(x) \longrightarrow z(x) + 1$, while: $\quad z(x-1) \longrightarrow z(x-1) - 1$.

If z exceeds z_c then the slope is too steep and the site is unstable. Then a grain moves to the next position, so that:

$$z(x) \longrightarrow z(x) - 2, \tag{2.2a}$$

$$z(x \pm 1) \longrightarrow z(x \pm 1) + 1. \tag{2.2b}$$

Now, if the site $x + 1$ becomes unstable after the transfer (with $z(x+1) > z_c$), then the process is iterated until either all locations are stable or the end of the grid is reached. For two-dimensional sandpile models, Bak *et al.* (1988) found the power spectral density for the Fourier transform of the time-series to be $S(f) \propto f^{-1.57}$; for three-dimensional models they obtained $S(f) \propto f^{-1.08}$.

2.1. Application to solar flares

The first astronomical application of this idea of which we are aware was the proposal by Lu & Hamilton (1991) that the solar coronal magnetic field is in a SOC state. They noted that observations of hard X-ray bursts in solar flares have a distribution that is a power-law in peak photon flux with logarithmic slope 1.8 over up to five decades (e.g.

Dennis 1985). This relation, $N(P) \propto P^{-1.8}$, is essentially independent of solar cycle phase.

The Lu & Hamilton model assumes that the flares are avalanches of smaller magnetic reconnection events and that all flares arise from the same physical process. Therefore the size of an observed flare depends on the number of small reconnection events of which it is comprised.

A lattice picture in three dimensions assumes the local field gradient is destroyed by reconnection when the local magnetic gradient is too large, as argued by Parker (1988). The magnetic gradient vector of a lattice point which exceeds this limit is eliminated at that particular location, but the gradient is shared with neighboring lattice points. In order for this model to work, changes in magnetic field must have a directionality (i.e., photospheric motions should on average increase B in a particular direction, a not unreasonable assumption over substantial areas).

Lu & Hamilton's computations were performed on a 30^3 grid. The results can be summarized as follows: energy release goes as $N(E) \propto E^{-1.4}$; peak flux goes as $N(P) \propto P^{-1.8}$ (as observed). Lu & Hamilton (1991) argue that these power-law indices are rather insensitive to the critical field gradient. While this picture is not clearly well motivated physically, it does incorporate some viable assumptions and does produce a way to understand an otherwise unexplained relation between the size and frequency of solar flares. Additional physical conditions and implications of this idea have been explored by Lu (1995).

2.2. *Application to gamma-ray bursts*

The possibility that a similar "avalanche" or "chain reaction" may play an important role in understanding details of gamma-ray bursts has been proposed by Stern & Svensson (1996). They note that while the time profiles of γ-ray bursts (GRBs) show a very wide range of behaviors, there are several statistical properties that have specific mathematical structures. In particular, the average peak-aligned profile and the autocorrelation function show stretched exponential behaviors. As for other SOC-type processes, this could imply that the wide range of GRB characteristics could arise from different random realizations of a simple stochastic process which is scale invariant in time.

The Stern & Svensson (1996) model is based upon a pulse avalanche, where a trigger event induces many additional flares by setting off a "chain reaction" in a near-critical regime. Such a basic picture was shown to fit both the diversity of GRB time-profiles and important average statistical properties of GRBs, such as their third-moment, their autocorrelation function and their duration distribution function. Similarly to Lu & Hamilton (1991), Stern & Svensson (1996) suggest that magnetic reconnection in a turbulent medium is a plausible underlying mechanism, one that can fit within an exploding fireball model for GRBs.

3. Applications to Accretion Disks

Mineshige, Ouchi & Nishimori (1994) first proposed that accretion disks around black holes could be in an SOC state. This could provide physical connections between bright spots (or determine the size thereof) and thereby engender coherent structures for interesting lengths of time. This was done in an attempt to produce the steeper than $\sim 1/f$ fluctuations seen in X-ray power-spectrum densities (PSD) for BH candidates such as Cyg X-1 (e.g. Miyamoto *et al.* 1993).

Their cellular automata model is based on the following prescription. The outer disk is smooth enough to not produce any variability; however, the inner disk is subject to some, not yet physically constrained, instability. (Mineshige *et al.* suggest this instability could be related to flares in the corona, presumed to lie above and below the disk, in analogy with Lu & Hamilton's picture.) The unstable inner disk is divided into zones (i, j) characterized by radial (r_i) and azimuthal (ϕ_j) coordinates. To model ordinary viscous accretion, the outer disk feeds a mass particle, m, to a zone in the outermost ring of the inner disk, r_1, with random ϕ. In lieu of more developed physics, they assume $M_{\mathrm{crit},i} \propto r$. If the mass contained in the $(i, j)^{th}$ zone, $M_{i,j} > M_{\mathrm{crit},i}$, they then assume enough mass is dumped from zone (i, j) to bring $M_{i,j} \leq M_{\mathrm{crit},i}$. This excreted mass is taken to spread to the three nearest zones one ring interior to the destabilized zone (for simulations corresponding to rigid rotation) or to three trailing interior zones (as a simple way of mimicking differential rotation).

Once the mass has been dumped from an overloaded zone into others closer to the central mass, one must test those adjacent interior zones for $M_{i+1,(j-1,j,j+1)} > M_{\mathrm{crit},i+1}$ (these second subscript choices pertain to rigid rotation). If this condition is fulfilled, one continues the avalanche by dumping mass into next ring inwards. One estimates the luminosity for each dumped m by:

$$L_x \simeq (GM_{BH}m)\left(\frac{1}{r_f} - \frac{1}{r_i}\right), \tag{3.3}$$

where M_{BH} is the mass of the central black hole, and r_i and r_f are the initial and final radii experienced by that particular mass unit during that particular flow inwards. Then one sums up for the energy released by all avalanching masses. Finally one reinjects mass at the outer edge of the inner disk and repeats the above procedures. For their canonical disk model, Mineshige *et al.* (1994a) discover approximate scalings where the number of events vary with their energy as $N(E) \propto E^{-1.35}$, the lifetime distribution varies as $N(\tau) \propto \tau^{-1.7}$, and the PSD follows $S(f) \propto f^{-1.8}$. They showed these power-law indices, or slopes, to be nearly independent of the amount of differential rotation artificially implemented in their simulations.

3.1. *Diffusion Plus Avalanches in Disks*

A modification of this picture was made by Mineshige, Takeuchi & Nishimori (1994), who included gradual mass inflow in every timestep, even if all zones were stable, with $M_{i,j} < M_{\mathrm{crit},i}$. By doing so Mineshige *et al.* (1994b) were able to model continuous accretion through viscosity in the disk, even in the absence of substantial mass flow (and energy release) produced through instabilities. In their standard model they transfered $3m$ if a zone is unstable (as before) and also $0.01m$ from one cell (randomly chosen) in each ring during each fundamental timestep.

The key result was that the PSD index β for $S(f) \propto f^{-\beta}$ was reduced from ~1.8 to ~1.6. They note that this is in better agreement with AGN (Lawrence & Papadakis 1993) and galactic BH candidate data (Miyamoto *et al.* 1993). Raising the diffusion rate substantially lowers β somewhat further. Other results from these models are that diffusion produces more small-scale avalanches in the inner regions, and decreases the chance for bigger flares. Small shots (energy releases) usually appear randomly in time since the mass is deposited into the outer ring at random points. On the other hand, big shots could have a non-random temporal distribution because a big event drains many cells below their critical values and therefore a longer time is needed to build back up to the critical state. Careful studies of the temporal distribution of various size flares could therefore, in principle, provide a test of this SOC-type model. However, this effect is reduced by diffusion, and we are unaware of additional studies that have performed this type of analysis on usefully large observational datasets.

3.2. *Extrinsic, relativistic modifications*

General relativistic effects, primarily Doppler boosts and gravitational light bending, also lower β (Abramowicz & Bao 1994). This work combined the basic idea of the bright-spot model (§ 1) with the SOC concept. The extent to which these extrinsic effects modify the PSD index is a strong function of the viewing angle, i, to the normal to the disk plane; larger values of i produce shallower slopes, with β reduced to $\sim$1.4 for $i = 80°$ from $\sim$1.8 for $i = 0°$ (Abramowicz & Bao 1994).

4. New variations on SOC in accretion disks

We have recently considered changes in various parameters for simple models such as those suggested by Mineshige *et al.* (1994a,b) to see if the light curves and power spectra for the fluctuations were actually as independent of initial and boundary conditions as they suggested. We examined models with different values of the:

- inner and outer radii of the unstable inner portion of the disk;
- initial deviations from critical density;
- accretion rate;
- amount of mass dumped when a cell exceeded the stability limit;
- amount of differential rotation;
- and diffusion to sideways and outward as well as inward zones.

We have also modified the gravitational potential from the Newtonian, with $\Phi \propto r^{-1}$, to the pseudo-Newtonian type, $\Phi \propto (r - 2)^{-1}$; the latter quite accurately reproduces many of the leading effects of the Schwarzschild geometry (Paczyński & Wiita 1980; Artimova, Björnsson & Novikov 1996). With these expressions we begin the use of notation where distances are given in units of $r_g = GM_{BH}/c^2$.

All of the results we display and discuss below illustrate only the intrinsic variability in the source rest frame. Nearly equivalently, one can say that these computations are restricted to the fluctuations triggered by avalanches as viewed face-on to the disk. One modification we have not yet performed is to incorporate Doppler and general relativistic effects (cf. Abramowicz & Bao 1994), which would allow for the additional extrinsic changes engendered by the observer's viewing angle; as noted above in § 3.2, this would tend to produce less steep PSD slopes.

Our first simulations used what appeared to be the standard model of Mineshige *et al.* (1994a), with 64 rings of unit radius extending inward from $r_1 = 100$ and with the individual zones starting just slightly below critical densities; given those parameters we do indeed reproduce a quasi-power law PSD with slope $\beta \approx -1.8$. These slopes depend very weakly upon *most* of the parameters needed to characterize the simple model. Of course, the early stages of the light curves are very sensitive to the initial deviations from the critical density. Only very small fluctuations are seen at early times if $m_{dump}/3 < M_{crit} - M_{init} \equiv \Delta M$; however, the disks eventually approach the critical values in most zones, and the later portions of the lightcurves appear similar to each other. We note that stable portions of the PSDs (those pertaining to higher frequencies) have very similar slopes even if there is a substantial delay before the critical state is reached throughout the disk (see Figs. 1 and 2).

Smaller values of the accretion rate tend to yield larger numbers of flares of specific energies; the light curves can retain a banded (or "quantized") appearance for significant times (see Fig. 3); nonetheless, values of β are quite independent of $\dot{M}$. In modeling differential rotation, not only did we consider simple extra zone shifts from ring to ring (as was done by Mineshige *et al.*) but we looked at models where the amount of trailing

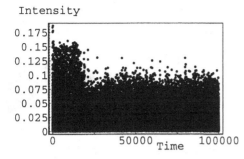
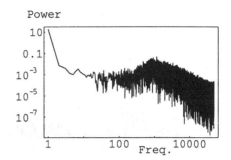

FIGURE 1. The left panel displays a light curve for a simulation with the following parameters: $r_1 = 200$; $r_N = 27.2$; $\Delta M = 0.5$; $\dot{M} = 1$; $m_{dump} = 2$. The right panel shows the corresponding PSD; the best fitting slope, $\beta = 1.54$.

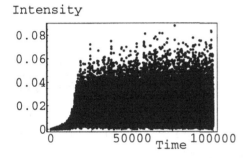
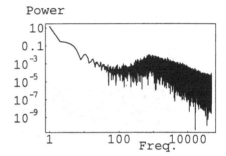

FIGURE 2. As in Fig. 1, with identical parameters except for a larger initial deviation from criticality, $\Delta M = 5.0$; for this case $\beta = 1.50$.

between the ith and $i+1^{th}$ rings was a function of r_i. This allowed for a more realistic approximation to Keplerian or sub-Keplerian rotation curves. Regardless of our approach, we found that changes in the degree of differential rotation also made for only very small differences ($\Delta\beta < 0.04$) in the PSDs, confirming their independence of the amount of rotation noted by Mineshige *et al.* (1994a). This is not surprising given that extrinsic orientation effects, which more strongly depend on differential rotation, have not been included.

In agreement with Mineshige *et al.* (1994b) we do find significant dependence of the power-law on diffusion. They showed that if the diffusion is only inward, then the slope of the PSD becomes shallower. This is a reasonable approximation to standard viscosity in disks, which transports matter inwards and angular momentum outwards. We also performed simulations which allow for a more general diffusion of energy, which might be more realistic if reconnection events provide the physical trigger. To allow for propagation of a disturbance partially sideways and back outwards as well as inwards (which was still presumed to dominate) we also considered cases where the mass outflow from an unstable zone would go 1/5 to each of three interior zones (60% inwards), 1/8 to each of the neighboring zones in the same ring (25% sideways) and 1/20 to each of the three exterior zones (15% outwards). We find that if the triggering zone can send some matter back outwards and sideways, then the slope can become steeper by up to $\Delta\beta \approx 0.15$–0.25 (compare Figs. 1 and 4).

We have found a substantial sensitivity of the power spectrum density index β to the location of the disk's inner radius (r_N) with respect to its outer radius (r_1). The power

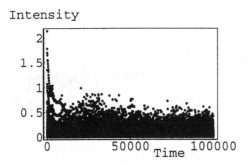

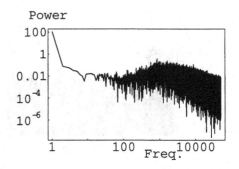

FIGURE 3. As in Fig. 1, with identical parameters except for: $r_1 = 198$; $r_N = 6.0$; here $\beta = 1.07$.

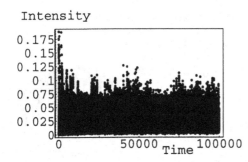

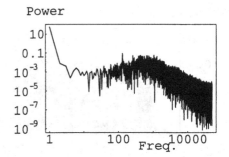

FIGURE 4. As in Fig. 1, with identical parameters; however, here the backwards and sideways diffusion discussed in the text has been included, and $\beta = 1.70$.

spectrum is steeper (higher values of β) if the inner unstable zones do not get too close to the BH. Even without including diffusion, $1.1 < \beta < 1.8$, is found, with the shallowest slopes corresponding to $r_N \approx 6$, i.e., at the innermost stable orbit for a non-rotating black hole, when the outer radius is fixed at $r_1 = 200$. The results obtained by fixing $r_N = 6.0$ and allowing r_1 to vary show that the dominant source of the variation in β is the ratio of r_1/r_N; these changes are illustrated in Figs. 3, 5 and 6. An analogy with advection dominated accretion flows (e.g., Chakrabarti 1996; Chen *et al.* 1995; Lasota, this volume; Narayan, this volume) may well exist here: if the flow loses its disk-like morphology within a larger inner radius (the "transition radius" for such advection dominated flows) then weaker fluctuations and higher values of β might be favored.

Relatively small changes are induced by changing the energy released in Eq. (3.3) from a pure Newtonian, and apparently scale free, r^{-1}, dependence to the pseudo-Newtonian, $(r - 2)^{-1}$, dependence. In cases where the inner radius was set at $r_N = 6$ so this effect would be maximized, we found the slope to become flatter by only $\Delta\beta \approx 0.1$ (compare Figs. 3 and 7). Therefore, we conclude that the differences in PSD slopes produced by the change in bounding radii are not due predominantly to the breaking of linear-scale invariance by the approximate inclusion of general relativistic effects, as one might at first have assumed. To further test this conclusion we performed simulations where the energy released was independent of r, but where the triggering of avalanches otherwise remained the same. In this toy model, which is closer to a standard sandpile picture, we found $\beta \approx 2.0$ with no dependence on the ratio r_1/r_N.

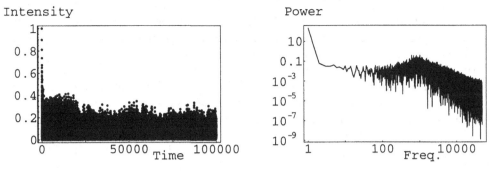

FIGURE 5. As in Fig. 3, with the only different parameter being: $r_1 = 31.6$; the resulting $\beta = 1.62$.

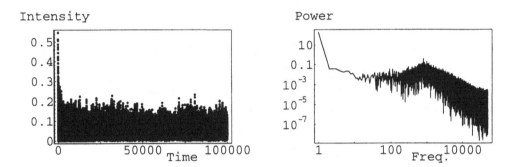

FIGURE 6. As in Fig. 3, with the only different parameter being: $r_1 = 18.8$; the resulting $\beta = 1.80$.

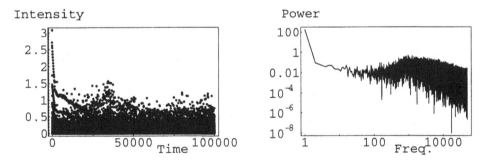

FIGURE 7. As in Fig. 3, with the only difference being the use of the pseudo-Newtonian potential instead of the standard Newtonian potential; $\beta = 0.96$.

5. Conclusions

Our simulations lead us to argue that a true state of self-organized criticality is not likely to be established in accretion disks because the fundamental units of energy release are probably not all the same; rather, energy released deeper in the gravitational well is more substantial for a given mass traversing a given distance. A realistic physical model might modify the statement made in the previous sentence, but need not do so. For example, consider magnetic flares based on reconnection events: if the energy density in the magnetic fields is close to equipartition with the pressure in the matter, then higher energy releases would still be expected closer to the inner edge of the disk. Therefore, even when the disk is operating right at the "hairy edge" of stability, so that flares of many different sizes can be randomly triggered by a single infalling mass unit (or reconnection event) at the outer edge of the unstable portion of the disk, thus implying that an approximate SOC state exists, we do not see all the properties of self-organized criticality. Rather, the key observables, i.e., the light curves and PSD, still depend on the region over which the instability operates; they are not independent of boundary conditions, as one might hope would be the case for a system that is a full-fledged SOC. Nonetheless, it is intriguing that these models seem capable of only reproducing variability power spectrum slopes between -1.0 and -2.0, as this also seems to span the range of observed PSD slopes seen in both galactic and extragalactic variable sources.

This work was supported by NASA Grant NAG 5-3098 and by Research Program Enhancement funds at Georgia State University.

REFERENCES

ABRAMOWICZ, M. A. & BAO, G. 1994 A comment on the paper by Mineshige et al. on the cellular automaton model for the variability of accretion disks around black holes. *Pub. Astron. Soc. Japan* **46**, 523–525.

ABRAMOWICZ, M. A., BAO, G., LANZA, A. & ZHANG, X.-H. 1989 X-ray AGN variability with hour to month timescales. In *Proc. 23rd ESLAB Symp. on Two Topics in X-Ray Astronomy* ed. J. Hunt & B. Battrick. ESA Sp-296, pp. 871–875. European Space Agency.

ABRAMOWICZ, M. A., BAO, G., LANZA, A. AND ZHANG, X.-H. 1991 On the variability power spectrum of AGN. *Astron. Astrophys.* **245**, 454–456.

ABRAMOWICZ, M. A., BAO, G., LARSSON, S. & WIITA, P. J. 1997 On the variability coherence observed in black hole candidates at different x-ray energies. *Astrophys. J.* **489**, 819–821

ABRAMOWICZ, M. A., LANZA, A., SPIEGEL, E. A., & SZUSZKIEWICZ, E. 1992 Vortices on accretion disks. *Nature*, **356**, 41–42.

ARTEMOVA, I. V., BJÖRNSSON, G., & NOVIKOV, I. D. 1996 Modified Newtonian potentials for the description of relativistic effects in accretion disks around black holes. *Astrophys. J.* **461**, 565–571.

BAK, P., TANG, C. & WIESENFELD, K. 1987 Self-organized criticality: an explanation of 1/f noise. *Phys. Rev. Lett.* **59**, 381–384.

BAK, P., TANG, C. & WIESENFELD, K. 1988 Self-organized criticality. *Phys. Rev. A.* **38**, 364–374.

BAO, G. & ABRAMOWICZ, M. A. 1996 Anticorrelation of variability amplitude with x-ray luminosity for active galactic nuclei. *Astrophys. J.* **465**, 646–649.

BAO, G., HADRAVA, P., WIITA, P. J., & XIONG, Y. 1997 Polarization variability of active galactic nuclei and x-ray binaries. *Astrophys. J.*, 20 Sept., in press.

BAO, G., WIITA, P. J., & HADRAVA, P. 1996 Energy-dependent polarization variability as a black hole signature. *Phys. Rev. Lett.* **77**, 12–15.

CHAKRABARTI, S. K. 1996 Accretion processes on a black hole. *Phys. Reports*, **266**, 229–390.

CHAKRABARTI, S. K. & WIITA, P. J. 1993 Spiral shocks in accretion disks as a contributor to variability in active galactic nuclei. *Astrophys. J.* **411**, 602–609.

CHEN, X., ABRAMOWICZ, M. A., LASOTA, J.-P., NARAYAN, R., AND YI, I. 1995 Unified description of accretion flows around black holes. *Astrophys. J.*, **443**, L61–L64.

DENNIS, B. R. 1985. Solar hard x-ray bursts. *Solar Physics* **100**, 465–490.

LAWRENCE, A. & PAPADAKIS, I. 1993 X-ray variability of active galactic nuclei: a universal power spectrum with luminosity-dependent amplitude. *Astrophys. J.* **414**, L85–L88.

LU, E. T. 1995 The statistical physics of solar active regions and the fundamental nature of solar flares. *Astrophys. J.* **446**, L109–L112.

LU, E. T. & HAMILTON, R. J. 1991 Avalanches and the distribution of solar flares. *Astrophys. J.* **380**, L89–L92.

MANGALAM, A. V. & WIITA, P. J. 1993 Accretion disk models for optical and ultraviolet microvariability in active galactic nuclei. *Astrophys. J.* **406**, 420–429.

MINESHIGE, S., OUCHI, N. B., & NISHIMORI, H. 1994a On the generation of 1/f fluctuations in x-rays from black-hole objects. *Pub. Astron. Soc. Japan* **46**, 97–105.

MINESHIGE, S., TAKEUCHI, M. & NISHIMORI, H. 1994b Is a black hole accretion disk in a self-organized critical state? *Astrophys. J.* **435**, L125–L128.

MIYAMOTO, S., IGA, S., KITAMOTO, S., & KAMADO, Y. 1993 Another canonical time variation of x-rays from black hole candidates in the very high flare state? *Astrophys. J.* **403**, L39–L42.

O'BRIEN, K., WU, L., & NAGEL, S. R. 1991 Avalanches in three and four dimensions. *Phys. Rev. A* **43**, 2052–2055.

PACZYŃSKI, B. & WIITA, P. J. 1980 Thick accretion disks and supercritical luminosities. *Astron. Astrophys.* **88**, 23–31.

PARKER, E. N. 1988 Nanoflares and the solar x-ray corona. *Astrophys. J.* **330**, 474–479.

PRESS, W. H. 1978 Flicker noises in astronomy and elsewhere. *Comments on Astrophys.* **7**, 103–120.

SAWADA, K., MATSUDA, T. & HACHISU, I. 1986 Accretion shocks in a close binary system. *Mon. Not. Royal Astr. Soc.* **221**, 679–686.

STERN, B. E. & SVENSSON, R. 1996 Evidence for "chain reaction" in the time profiles of gamma-ray bursts. *Astrophys. J.* **469**, L109–L113.

VAUGHN, B. A. & NOWAK, M. A. 1997 X-ray variability coherence: how to compute it, what it means, and how it constrains models of GX 339−4 and Cygnus X-1. *Astrophys. J.* **474**, L43–L46.

WIITA, P. J., MILLER, H. R., CARINI, M. T. & ROSEN, A. 1991 Microvariability in blazars via accretion disk instabilities. In *Structure and Emission Properties of Accretion Disks.*, ed. C. Bertout, S. Collin-Souffrin, J.P. Lasota and J. Tran Thanh Van, 6th I.A.P. Astrophysics Meeting / I.A.U. Colloq. 129, pp. 557–558. Editions Frontiéres, Gif-sur-Yvette.

ZHANG, X.-H. & BAO, G. 1991 The rotation of accretion disk and the power spectrum of x-ray flickering. *Astron. Astrophys.* **246**, 21–31.

Old and new advances in black hole accretion disk theory

By ROLAND SVENSSON

Stockholm Observatory, S-133 36 Saltsjöbaden, Stockholm, Sweden

A summary is given of the high-lights during the Reykjavik Midsummer Symposium on Non-Linear Phenomena in Accretion Disks around Black Holes. Such high-lights include the recent advances on understanding: 1) the accretion disk solution branch dominated by advection (i.e., advection dominated accretion flows, ADAFs), 2) the importance of magnetic fields in many different respects, most importantly being responsible for the self-sustained MHD-turbulence giving rise to the disk viscosity, and 3) the details of the radiation processes giving rise to the X/γ-ray continuum originating close to the black hole.

Some old advances are unfortunately also necessary to discuss here. It is pointed out to the accretion disk research community, that many of the research papers published on ADAFs 1994-1997 do not accurately present the history of ADAF research. Some of the results that were presented as new and original actually appeared in a paper by Ichimaru already in 1977. Also not quoted are the papers from the 1970's and 1980's calculating the temperature structure of and the spectra from quasi-spherical accretion flows onto black holes. As the ADAFs are close to being quasi-spherical, the resulting spectra of ADAFs and quasi-spherical flows are almost identical. Further of the recent ADAF results are therefore not new results as is sometimes claimed.

In spite of all the recent progress of various aspects of accretion flows around black holes, many of the research lines have still not been merged providing potential for further dramatic progress in coming years.

1. Why Iceland?

This is the very first conference in astrophysics that has ever taken place on Iceland, a country with only about 2-3 astronomers. It is natural to ask the question: Why was it organized on Iceland? This question fortunately has several reasonable answers:

First, the astronomy population on Iceland is not that small as it may seem at first sight. It is approximately similar to that in other Western countries, i.e., about 1 astronomer per 100 000 in population. Iceland's 250 000 citizens imply a total of $2.5 \pm \sqrt{2.5}$ astronomers, in good agreement with the actual number.

Second, one of these 2.5 astronomers works on accretion disks which makes Iceland probably the country with the world's largest fraction of astronomers working in this field.

Third, in the past, at least according to classic literature, there has been at least two historic studies of "black hole interiors" originating on Iceland. In Jules Verne's *Voyage au Centre de la Terre* (1864), the research team led by Professor Otto Lidenbrock from Hamburg entered a hole (most likely being black) on the bottom of the crater of Snaefell in 1863. Here, they followed in the footsteps of the 16th century Icelandic scientist Arne Saknussemm who had explored this black hole before them as described in Jules Verne's novel.

Fourth, and maybe the most important reason is that the research area of accretion disks has grown rapidly, not only worldwide, but also in most of the Nordic countries (Denmark, Finland, Iceland, Norway, Sweden) over the last 10 years. Because of this, all the support for this conference originated from these countries directly or indirectly with the requirement, of course, that the conference had to take place in one of these countries.

The summary talk summarized most of the talks at the Midsummer Symposium. However, as the chapters in this monograph were written up to a year after the symposium, their content are often different from that of the symposium talks occasionally being influenced by the summary talk. This summary chapter still summarizes the symposia talks, but also includes some comments on the material in the written reviews.

2. Accretion disks

One of the main topics of this Midsummer Symposium is the recent research bandwagon on advection dominated accretion flows (ADAFs) starting in 1994 with probably of the order of 100 papers since then. Many of the results are covered in Abramowicz (this volume), Björnsson (this volume), Narayan, Mahadevan, & Quataert (this volume), and Lasota (this volume). The problem is that in all the papers written before the summary talk of this symposium, the history of research on ADAFs has not been accurately presented. In my summary talk, I wanted to take the opportunity to set the record straight. I therefore here quote a few paragraphs from the section on the history of theoretical accretion disk research in my chapter (Svensson 1997) in *Relativistic Astrophysics: A Conference in Honour of Prof. I.D. Novikov's 60th birthday*:

2.1. *Theoretical history (from Svensson 1997)*

The underlying framework for almost all efforts to understand active galactic nuclei (AGN) is the accretion disk picture described in the classical papers by Novikov & Thorne (1973) and Shakura & Sunyaev (1973). This original picture was partly inspired by the extreme optical AGN luminosities. Here, effectively optically thick, rather cold matter forms a geometrically thin, differentially rotating Keplerian disk around a supermassive black hole. The differential motion causes viscous dissipation of gravitational binding energy resulting in outward transportation of angular momentum and inward transportation of matter. The dissipated energy diffuses vertically and emerges as black body radiation, mostly in the optical-UV spectral range for the case of AGN.

Observations have also been the driving force in the discovery of two other solution branches.

a) Hard X-rays from the galactic black hole candidate, Cyg X-1, as well as the discovery of strong X-ray emission from most AGN led to the need for a hot accretion disk solution. Shapiro, Lightman & Eardley (1976) (SLE) found a hot, effectively optically thin, rather geometrically thin solution branch, where the ions and the electrons are in energy balance, with the ions being heated by dissipation and cooled through Coulomb exchange, leading to ion temperatures of order 10^{11} - 10^{12} K. The efficient cooling of electrons through a variety of mechanisms above 10^9 K, leads to electron temperatures being locked around 10^9 - 10^{10} K. The SLE-solution is thermally unstable.

b) Observations of Cyg X-1 and of radio galaxies led to two independent discoveries of the third solution branch. Ichimaru (1977) developed a model to explain the soft high and hard low state of Cyg X-1. The soft high state is due to the disk being in the optically thick cold state, and the hard low state occurs when the disk develops into a very hot, optically thin state. This solution branch is similar to the SLE-solution, except that now the ions are not in local energy balance. It was found that if the ions were sufficiently hot they would not cool on an inflow time scale, but rather the ions would heat up both by adiabatic compression and viscous dissipation and would carry most of that energy with them into the black hole. Only a small fraction would be transferred to the electrons, so the efficiency of the accretion is much less than the normal $\sim 10\%$. As the ion temperature was found to be close to virial, these disks are geometrically thick. Just as for the SLE-case above, the electrons decouple to be locked at 10^9 - 10^{10} K. The flows resemble the quasi-spherical dissipative flows studied by, e.g., Mészároz (1975) and Maraschi, Roasio & Treves (1982). Independently, Rees *et al.* (1982) and Phinney (1983) proposed that a similar inefficient accretion disk solution is responsible for the low nuclear luminosities in radio galaxies with large radio lobes and thus quite massive black holes. These geometrically thick disks were named *ion tori*. Rees *et al.* (1982) specified more clearly than Ichimaru (1977) the critical accretion rate above which the flow is dense enough for the ions to cool on an inflow time scale and the ion tori-branch would not exist. Further considerations of ion tori were made by Begelman, Sikora & Rees (1987). Ichimaru (1977) emphasized that his version of ion tori is thermally stable.

The ion-tori branch has been extensively studied and applied over the last two years (1995-1996) with more than 30 papers by Narayan and co-workers, Abramowicz and co-workers, as well as many others. These studies confirm and extend the original results, although some papers do not quote or recognize the original results. New terminology has been introduced based on an Eulerian viewpoint rather than a Lagrangian. Instead of the ions not cooling, it is said that the local volume is "advectively cooled" due to the ions carrying away their energy. The *ion tori* are therefore renamed as *advection-dominated disks*.

2.2. *Why did history go wrong?*

From the above, it is clear that just 4 years after the start of research on modern accretion disk theory in 1973, the pioneering work on three of the four major accretion disk branches had been done (see Narayan *et al.*, this volume for a description of the four branches). As a graduate student with a side interest in accretion disks, I myself read the paper by Ichimaru back when it was published and a few times since then, but did not realize that Ichimaru had discovered a fundamentally new accretion disk branch. When the ADAF research bandwagon got going in 1995, I spent time reading those papers and eventually realized that Ichimaru, although not quoted, actually had made the original discovery of the ADAF-solution. Instead, in the literature the ADAF-solution was presented as a new discovery. Even the issue of whether the ADAF-solution was new or not was discussed in some papers. Let me give a few quotations: "Recently, a new class of two-temperature advection-dominated solutions has been discovered"; "We have found new types of optically thin disk solutions where cooling is dominated by the radial advection of heat"; "Is our new solution really new? Certainly, to the extent that we have for the first time included advection and treated the dynamics of the flow consistently, the advection-dominated solution... is new. But even more fundamentally, it is our impression that the existence of *two* hot solutions was not appreciated until now". In order to correct the history of ADAF research, I wrote the paragraphs above in June 1996 as part of a review on X-rays and gamma-rays from AGNs (Svensson 1997), submitted it to the pre-print archives, sent preprints to about 150 astronomical libraries, and spread 100s of copies at conferences. But even so, the information did not penetrate the research community. So, as a summarizer of this accretion disk symposium, I took the liberty to show that the most important results of recent ADAF-research were already obtained by Ichimaru in 1977. Although Ichimaru's work was not mentioned in any of the talks at this symposium, it is gratifying to notice that he is quoted in three of the chapters. Furthermore, Ichimaru is now regularly quoted in ADAF-papers appearing since this symposium.

Why was Ichimaru's 1977 paper forgotten by almost everybody? It may depend on the way literature searches are done. One searches a few years back, and assumes that the major reviews cover the most important old papers. And once the ADAF history was written, it became adopted by the research community. But it does not explain why the Ichimaru-paper was ignored already in the late 1970s. It was quoted in the Cygnus X-1 review by Liang & Nolan (1984) but not for providing a new accretion disk solution, or for its explanation of the two spectral states of Cyg X-1.

2.3. *ADAF principles*

Much of the recent work on different accretion disk solutions and on ADAFs, in particular, is summarized in Abramowicz (this volume), Björnsson (this volume), and Narayan, Mahavedan, & Quataert (this volume). The recent work has among other things developed self-consistent solutions for the dynamics of ADAFs, has elucidated the connections between the different solutions branches, and has calculated detailed spectra from such flows.

One of the outstanding questions is which of the solution branches (see Figure 1 in Abramowicz or Björnsson, this volume) a disk chooses. Narayan & Yi (1995) discussed three possibilities. In the summary talk, I provided names for these three options (and thank Andy Fabian for proposing the "strong" and the "weak" names).

• The Strong ADAF Principle: The disk always chooses the ADAF branch if it exists.

• The Weak ADAF Principle: The disk chooses an ADAF branch when no other stable branch is available.

• The Initial Condition Principle: The initial conditions determine which branch is chosen.

One should note here that already Ichimaru (1977) employed the weak and the initial condition principles in his scenario for the two states of Cyg X-1. Depending on the initial conditions at large radii, the disk gas either goes to the ADAF branch (hard state) or cools down to the gas-pressure dominated standard Shakura-Sunyaev branch (soft state). At some smaller radius in the latter case, radiation pressure starts dominating, the standard branch becomes unstable, and the weak principle says that the disk then makes a transition to the ADAF-branch. Furthermore, Ichimaru (1977) also obtained the result that there is a maximum accretion rate above which the optically thin ADAF branch does not exist, a fact not recognized in recent ADAF-papers where this critical accretion rate was rediscovered (for this accretion rate the cooling time scale becomes equal to the inflow time scale and the disk cannot remain in an ADAF state, see §3.2.2 in Narayan *et al.*, this volume).

How unique are the ADAF-models and scenarios that are proposed for various phenomena? Depending on which ADAF-principle one uses, one gets different models and scenarios, which gives some latitude for the model-builder. It is therefore important to develop a physical understanding of the transitions (with changing radius or accretion rate) between the branches and to determine which principle applies. Furthermore, the whole solution structure with its different branches (see Figure 1 in Abramowicz, this volume, or in Björnsson, this volume) depends strongly on the type of viscosity prescription that is used. The prescription mostly used is $t_{r\phi} = \alpha P_{\mathrm{tot}}$. Other prescriptions such as $t_{r\phi} = \alpha P_{\mathrm{gas}}$ or $t_{r\phi} = \alpha \sqrt{P_{\mathrm{rad}} P_{\mathrm{gas}}}$ are likely to give very different solution structure and scenarios. This is rarely discussed in the ADAF-literature. Note that the solution structure in the Figure quoted only includes bremsstrahlung as a soft photon source for Comptonization. The more realistic case of unsaturated thermal Comptonization (without specifying the soft photon source) was considered by Zdziarski (1998).

2.4. *Similarities between quasi-spherical accretions and ADAFs*

One of the ADAF-results is that the accretion is almost spherical. As the proton temperature is close to virial, the vertical scale height is close to the radius, R. All physical parameters then scale with radius approximately as for free fall spherical accretion (see eqs. 3.15 in Narayan *et al.*, this volume). The coefficients may be different depending upon the viscosity parameter, α, but for α of order unity, even the coefficients are approximately the same. This means that the scenarios of dissipative quasi-spherical accretion of the late 1970s (e.g., Mészároz (1975) and Maraschi, Roasio & Treves (1982)) and ADAFs are essentially identical, the difference being that the workers of the late 1970s did not consider the dynamics in detail but rather used intuition to conclude that the flow scaled as free fall flows. Many of the resulting conclusions are the same. It was, e.g., noted in some early (quasi-)spherical accretion papers that the solutions depend on the black hole mass mainly through the combination $\dot{m} \equiv \dot{M}/\dot{M}_{\rm Edd} \propto \dot{M}/M$, and that therefore the solutions are similar for both galactic black holes and for super massive AGN black holes. Other results were that a *two-temperature* structure develops in the inner, say, 100 Schwarzschild radii; that the proton temperature remains close to virial, while the electron temperature saturates at about a few times 10^9 K; and that the luminosity scales as $\dot{M}^2$. Some of the radiation processes that were included were: self-absorbed cyclo-synchrotron emission, Compton scattering, and sometimes pion production in proton-proton collisions. The accretion rates, $\dot{m}$, considered were of order unity as the purpose was to explain the luminous AGNs.

These results again appear in the ADAF literature (see, e.g., Figures 4, 6, and 7 in Narayan *et al.*, this volume). One important difference now is that also small $\dot{m}$ are considered giving rise to very different spectra (see Figure 6 in the chapter of Narayan *et al.*, this volume).

2.5. *Applications of ADAFs*

The ADAFs have been included in scenarios and models for several astrophysical phenomena as described by Narayan *et al.* (this volume) and by Lasota (this volume). Such phenomena include explaining the quiescent state of three black hole X-ray transients, the different spectral states and spectral transitions of soft X-ray transients, low-luminosity galactic nuclei such as Sgr A* in the the Galactic Center, the LINER NGC 4258, and possible "dead quasars" such as M87 and M60. Lasota (this volume) discusses in greater detail how the spectral properties of the outbursts of the soft X-ray transient GRO J1655-40 can be explained within a scenario with an inner ADAF + an outer cold disk.

As mentioned above, applying different ADAF principles give rise to different scenarios. One such case is the efforts to explain the different spectral states of soft X-ray transients. Chen & Taam (1996) apply the weak principle and finds the transition radius between the outer cold disk and the ADAF to increase with accretion rate. Esin, McClintock, & Narayan (1997), on the other hand, in their detailed work apply the strong ADAF principle and finds the transition radius to decrease with increasing accretion rate (see Figure 11 in Narayan *et al.*, this volume)

Another case is the LINER NGC 4258 (Narayan *et al.*, this volume), where the strong ADAF principle gives rise to an ADAF inside the masering molecular disk, while the weak ADAF principle gives rise to a standard thin disk (Neufeld & Maloney 1995). Narayan *et al.* (this volume) argue that the latter scenario has a too low accretion rate to explain the observed emission.

3. Evidence for the existence of black holes and surrounding accretion disks

These topics were covered by Andy Fabian in his talk. In this volume, most of the evidence is, however, discussed in the two observational chapters by Charles (Galactic black holes) and by Madejski (supermassive black holes), while the review chapter by Fabian is limited to the broad Fe emission lines generated by the inner parts of a cold thin accretion disk.

The evidence for supermassive black holes in galactic nuclei has in the past been indirect. X-ray variability has set an upper limit to the size of the emitting region and thus to the mass. The luminosity has provided an estimate of the mass assuming the object to be radiating at the Eddington luminosity. Measuring the velocity fields close to the nuclei of nearby galaxies has also provided an estimate of the mass enclosed within that region. Recently, there has, however, been dramatic progress as described by Madejski (this volume). The VLBA mega-masers in the LINER NGC 4258 showed a Keplerian velocity profile indicating a "point" mass of $3.6 \times 10^7 M_\odot$. And *ASCA* observations showed Fe-line profiles broadened and distorted in precisely the way expected for emission from a rotating disk just outside a black hole (see Figure 4 in Madejski, this volume, or Figures 5 and 6 in Fabian, this volume). Some observations even sets constraints on the rotation of the black hole. Future space missions with observing capability of the Fe-line will provide ample opportunities to explore the strong gravitational field close to the event horizon of black holes. The Fe-line at the same time provides evidence for the existence and the properties of a cold reflecting disk close to the black hole.

There has been similar progress regarding determining the dynamical mass of the compact object in several galactic soft X-ray transients (see Table 3 in Charles, this volume). Again, the progress is mainly observational depending on the X-ray satellites providing the discovery of the transients, and the very large ground-based telescopes providing the dynamical mass-determinations. Present and future X-ray missions with all-sky monitors will discover new transients increasing the statistics and possibly broadening the range of properties of galactic black holes.

In this context, one should note the exotic but qualitative contribution by Novikov (this volume) on the physics just outside and inside the event horizon. Of particular interest is the qualitative discussion of the tremendous growth of the internal mass (mass inflation) of the black hole interior during the formation process.

4. Radiation processes

While the classical rates for bremsstrahlung, cyclo/synchrotron radiation, and Compton scattering in the nonrelativistic and relativistic limits were sufficient in the early disk models, the electron temperatures of 10^9 - 10^{10} K indicated by both observations and theory required the calculation of transrelativistic rates of the above processes as well as for pair processes that becomes important at these temperatures. Rate calculations as well as exploring the properties of pair and energy balance in hot plasma clouds were done in the 1980s by Lightman, Svensson, Zdziarski and others. The Compton scattering kernel probably received its definite treatment in Nagirner & Poutanen (1994). Recent improvements in some rates were obtained by Mahadevan, Quataert, and others. Much of this microphysics have been included both into the SLE-solutions and into the ADAF-models. One problem has been to determine the importance of electron-positron pairs in hot accretion flows. The most detailed considerations so far show that pairs have at most a moderate influence in a very limited region of parameter space (see Björnsson, this volume).

To obtain approximate spectra, the Kompaneets equation with relativistic corrections and with a simple escape probability replacing the radiative transfer is sufficient (Lightman & Zdziarski 1987). However, if one want to obtain constraints on the geometry from detailed observed spectra, then methods to obtain exact radiative transfer/Comptonization solutions in different accretion geometries must be developed. Such methods were developed by Haardt (1993) (approximate treatment of the Compton scattering) and Poutanen & Svensson (1996) (exact treatment) who solved the radiative transfer for each scattering order separately (the iterative scattering method). These codes are fast enough to be implemented in XSPEC, the standard X-ray spectral fitting package, and exactly computed model spectra can now be used when interpreting the observations. The codes, furthermore, includes the reprocessing (both absorption, reflection, and transmission) of X-rays by the cold matter. These methods have mostly been used to study radiative transfer in two-phase media consisting of cold and hot gas with simple geometries, but they have also recently been integrated into the ADAF models by Narayan and co-workers.

Poutanen (this volume) concludes that the galactic sources with their smaller reflection are best fit with a hot inner disc surrounded by a cold outer disc, while the Seyfert galaxies with their larger reflection are best fit with a geometry where the X-ray emission originates from active regions (magnetic flares?) atop of a cold disk (extending all the way in to the black hole as the broad, distorted Fe-lines indicates). The spectral predictions of ADAF models have not been tested against the detailed spectral observations using χ^2 fittings. One should therefore at the present moment have greater confidence in the conclusions of the work doing detailed radiative transfer and spectral modelling and fittings in simple geometries.

Poutanen also describes how the *detailed* spectra of the spectral transition between the hard and the soft states of Cyg X-1 can be described by a simple hybrid pair model (see Figure 9 of Poutanen, this volume) in a geometry with a hot inner disk surrounded by an outer cold disk and where the transition radius changes during the transition. The geometry is similar to the ADAF scenario suggested by Esin *et al.* (see Figure 14 in Narayan *et al.* this volume).

It is clear that the natural evolution is to merge the detailed spectral models with various accretion disk scenarios. One weakness of the ADAF-literature is that the spectral predictions of the (broad band) ADAF model is normally only compared with the standard (narrow band) disk model of Shakura & Sunyaev (1973). Other alternative scenarios, such as the two phase scenario where cold clouds are submerged in a hot medium (e.g., Krolik 1998 for a recent work on this scenario), are normally not discussed.

Another problem is the determination of the detailed spectral predictions from the cold disk, where the expected spectrum certainly is not that of a black body (Krolik, this volume).

5. The effects of magnetic fields

Magnetic fields are important in several respects in accretion disks around black holes. Several of these were discussed during the symposium and here some of them are listed:

• The most important influence of magnetic fields in an accretion disk is probably as being the agent generating self-sustained MHD-turbulence in a differentially rotating disk and thereby providing the necessary anomalous viscosity needed in black hole accretion disks. Brandenburg (this volume) describes the results of 3-D numerical simulations schematically in Figure 2. The Keplerian shear gives rise to large scale magnetic fields that in its turn generates turbulence through the Balbus-Hawley and Parker instabilities. Finally, the turbulence regenerates the magnetic field through the dynamo effect. The energy flow is shown in Figure 5 (Brandenburg, this volume). Approximately the same power is released in Joule heating (of electrons and ions) as in viscous heating (of ions mainly). One of the most important results is the magnitude of the Shakura-Sunyaev viscosity parameter, α, and the realization that α is not a constant.

• The magnetic field generates vertical stratification. There will be two different scale heights for the magnetic and the gaseous pressures. The magnetic field is more uniformly distributed vertically than the matter.

• The magnetic field plays a crucial role in the generation of a corona around the disk. The hope is to use a short time sequence from a simulation run of MHD-turbulence in the disk itself. This time sequence will form the driving background for a corona simulation. Nordlund speculated that spontaneous magnetic dissipation may generate self-organized criticality in the corona, similar to what has been found in simulations for the solar corona. Here, one should also note the partly unsuccessful efforts described by Wiita (this volume) in obtaining a self-organized critical state of the accretion disk itself.

• The magnetic field may act as a confining agent for cold gas (in the scenario where cold gas clouds coexist with a hot tenuous medium) as discussed by Celotti & Rees (this volume). The field may also be responsible for confining the matter or pairs in the active regions discussed by Poutanen (this volume).

• The magnetic field plays an important role as electrons (and positrons) in its presence generates cyclo-synchrotron photons in the radio to IR-range. These photons serve as seed photons in the Comptonization process, and is a crucial component in the ADAF models (Narayan *et al.*, this volume).

• Under certain conditions, synchrotron self-absorption dominates over Coulomb scattering as a thermalizing mechanism. The electrons emit and absorb cyclo-synchrotron photons resulting in a Rayleigh-Jeans self-absorbed photon spectrum and a Maxwellian distribution for the electrons (Ghisellini, Guilbert, & Svensson 1988; Ghisellini & Svensson 1989; Ghisellini, Haardt, & Svensson 1998). The thermalization time scale is just a few synchrotron cooling times.

6. Larger scale phenomena

Some phenomena on larger scales were discussed during the symposium. In an interesting talk, Pringle showed how the radiation from the central source may cause radiation-driven warping of the disk on larger scales. In certain cases, the inner disk may even turn upside down relative the outer disk resulting in self-shadowing of the radiation from the central source. The resulting ionization cone may not necessarily be aligned with the central source.

Merging of galaxies containing supermassive black holes may lead to a supermassive binary black hole in the resulting galaxy. The question is: How does the binary black hole evolve? Artymowicz (this volume) discusses the history of the efforts trying to solve this problem. After some 20 years of research it now seems that dynamical friction does not cause the eccentricity of the black hole orbits to grow. The interaction of a binary black hole with a common disk may resolve the difficulty of getting black hole-black hole merging to occur in less than a Hubble time. In the process, the black holes are fed, and a periodic light curve may be observed. One such example is OJ 287 with a period of about 13 years. A black hole circling a primary black hole with an accretion disk may cause warping of the disk as shown by Papaloizou *et al.* (this volume). The obvious application is the observed warping of the masering disk in the LINER NGC 4258.

Another issue discussed was the survival of vortices in accretion disks. The common notion is that Keplerian shear would kill such vortices on a few rotation time scales. In Spiegel's talk (and in Bracco *et al.*, this volume) it was shown that coherent structures indeed form. On the other hand, in the local simulations by Brandenburg (this volume) including magnetic fields, vortices do not form. The issue of vortices, their survival and influence is not yet settled.

7. Conclusion

This was a most rewarding symposium where leading scientists presented the most recent dramatic developments regarding several of the physics areas needed for realistic modelling of black hole accretion flows. These areas include radiative processes in hot plasmas, radiation transfer in hot and cold plasmas, MHD in differentially rotating gas, the origin of disk viscosity, magnetic flares, gas or MHD-simulations of flows, and so on. Some of these research lines have already merged. As each subproblem is understood, further merging provides the potential for dramatic progress in coming years.

The author acknowledges support from the Swedish Natural Science Research Council and the Swedish National Space Board.

REFERENCES

BEGELMAN, M. C., SIKORA, M. & REES, M. J. 1987 Thermal and dynamical effects of pair production on two temperature accretion flows. *Astrophys. J.* **313**, 689–698.

CHEN, X. & TAAM, R. E. 1996 The spectral states of black hole X-ray binary sources. *Astrophys. J.* **466**, 404–409.

ESIN, A. A., MC CLINTOCK, J. E., & NARAYAN, R. 1997 Advection-dominated accretion and the spectral states of black hole X-ray binaries: Application to Novae Muscae *Astrophys. J.* **489**, 865–889.

GHISELLINI, G. & SVENSSON, R. 1990 Synchrotron self-absorption as a thermalizing mechanism. in *Physical Processes in Hot Cosmic Plasmas* (eds. W. Brinkmann, A. C. Fabian & F. Giovannelli) pp. 395–400. Kluwer.

GHISELLINI, G., HAARDT, F. & SVENSSON, R. 1998 Thermalization by synchrotron absorption in compact sources: electron and photon distributions. *Monthly Not. Roy. Astr. Soc.* , **297**, 348–354.

GHISELLINI, G., GUILBERT, P. W. & SVENSSON, R. 1988 The synchrotron boiler. *Astrophys. J. Letters* **335**, L5–L8.

HAARDT, F. 1993 Anisotropic Comptonization in thermal plasmas: spectral distribution in plane parallel geometry. *Astrophys. J.* **413**, 680–693.

ICHIMARU, S. 1977 Bimodal behavior of accretion disks: theory and application to Cyg X-1 transitions. *Astrophys. J.* **214**, 840–855.

KROLIK, J. H. 1998 A new equilibrium for accretion disks around black holes *Astrophys. J. Letters* **498**, L13–L16.

LIANG, E. P. & NOLAN, P. L. 1984 Cygnus X-1 revisited *Space Science Reviews* **38**, 353–384.

LIGHTMAN, A. P. & ZDZIARSKI, A. A. 1987 Pair production and Compton scattering in compact sources and comparison to observations of active galactic nuclei. *Astrophys. J.* **319**, 643–661.

MARASCHI, L., ROASIO, R. & TREVES, A. 1982 The effect of multiple Compton scattering on the temperature and emission spectra of accreting black holes. *Astrophys. J.* **253**, 312–317.

MÉSZÁROZ, P. 1975 Radiation from spherical accretion onto black holes. *Astr. Astroph.* **44**, 59–68.

NAGIRNER, D. J. & POUTANEN, J. 1994 Single Compton scattering. *Astrophys. Space Phys. Reviews* **9**, 1–83.

NARAYAN, R. & YI, I. 1995 Advection-dominated accretion: Underfed black holes and neutron stars. *Astrophys. J.* **452**, 710–735.

NEUFELD, D. A. & MALONEY, P. R. 1995 The mass accretion rate through the masing molecular disk in the active galactic galaxy NGC 4258. *Astrophys. J. Letters* **447**, L17–L20.

NOVIKOV, I. D. & THORNE, K. S. 1973 Astrophysics of black holes. In *Black Holes, Les Houches* (ed. C. De Witt & B. DeWitt) pp. 343–450. Gordon & Breach.

PHINNEY, E. S. 1983 A theory of radio sources. PhD dissertation, University of Cambridge.

POUTANEN, J. & SVENSSON, R. 1996 The two-phase pair corona model for active galactic nuclei and X-ray binaries: How to obtain exact solutions. *Astrophys. J.* **470**, 249–268.

REES, M. J., BEGELMAN, M. C., BLANDFORD, R. D. & PHINNEY, E. S. 1982 Ion-supported tori and the origin of radio jets. *Nature* **295**, 17–21.

SHAKURA, N. I. & SUNYAEV, R. A. 1973 Black holes in binary systems. Observational appearance. *Astr. Astroph.* **24**, 337–355.

SHAPIRO, S. L., LIGHTMAN, A. P. & EARDLEY D. N. 1976 A two-temperature accretion disk model for Cygnus X-1: structure and spectrum. *Astrophys. J.* **204**, 187–199.

SVENSSON, R. 1997 X-rays and Gamma Rays from Active Galactic Nuclei in *Relativistic Astrophysics: A Conference in Honour of Professor I. D. Novikov's 60th Birthday* (ed. B. J. T. Jones & D. Marković), pp. 235–249, Cambridge University Press

ZDZIARSKI, A. A. 1998 Hot accretion discs with thermal Comptonization and advection in luminous black hole sources *Monthly Not. Roy. Astr. Soc. (Letters)* , **296**, L51–L55.